Product Management

Product Management

Value, Quality, Cost, Price, Profits, and Organization

H. E. Cook

C. J. Gauthier Professor
Department of Mechanical and Industrial Engineering
University of Illinois at Urbana-Champaign

CHAPMAN & HALL

London · Weinheim · New York · Tokyo · Melbourne · Madras

First published in 1997 by
Chapman and Hall Ltd.

ISBN 0-412-31970-5

A catalogue record for this book is available from the British Library

Library of Congress Catalog Card Number: 97-067487

Cook: Product Management

First Indian Reprint 2011

ISBN 978-81-8489-841-5

This edition is published by Springer (India) Private Limited, (a part of Springer Science+Business Media), Registered Office: 3rd Floor, Gandharva Mahavidyalaya, 212, Deen Dayal Upadhyaya Marg, New Delhi – 110 002, India

Printed in India by Rashtriya Printers, Delhi.

To my three dearest people: Valle, Maegan, and Adam, in order of appearance.

Contents

Preface

This book is about managing the development of new products for highly competitive markets. Particular attention has been given to tightly integrating current tools and concepts within a common framework including value engineering, market research, total quality management, quality function deployment (QFD), Taguchi methods, activity-based costing, statistical process control (SPC), product planning, pricing, and systems engineering. The view taken here is that the enterprise develops the profound knowledge and proprietary insight that it needs to favorably differentiate its products in the marketplace by continually generating fresh ideas, testing them through controlled but limited experimentation, and rapidly adopting those ideas that work. A strong attempt has been made to find the right balance between simplicity, empiricism, and rigor so that diverse groups within the enterprise can use and trust the structured methodology developed within its bounds of applicability.

A model of the firm is used throughout that is no more complex than necessary to connect the means of the enterprise to its ends. Based upon the most elementary yet powerful arguments from microeconomic theory, the demand for a product is developed in terms of its price, its value to the customer, and the values and prices of the products competing against it. The deployment of strategic decision making is emphasized because, in an intense competitive environment, every proposed action at every level of the enterprise should be examined for its strategic significance, not just at the level of highest management. For strategic significance to be used at all levels of decision making, the systems viewpoint must be understood and practiced throughout the enterprise. Consequently, the systems viewpoint is developed in detail.

Revisions to certain tools and concepts were required to integrate them into a common framework: for example, the definition of value championed by value engineers – product function divided by cost – for making subsystem and component level cost versus benefit trade-offs is not used here as it does not facilitate making trade-offs between different subsystems at the critical system level of design, the reason being that subsystem values would not be defined in a common set of units. Instead the value associated with an attribute change of a product at any design level is defined in terms of the change in product demand that the attribute change generates divided by an appropriate constant obtained from the price elasticity of demand. Classical conjoint analysis is modified by asking the respondents to the market research survey to make choices between a fixed baseline at a fixed price and product alternatives at different prices. The need for the fixed baseline when making simulated choices comes from the prospect theory findings of Tversky and Kahneman regarding the value of gains versus losses. The total quality of a product is defined as its net value to all of society including the customer, the enterprise, and the rest of society. The QFD process is only used conceptually here to set up a more direct and quantitative process for forecasting the impact of product design and manufacturing alternatives on demand, profit, and total quality. A new procedure is introduced for testing the significance of the experimentally measured signal-to-noise ratios championed by Taguchi for parameter design. Because just-in-time logistics are reducing the length of production runs, short-run SPC is covered along with classical, long-run SPC.

Although of considerable importance today, the relationships between the attributes of a product and its value to the customer will take on added weight in the future as firms replace prototype hardware with computer simulations to reduce product development time and costs. For such plans to be effective, however, engineers will need to learn how to go beyond the traditional end points of their calculations and project the dollars that changes in power, weight, speed, stress, noise level, sound, vibration, roominess, style, and so on generate in customer value. Consequently, the problem of how to translate proposed attribute changes for future products into forecast value changes is explored in depth.

The overall problem of how to make accurate computer simulations of the product attributes of importance to the customer, and to accurately project how those attribute changes will affect value to the customer and manufacturing costs, represents a research problem of considerable magnitude and significance requiring a cross-section of disciplines to properly address it. Issues raised in the text which are also important areas for future work include: (1) the need to explore the impact that a prospect theory kink in the value or loss function curve has on the loss of quality as a result of variation in the attribute and (2) the need to explore new accounting practices that are better geared to measuring the strategic impact that operational units make to the success of the enterprise. The more analytical approach taken here, focusing upon the fundamental metrics that lead to sustained profitability, is in contrast to the comparative analysis techniques (case studies) used in most treatments of competitiveness and business strategy. There is a danger, however, in drawing exclusively upon comparative studies to guide the actions of any enterprise. First of all, there can be large cultural differences between firms which will let one thing work well in one and not so well in another. Secondly, the interest is naturally in the best practices of the market leader, but by the time you grasp, digest, and implement one of the leader's best practices, it may have already been replaced by a better practice. Consistent leaders do not rest on their laurels. It is difficult, therefore, to catch a leader using only a strategy of imitation. Toyota's practices, for example, are not static.

As we move into the future, it is likely that the bottom line for evaluating the strategic impact of product decisions will more and more become based upon the impact that the decisions make upon society as a whole. In keeping with this viewpoint, the definition of total quality used here is the net value of the product to all of society, being the sum of its net value to the enterprise, its customers, and environmental quality. This should not, however, be in serious conflict with the traditional bottom line of profit because the best measure today of how well an enterprise is managing total quality is its level of sustained profitability provided that it is meeting the environmental standards set by governmental regulations.

Applications of the variety of tools described in this book are greatly facilitated by contemporary spreadsheet programs and graphics packages because of their wide flexibility. For example, the extensive array of look-up tables needed to support an in-depth discussion of statistics is no longer needed as most data can be generated on demand using spreadsheet functions. Limited statistical tables are provided in the text for illustration purposes and for occasionally cross-checking that the spreadsheet functions are being used correctly. The generality, power, and simplicity provided by matrix algebra in the design and analysis of experiments can now be fully and readily used in practice because of the matrix operations provided by the spreadsheet programs. Monte Carlo calculations are also readily made using the random number generator provided by the spreadsheet. An attempt has been made to make the book self-contained, extensive help being provided in Appendices A and B for readers not familiar with statistics and matrix algebra or desiring a review. The particular spreadsheet program used for the computations given in the book was Microsoft Excel® version 5.0. The graphics package used was KaleidaGraph® version 3.05.

It is hoped that the wide use of automotive examples has pedagogical value because most persons are familiar with automobiles. The methodology developed, nevertheless, is general and can be applied to any product or, in fact, to any service although the latter is not developed here with examples.

Harry E. Cook

Acknowledgments

The research effort which has led to the methodology described in this book was supported largely by industrial sponsors and by the Grayce Wicall and C.J. Gauthier Laboratory Professorships endowed by Mr C.J. Gauthier. Very special thanks go to Jim Bateman, Dan Benz, Jonathan Brill, Joe Donndelinger, Rich Ingram, Chris Magee, and Joe Sparks for their strong interest in the research and the sharing of their fresh ideas about important directions. The students who have participated in this research and I are deeply grateful to them and for the financial support provided by their companies. I am also grateful to Murat Aycin, Felicia Moss, and Svetlana Shinkareva for their review and critique of a portion of the text. I have benefited greatly from timely and insightful discussions with Dick DeVor, David Goldberg, Doug Simpson, and Aloz Sluga. I remain eternally grateful to Amanda Horner for the important administrative support provided over the last six years which has allowed me to devote more time to the book than I had imagined I would find at the outset. Finally, I have been fortunate to have worked closely with many delightful and imaginative graduate students as part of an ongoing product management research program. These include Shandon Alderson, Hussein Ali, Sheldon Bailiff, Curtis Bush, Elizabeth Cowan, Edwin Dair, Joe Donndelinger, Andy Elsbury, Mike Gill, Prakash Kolli, Mike Lee, Greg McConville, Eric Monroe, Terrence Mosely, Matt Neidlinger, Mike Pozar, Mark Rimkus, John Runnion, Torsten Schildt, Orlando Sellers, Becky Silver, Mark Simek, Luana Weiss, Kimberly Williams, Tony Woods, and Andrew Wu.

1 Product realization in the global marketplace

1.1 Structuring the unstructured problem

Product realization refers to the complex, unstructured problem of bringing a new product from its conceptual beginnings into and through its full life cycle. The major steps include idea generation, research, development, manufacturing, sales, service, and recycle/disposal (Figure 1.1). The problem is complex because of the need to coordinate and synchronize efforts across the organizational units within the enterprise and because three important stakeholders outside of the enterprise – customers, stockholders, and governmental regulatory agencies – must be more than satisfied by the outcome. The problem is unstructured because many different design approaches are feasible and myriad possibilities exist for developing and manufacturing the product. Product management is the process administering and guiding product realization by giving it structure and reliable tools for reducing its complexity and making good, timely decisions.

This book develops a structured methodology for guiding the enterprise in developing new products. The focus is on making strategic decisions – decisions which will significantly impact profitability – in a timely and consistent fashion at every level of the enterprise. It requires that we become familiar with ele-

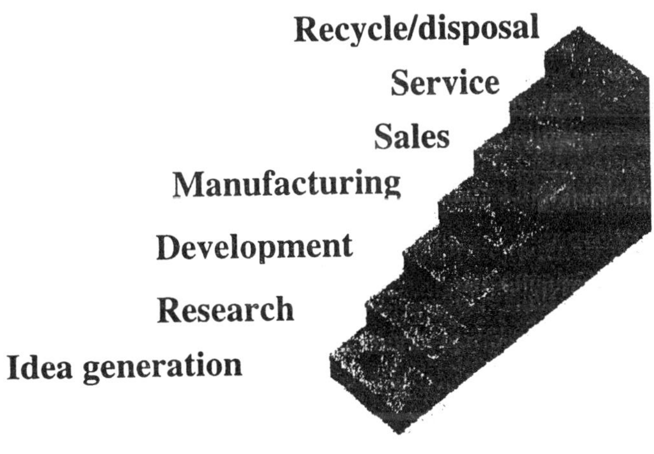

Figure 1.1 Steps in product realization.

mentary but fundamental economic, statistical, financial, organizational, and behavioral concepts and integrate them into a comprehensive toolset for managing the product realization process.

1.2 A product management case study

The range of issues faced by those involved in product management – assessing product value, managing trade-off decisions, facing-off against tough competitors, and dealing with uncertainty – can be illustrated by considering the plight in the mid-1980s of Planes Inc., a hypothetical commercial aircraft company experiencing declining market share and profits. It wants to reverse its slide and become the dominant producer of aircraft for transporting passengers on long, mainly international, routes by making a major breakthrough in aircraft design and technology.

On its drawing boards are the conceptual designs for two new aircraft (Figures 1.2 and 1.3) either of which might possibly lead to its hoped-for turnaround. One, called the LST, is a large subsonic transport having a speed of mach 0.84. The other, called the SST, is a supersonic transport having a speed of mach 2.2. (Assume that the 747 and Concorde do not exist.) Planes Inc. cannot afford to do the two aircraft simultaneously as each alone will cost $8 billion to develop. Facing the company are three crucial questions:

1. Which of the two aircraft should it develop?
2. Once that decision is made, how should it manage the complex development program required by either plane?
3. What happens if it pursues one alternative and one of its competitors pursues the other or if one or more competitors also pursue the alternative it has chosen?

To answer these questions, the company must weigh the opportunities and risks of each alternative. Assessments are needed ranging from the right price of a ticket to the price of a plane and from the size of the market to the size of the investment. Once these have been made, the profitability of each alternative must then be forecast.

The risk to the bottom line arises from the degree of uncertainty in the profit forecast. This results from the lack of a firm understanding of what the actions of competitors might or might not be, what government regulations might or might not be in place, and what unforeseen pitfalls exist in pursuing and implementing the potentials of a new technology.

1.2.1 AS-IS VERSUS TO-BE

The starting point for the analysis is an assessment of the **as-is** condition of the current marketplace. Aircraft currently in use for long hauls carry 250 persons

Figure 1.2 Large subsonic transport (LST) concept.

Figure 1.3 Supersonic transport (SST) concept.

and current ticket prices are $816 for economy and $1630 for first class. The **to-be** market will be greatly influenced by which of the two new planes is built. The unique feature of the proposed LST is its size which will give it the ability to carry more than 360 passengers. For the average overseas trip, the LST operating costs of $60 000, which includes depreciation, projects to a per passenger cost of $166 and represents a $74 reduction in operating costs per passenger from the current $240. The entire as-is market could be absorbed by the LST.

Although the SST will carry only 120 passengers, which more than doubles its operating costs per passenger, a market niche is targeted for this plane of persons who value their time very highly and who Planes Inc. believes will pay a higher ticket price for markedly reduced travel time. For the average overseas trip, the SST's operating costs of $60 000 including depreciation projects to a per passenger operating cost of $500 which is $260 higher than current operating costs.

1.2.2 MARKET RESEARCH

Planes Inc. considered immediately approaching its customers, the airline companies, and asking them which alternative they preferred after describing the capabilities of each plane including passenger capacity, range, operating costs, operating lifetime, and price. However, before answering Planes Inc., the airline companies would almost certainly take the information they were given to their customers, the potential passengers for the two aircraft, to get their reaction. After analyzing the results, the airlines' decision would be based upon which of the two planes would make the most sense from their business perspective.

It was possible, of course, that the airlines might do their market research inadequately and make a decision which would prove later to be incorrect, or they could decide that neither plane is needed. An incorrect decision as to the right alternative would be a catastrophe for Planes Inc. because only a few of the new aircraft would be sold and the losses from the unrecovered development costs would bankrupt the company. Also the decision not to build either could ultimately be fatal to Planes Inc. fortunes because, without the type of breakthrough envisioned, its position as a manufacturer would continue to deteriorate as a result of the cost and performance leadership enjoyed by its competitors in conventional aircraft. Finally, before the project could begin, the banks would need to be convinced that the required technological breakthroughs could be made and that the necessary customers would be there.

Although the banks would require the airlines to convincingly state their intent with regard to the proposed new plane through progress payments to Planes Inc., the major burden of convincing the banks to grant the loan for developing the new aircraft would be on Planes Inc.'s shoulders not the airlines'. Thus an in-depth understanding of the market would be invaluable in presenting its case. Consequently, Planes Inc. decided to independently pursue preference studies of potential passengers in regard to the two aircraft before approaching the airline companies. The results of its market research are shown in Table 1.1.

1.2.3 VALUE OF ATTRIBUTE CHANGES

Planes Inc. also found that the value to the ticket-buying customer in reducing trip time t could be approximated by a simple expression of the form:

$$V = V_0 + k[t_0 - t]$$

Table 1.1 Market study (hypothetical) of large subsonic transport (LST) versus supersonic transport (SST)

Factor	*Units*	*Current ST*	*Large ST*	*SST*
Passenger capacity	[# Persons]	250	360	120
Speed	[Mach #]	0.84	0.84	2.2
Economy ticket	[$]	816	742	N/A
First-class ticket	[$]	1630	1556	1890
Investment	[10^9 $]	0	8	8
Operating costs	[$/Trip]	60 000	60 000	60 000
Operating costs	[$/Passenger]	240	166	500
Market size	[# Planes]	720	590	77
Forecast price	[10^6 $/Plane]	N/A	150	150
Forecast variable cost	[10^6 $/Plane]	N/A	75	75
Break-even	[# Planes]	N/A	106	106
Forecast profits	[10^9 $]	N/A	36.25	–2.225
Build rate	[# Planes/Month]	N/A	8	8
Time to break even	[Months]	0	13.25	Never

where V_0 is the value of the trip for the time t_0 currently required and k is the value that the customer places on his or her time. For the average economy class customer, k was $30/hr but for the average first-class passenger it was much higher at $100/hr. A reduction in trip time from 8 to 3 hours would result in $150 added value to each economy passenger and $500 to the first-class passenger. For the average first-class customer, the added amount of $260 to the price of a first-class ticket was well below the value added of $500 representing a net value gain of $240. For the average economy-class customer, however, the inclination to consider the amenities of higher-price, first-class travel was even less enticing than before, the first-class ticket increasing in price by $260 for a gain in value of only $150.

1.2.4 PRICE ELASTICITY AND MARKET SIZE

With this information, the market research department examined previous demand changes with price and discovered that a 10% price reduction on average resulted in a 20% increase in the number of people flying. This suggested to them that a net value increase of $240 would increase the number of first-class passengers by 29%. Since the current market share of first-class passengers was 10% of all passengers, the size of the segment for the SST was optimistically forecast at 12.9% of the current total number of passengers (economy and first class). The optimism was based upon the assumption that all first-class passengers would choose to take the SST.

This assumption, however, was strongly challenged by the financial analysts. Marketing reconsidered and projected that the SST would take 50% of the existing first-class market plus all of the forecast growth in the number of SST passengers. This resulted in an estimate that the to-be SST market would be 7.8% of the as-is economy plus first-class market.

1.2.5 ACTIONS OF COMPETITORS

But what if a competitor simultaneously developed an LST similar to the one on Planes Inc. drawing boards? This could lead to the price of a first-class ticket for the LST being reduced by $74 versus the current price. This action would by itself lead to a 9% increase in first-class passengers and capture sales from persons who might otherwise use the SST. Thus, with competition from an LST, the SST might only realize a share equal to 7.4%. Further discounting of the forecast SST gains would come from the fact that the LSTs would be able to offer much better flight availability, on the order of 10 LST flights to each SST flight. With all of these considerations in mind, the number of SSTs required to service this market was forecast finally at 77 planes.

Although the LST would offer no improvement in value by reducing trip time, its cost savings per passenger would allow the price to be reduced from $816 to $742. Thus the 9% net price reduction would increase, according to the price elasticity of two, the size of the long-range aircraft market by 18% resulting in a need for 590 LST planes seating 360 each versus today's 720 planes seating 250 each. This translated into the LST's projected ultimate market size being 7.7 times that for the SST.

1.2.6 BREAK-EVEN TIME

The estimated variable cost to build one LST or one SST was the same at $75 million. Thus at the planned selling price of $150 million per plane for both the LST and SST, either alternative should break even (recoup its initial investment through the net revenue generated by each plane sold) at 106 planes. If only Planes Inc. built the LST, this would occur after thirteen months at its projected build rate of eight planes per month and result in a long-term profitability of over $36 billion. But it was not realistic to assume that Planes Inc.'s two major competitors would remain outside such a lucrative market. With the LST, the forecast demand was large enough for each of the three to make a long-term profit between $7 and $10 billion. However, the SST was not forecast as being profitable even for a monopoly because its 106 break-even number was higher than its projected market of 77 planes.

Although not factored directly into the financial analysis, Planes Inc. along with the airline companies was also concerned that the SST might be restricted to operate below supersonic speeds to reduce noise over populated areas. This would reduce its speed advantage and increase its operating costs because the SST's engines were designed to operate most efficiently at supersonic speeds. The skin of the SST would be hot on landing which would increase the aircraft's turn-around time because it would have to be cooled before refueling and inspection could begin. Potential unforeseen weight increases, which always occurred in the development of new aircraft in the past, would also present acute problems to the SST because the operating life of the stressed hot skin was very sensitive to added weight. Not only were the numbers not favorable but the uncertainties and resulting risks were much higher with the SST.

1.2.7 MAKING THE DECISION

Thus, Planes Inc. saw wisdom in pursuing the LST versus the SST alternative and the airline companies agreed with this assessment after performing their own research. But choosing the right alternative on paper at the outset of a new product program does not assure that the new plane will be developed as planned. How does Planes Inc. continue to consistently make the right decisions as the development of this new aircraft unfolds? Does it mandate that all decisions be made from the top? If not, how much should be delegated and how should it be delegated? How should it organize to most effectively develop the plane? Which components should be made inside and which should be purchased from outside?

1.2.8 WHAT THE CASE SHOWS

This case highlights five general needs that must be met to assure the successful development and sale of any new product:

1. **The need for routinely but systematically weighing alternatives properly and selecting the one most appropriate:** the major decision in the case was whether to pursue the LST or the SST. On choosing the LST, selection among competing alternatives, however, has not ended. Which engine, the one from supplier A or the one from supplier B, should be used? Should graphite composites be used in place of aluminum? If so where? How carefully should the temperature and humidity be controlled in the passenger compartment? What should the leg room be between the seats? What level of avionics should be used? Fly by wire? Panel assembly by rivets or glue? How much manufacturing automation should be used? What criteria should be used to trade off cargo space, passenger space, and fuel volume?
2. **The need to determine the joint value of multiple product attributes:** two product issues were considered in the above case – trip time and operating cost per passenger mile. Trip time affected the value of the trip to the customer and operating costs ultimately affected the price of the ticket. Products will generally have many attributes and they will often be conflicting. How does the airline make the trade-off between the added value of increasing knee room and the added operating cost resulting from less capacity for passengers? This begs the questions: What is the value of added knee room? What is the value of added head room for egress and ingress versus more overhead space for carry-on bags? How does the marketing

department best communicate the value of product attributes to the engineers who ultimately will be responsible for making most of the trade-off decisions? The externalities of a product, for example the loud noise generated by the SST at supersonic speeds, that affect all of society and not just the buyer are important attributes that can make or break a product depending upon how they are addressed.

3. **The need to understand how demand for a product changes with its value and price and with changes in the value and price of competitive products:** a major factor in the decision made by Planes Inc. was the relationship between product demand and changes in product value and price. How would the change in value or price of a competing product affect the demand of Planes Inc.'s proposed new product?
4. **The need to manage the product realization process systematically so that good responsive decision making occurs at all levels of the organization:** although Planes Inc. felt that its decision to pursue the LST was correct, many more decisions must be made as to which alternative to pursue at the subsystem and component level before the LST is developed. If these decisions are not made smartly, the performance, demand, and profits expected for the LST may fall way short of objectives. The use of a structured methodology would assure that decisions were made consistently across all elements of the enterprise. The idea is to make decisions as wisely on the factory floor as in the Board Room and vice versa!
5. **The need to manage uncertainty:** in the decisions made by Planes Inc., none of the factors involved were known precisely yet management was willing to make a 'bet the company' decision involving an $8 billion investment. Risk taking is a major element in the development of all new products. It arises from two competing factors. First of all, a relationship is at work regarding uncertainty and time. Simply stated, if you want to know something more accurately you will need to spend more time to get it. This uncertainty appears in answering the question, 'How much better is the new product than the existing product?' However, as you take more time to reduce the uncertainty relative to the existing product, you increase the risk of a competitor beating you to the market with a better product. Therefore, a point in time always exists where a decision should be made based upon the amount of information already collected, the rate at which additional information is being gathered, and the forecast opportunities being lost by having not made the decision. Fortunately, statisticians have developed techniques for acquiring information effectively and efficiently. By using these techniques risk can be reduced but never eliminated. An important part of our effort here will be applying what the statisticians have developed to the product realization problem.

1.2.9 MARKET REACTION TO THE REAL CONCORDE, A SYNOPSIS

The above hypothetical case begs the question: 'What has been the market reaction to the real Concorde?' Feldman has described the development of the plane from a political and sociological perspective in his analysis of high technology failures [1]. Concorde was planned and developed with political objectives in mind – standing up to the United States in the 'technology race,' aiding England's entry into the Common Market, and jobs – rather than profits.

Market research was abysmal. In 1969, sales between 250 and 1250 planes were predicted by 1978. Earlier (1961), a French study projected 185 planes would be needed on French domestic routes alone! Many decisions were secret even though public funds were used. The plane had limited range and capacity and it was very expensive to operate and not allowed to fly supersonic over most populated areas. The 747SP could actually reach Sydney faster from London because of Concorde having to stop and refuel. Not surprisingly, the aircraft has been a commercial failure and it failed before the energy crisis. Only fourteen were built beyond the original two prototypes. The national airlines in England and France are subsidized by their governments to cover the losses generated by the Concordes still being flown.

In the hypothetical case, we made the decision for the LST because we were using profits as the objective. These questions remain for Concorde: Did it achieve its political objectives? If so were the costs worth it? Or were there other causes that could have better used public funds? Was this effort a net benefit or net loss to society? If a new SST is attempted, what should its specifications be?

1.3 The global marketplace

In the not-too-distant past, we saw the product realization process in terms of the individual units within the enterprise operating sequentially over time to develop the product (Figure 1.4a). When one part of the job was finished, the results were 'tossed over the wall' to the next person or operation in the chain. Now we understand product realization from a systems perspective, a system which operates successfully as a result of the close integration of these disciplines operating concurrently at the outset of the process (Figure 1.4b).

The intense competition in global markets was the driving force shifting our viewpoint of product realization from a sequential to a concurrent process. The U.S. market has been a major 'battleground' for the competitive wars (Figure 1.5).

At the outset of intense global competition in the late 1970s and 1980s, many industries in the U.S. were not highly successful as shown by the losses suffered in the balance of trade by major U.S. manufacturing sectors. The trade imbalance in automobiles (Figure 1.6) grew significantly from 1976 to 1988.

A sizable, negative trade balance also appeared in semiconductors, machine tools and consumer electronics. In looking for an explanation for this poor performance, the MIT study [2] attributed it to six things:

1. outdated strategies;
2. short-term horizons;
3. weaknesses in development and production;

Figure 1.4a In the past, product realization was a sequential and slow process. When one group finished its work, it 'threw it over the wall' to the next.

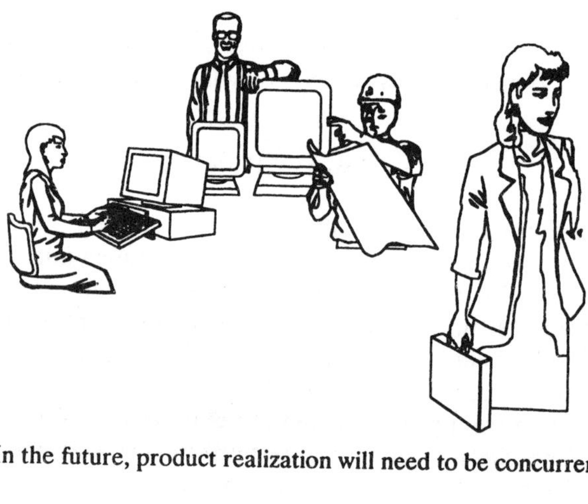

Figure 1.4b In the future, product realization will need to be concurrent and fast.

Figure 1.5 The United States has been the primary battleground for competitive wars.

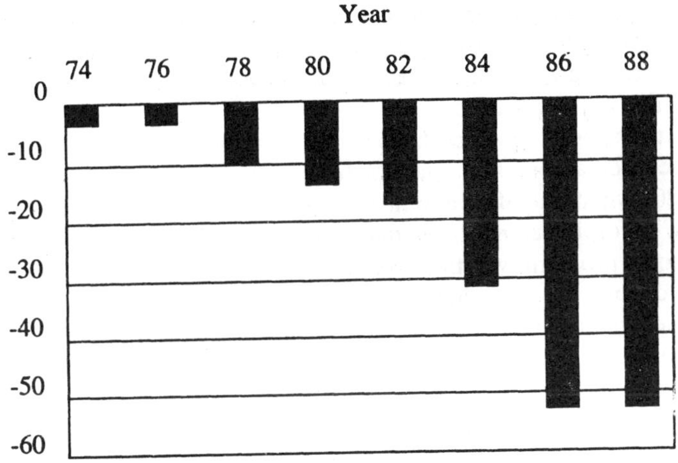

Figure 1.6 U.S. trade imbalance in automobiles, 1974–1988. (Source: U.S. Department of Commerce, International Trade Administration, Office of Trade and Information Analysis, as summarized by Dertouzos, Lester and Solow [2].)

4. neglect of human resources;
5. failures of cooperation;
6. government and industry at cross purposes.

The major outdated strategy identified was 'mass production of standard commodity goods.' When the automobile was a new technology replacing the horse and buggy, the mass production strategy of Ford Motor Company under its founder worked well. The automobile became affordable to a large percentage of the population and Ford was very profitable. The initial strategy was to build virtually every car the same, even the same color, black, and produce them forever. The flaw in over-standardization and long production runs became apparent to Alfred P. Sloan and he positioned General Motors to go head-to-head with Ford by offering greater diversity in the product. GM came to dominate the U.S. automotive business with Sloan's approach as Henry Ford held onto his strategy of a single, low-cost product. Ford eventually had to adopt GM's practices of product differentiation to stay in business. Ironically GM's market share relative to Ford only began to recede in the mid 1980s as GM's brands – Chevrolet, Pontiac, Oldsmobile, Buick, and Cadillac – became less distinctive through the use of common platforms and exterior stampings that reduced product differentiation. Today the strategy is once again toward more product diversity to better reach the wide variety of customers in the market. The approach is to discover flexible ways of manufacturing and assembly that yield high production rates of differentiated products coming down the same assembly line while maintaining relatively low costs for tooling the different products.

The competitive spirit that propelled U.S. industry in the first half of the 20th century began to wane at the mid-point of the second half of the century for several reasons. One was that the second world war left much of the industrialized world in the northern hemisphere, except for North America, in shambles. In such an atmosphere, it was perhaps natural that an over-confidence in U.S. practices developed among U.S. manufacturers who were not truly shaken until the early 1980s when one after another of their products began to lose major market share to foreign competitors that were making better products at less cost.

A second problem was that the U.S. businesses that were successful competitors during the first half of the century had grown to be very large. Companies that retained a strong central management structure were no longer as responsive to meeting the challenge of rapidly changing markets and less likely to take the kind of risks that led to their original growth. However, companies such as General Electric and Hewlett Packard, which rested considerable autonomy in their divisions and continue to do so, have prospered during this same period.

The problem of focusing on the short term at the expense of the long term was the second factor identified in the MIT study. Focus on the longer term is facilitated by having access to capital at low interest rates and ownership by institutions such as banks that have a longer view. Because the cost of borrowing can differ appreciably between nations, the enterprises in a nation with low rates can take a longer view than those in another with higher rates. It is easy for institutions having short-term objectives such as mutual funds to move their large investments from one company's stock to another at relatively low transaction costs and thereby seek each incremental gain from wherever the opportunity is at the moment. It is natural for reward systems within companies whose stock is subject to markets of this type to reflect short-term goals.

The weaknesses identified in product development and production were not all of technical origin. The management practice of sequential product development from marketing to planning to engineering to purchasing to manufacturing to assembly and finally to sales and service has proven to be slow and costly. Not understanding the importance of manufacturing excellence to the strength and competitiveness of the enterprise has also hurt many U.S. manufacturers [3].

One early tenet of mass production was to design jobs such that persons with low skills and low wage rates could perform the tasks. Tasks were structured to be simple and repetitive. This strategy served the needs of many nations reasonably well as there was a large pool of unskilled labor during the first half of the 20th century. This practice, however, caused many companies to not grasp the importance of its human resources in the global competitive manufacturing environment that had emerged by the 1980s. The intensity of competition today mandates that each company use all of its resources at their fullest, if it is to be successful.

The traditional view of competitive advantage was that it was generated by special access to raw materials, capital, and technology. Today raw materials are sold on the world market, many capital markets (but not all) are open with funds moving at electronic speeds and knowledge and technology can be transferred quickly through reverse engineering and improved upon by many firms. What remains are the people within the enterprise and they represent the most secure foundation for generating a sustained competitive edge in today's global markets [4]. The companies that best use their entire workforce in designing and manufacturing products for demanding customers will generate the competitive edge of low costs and a high rate of performance improvement.

Failures of cooperation between those working within manufacturing concerns and also between those in private enterprise and government were the last two general problems identified in the MIT study. The conflicts within the firm were not just between labor and management but between those in manufacturing management and engineering management and between the finance staff and everybody else. There is a tendency for each organizational unit to distrust the others. In the absence of a strong competitive environment there was not a common enemy outside so the enemies were within. The political parties also reflected the battle between labor and management with the Democrats favoring labor and the Republicans favoring management.

1.4 Performance hill race

Although manufacturers talk about improving their competitiveness in terms of an analogy to baseball or football – need to hit a home run, better blocking and tackling – the analogies are misleading [5]. There are no time-outs in the global marketplace. There are no umpires to call the close plays nor is the playing field limited to familiar ball parks. The game played is not seasonal with periods for rest and recuperation. Also if you did not do well last year, you do not get first choice of the upcoming talent from the college draft. You also do not take on one opponent at a time on days scheduled well in advance.

Instead global competition is like a foot race between manufacturers up a performance hill (Figure 1.7). The height of a manufacturer's position on the hill at any point in time is a measure of the performance of its product. The runners are sustained in this effort by sales to customers that they encounter at various points on the hill. The racers are handicapped by weights proportional to their costs. If some run faster and have lower costs, few customers will be left for those that lag behind. Without this nourishment from sales, the laggards will have to drop out of the race.

Competitive advantage in the performance hill race is generated by low costs and moving quickly. This is typical of companies that have been found to be successful in highly competitive global markets. These companies have lower costs, higher quality, and greater speed in delivering their products to their customers than companies not so successful [6]. They can correctly be referred to as 'lean' in that they tend to do a better job with fewer resources than many of their com-

Figure 1.7 Performance hill race is a good sports analogy for business in a competitive environment.

petitors [7]. The questions that we need to address are: What must a company do to assure that its workforce can design and manufacture winning products year in and year out over the long term? How should a manufacturing enterprise operate on a day-to-day basis? What are the rules, goals, procedures and disciplines that should govern its actions?

1.5 Paradigms

There are rules that we accept without question called paradigms [8]. Newton's laws of motion are an example of a paradigm that is eminently correct within the environment of velocities associated with normal, day-to-day human encounters. We do not work downwards from relativity theory to arrive at the equations of motion for an automobile or an aircraft. In this way, paradigms are useful because we don't need to take the time to solve each new problem starting with a review and application of first principles.

But the paradigms that we employ in our daily routines and business do not have the exactitude of Newton's laws of motions and most are highly sensitive to the environment that supports them. Consider the example of an international business trip. On leaving the airport, you discover as you precariously maneuver in a driving rainstorm that the windshield wiper on the rented car has its control somewhere other than the first three places you reached. Your routine drive home is replaced by pondering signs and directions in an unfamiliar city. On finding the hotel and poorly speaking the language, you discover that the non-smoking room which you thought you had requested was last occupied by a chain smoker. At dinner, the waiter serves you something quite different from what you imagined you asked for. Once settled and asleep, you wake up and walk into the wall on your way to the bathroom. When your system tells you it is one o'clock in the morning, dawn appears. Behavior which was routine and adequate in our familiar environment becomes awkward and inadequate in the unfamiliar environment.

Just as the social and cultural environment for the businessperson on a trip changes as described above and interrupts his or her routine, the competitive environment governing the success of an enterprise changes and the enterprise must react properly to it to remain successful. However, changes in the competitive environment will often not be signaled by the strong cue that landing in London, Paris, or Tokyo gives to the immediate changes coming in the social and cultural environment. Only after many failures in succession might it become clear to an enterprise that the competitive environment which once supported one of its cherished standard operating procedures has changed.

An important reason outdated business paradigms remain in vogue is that society institutionalizes them. Everybody jumps on the bandwagon – universities, consultants, foundations, government agencies, and industry – to support the conventional wisdom and in doing so the paradigm becomes entrenched well past its limits of applicability. The organization of product realization is an example of a paradigm that is now shifting from a functional to a team-based process. The shift is perhaps further along today in our manufacturing enterprises than in our colleges and universities where functionalism has been institutionalized through the extensive departmentalization of education into highly specialized units. There are several other manufacturing paradigms that were once strongly held but are now outdated:

- Quality improvements increase costs.
- You push production through a plant.
- You need significant in-process inventory to maintain production.
- You need to increase product development lead time to improve quality.
- You need strong functional divisions that operate sequentially to develop new products.

Only one manufacturing paradigm remains sacrosanct, the Law of Commerce: satisfy the customer, keeping costs low!

Entrepreneurial enterprises use 'creative destruction' to generate paradigm shifts [9] and Japanese manufacturers have shifted many paradigms 180°. Most shifts came about by observing and learning on the factory floor. Instead of studying the production process using sophisticated mathematical techniques, straightforward but fundamental questions were asked: inventory represents a cost not an asset, how can it be reduced? The environment that generates high inventory – low quality and slow tool changes – had to be changed if low inventory was to be achieved. High quality was also believed to result in high costs so this paradigm had to be shifted before inventory could be reduced.

The concept of having components arriving in small batches (low inventory) just in time for assembly or for the next operation occurred to Mr Ohno of Toyota when he visited an American supermarket in the 1950s and saw how fresh produce and vegetables were maintained by frequent delivery. Two major paradigm shifts that were developed by Toyota under Mr Ohno's direction are summarized in Table 1.2 along with the necessary environmental changes.

Table 1.2 Toyota paradigm shifts

Paradigm shift	*Environmental shifts required*
Large to small inventory	From low to high quality and from slow to rapid tool changes
Quality increases costs to quality decreases costs	From emphasis on on-line quality control methods to emphasis on off-line quality methods

1.6 Concurrent engineering

The paradigm shift under way from sequential to concurrent product realization is of considerable importance. The reason for this shift is that, as a rule of thumb, 80% of the actual cost of a new product is committed during the first 20% of the development period (Figure 1.8). Consequently, if the manufacturing engineer has not been consulted, unnecessary manufacturing costs, unforeseen by the designer, could be dictated by the design that give no appreciable added value to the customer in terms of product performance. This situation can be avoided when there is early and close interaction between the design and manufacturing engineers. Overall excellence in product management also requires close collaboration between marketing, purchasing, design, manufacturing, and service from the outset of the product realization process.

Another feature of concurrent or simultaneous engineering is that fewer design changes are required (Figure 1.9) and those that are needed occur early in the process when the cost of making the change involves simply a reordering of information or intent and not the wholesale scrapping or rework of tools, dies and facilities which can generate enormous added costs and delays in the product introduction date.

Concurrent product realization teams are formed from members from the separate elements of the organization. One type of organization frequently

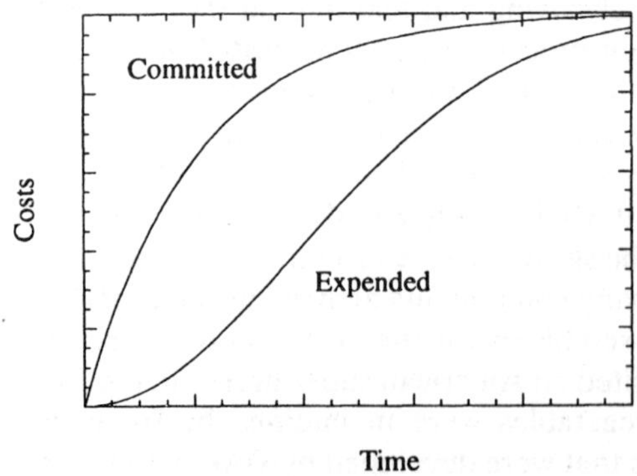

Figure 1.8 Typical behaviour of funds committed and funds expended over time for a new product.

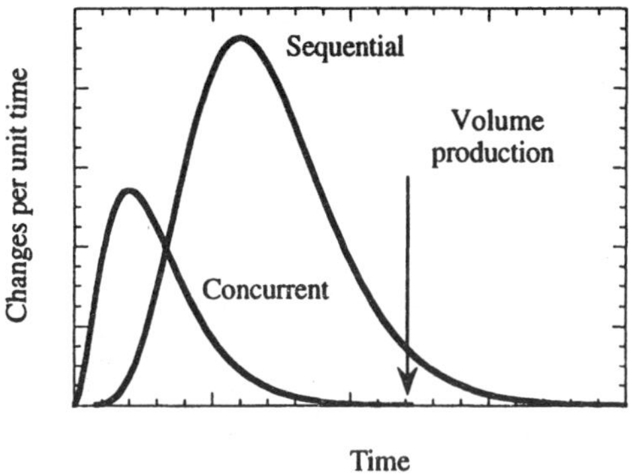

Figure 1.9 Distribution of product design changes versus time.

used by manufacturing enterprises is the functional organization (Figure 1.10). Many feedback and control channels are not displayed in this figure for the sake of clarity. Also the important boxes for finance and legal staffs as well as the management hierarchy have not been shown. In the conventional wisdom of manufacturing organizations, vice-presidents would be in charge of the functions shown.

Although the figure indicates that only the marketing function communicates with the customer, the other groups will also generally have customer contacts so that they can better understand how certain aspects of the design and manufacturing process that they are responsible for are being accepted by the customers. Responsibility for translating customer needs into product specifications is shared to a considerable degree by marketing, planning, and engineering and to a lesser extent, generally, by purchasing and manufacturing.

When a product realization team is formed, responsibility for system design is generally transferred to the team. Nevertheless, problems in team leadership can arise as a result of (1) divided loyalties between the team and the home organization of the team leader, (2) the leader being given the responsibility but

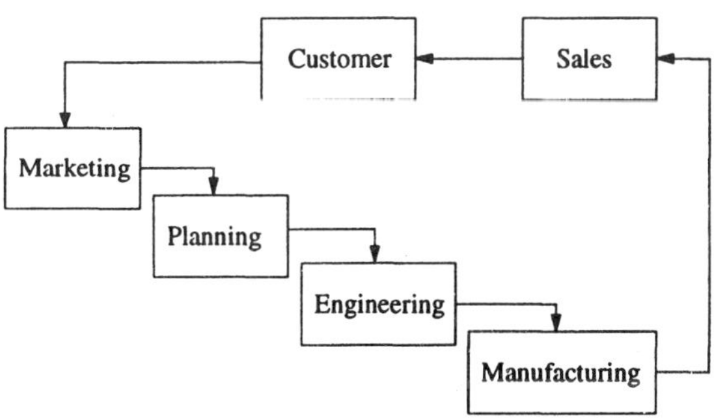

Figure 1.10 The functional organization with the customer shown in the loop.

not full authority for the effort and (3) leaders being given new jobs before the process is completed. Having an empowered team leader who will be there for the duration is believed to be very important [10] in realizing a timely product that simultaneously satisfies customers needs and generates profits for the manufacturer.

1.7 Systems viewpoint

An important perspective of product realization is obtained by drawing a diagram of the process flow that begins with the manufacture of components which in turn are assembled into subsystems that are combined to generate the system that is sold to the customer (Figure 1.11).

Purchase requests for the basic raw materials needed to start the process flow are rather simple in terms of their information content, e.g. bulk orders of commodities such as steel and plastic feedstocks. The information content of a finished product, however, such as a plane or car or computer is enormous because a large number of parts of different materials and geometry are packaged together in a highly regimented manner. This large increase in information content in the product as it advances through the manufacturing process is associated with an increase in **product order** (or a decrease in the **product entropy**) and is achieved only through considerable work on the part of the enterprise and its suppliers. The entropy of the overall process, of course, must increase according to the second law of thermodynamics.

Once the product is assembled and sold, the order achieved in the product starts to unravel as a result of the dissipative forces of nature – corrosion, erosion, diffusion, evaporation, wear, fatigue, irradiation, ionization, and impact (Figure 1.12). With time, entropy reduces design order and the product no longer functions or functions poorly and must either be repaired or replaced. Nevertheless, the overall birth and death process has a silver lining in a competitive market as the consumer can reasonably expect that the product chosen to replace the old one will be better.

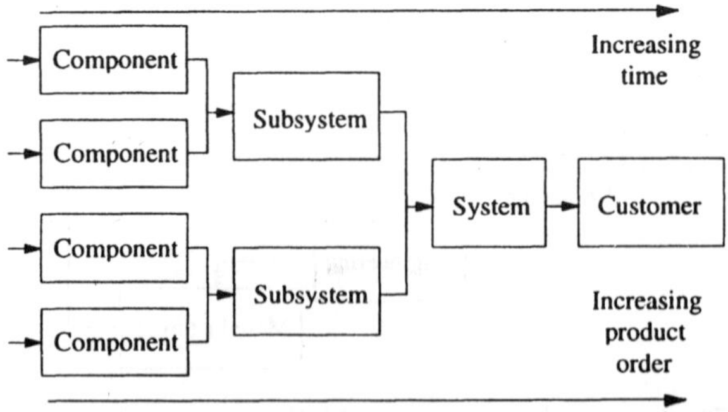

Figure 1.11 Increase of product order with time as a result of manufacture and assembly.

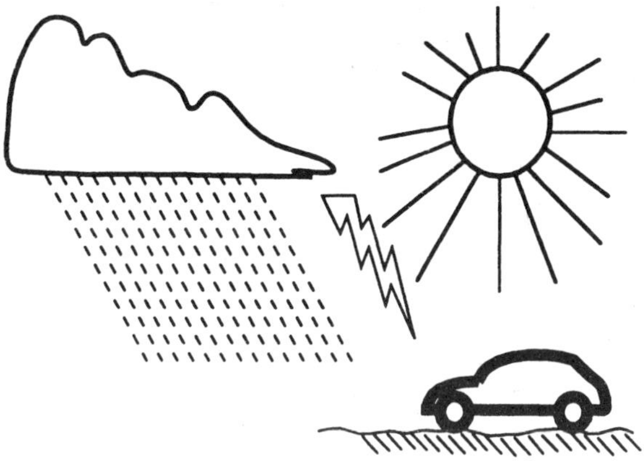

Figure 1.12 The dissapative forces of nature undo the order created in the product.

The inevitability of the demise of every product makes it important to design products with materials that can be recycled. Moreover, major emphasis must be given to the total impact of the product over its entire life cycle on all of society during the planning and development of the product. Thus, if we desire to improve product management, we need to view the process from the perspective of all the stakeholders: the manufacturing enterprise, its customers, its competitors, and the rest of society (Figure 1.13). Each of the four stakeholders has a pivotal role in product realization over its life cycle. The customer is the arbiter of product value and exercises this power when making the choice between purchasing a product from a given enterprise or from its competitor. The enterprise is responsible for the costs of its product. Peter Drucker [11] has expressed the arbiters of value and the sources of cost clearly and succinctly:

> Finally the most important single thing to remember about any enterprise is that results exist only on the outside. The result of a business is a satisfied customer ... Inside an enterprise there are only costs.

The pace of innovation needed by the enterprise to survive, which in terms of our earlier analogy is the minimum speed at which it must run up the performance hill, is set by the pace of its strongest competitors. Society at large is also involved in the transactions between the enterprise and its customers as a result of the impact that product realization has on environmental quality or what economists call **externalities**. Your radio may generate music to your ears but it may be noise to your neighbor. The mine that yielded the ore for the steel to build your factory may have dirtied the stream that was the favorite fishing spot for many. As the world's industrial output and population has grown rapidly over the last century, management of the value added to society of useful products versus the losses in environmental quality generated from their manufacture, use, and disposal has become a major but still poorly researched and understood factor in the product realization process.

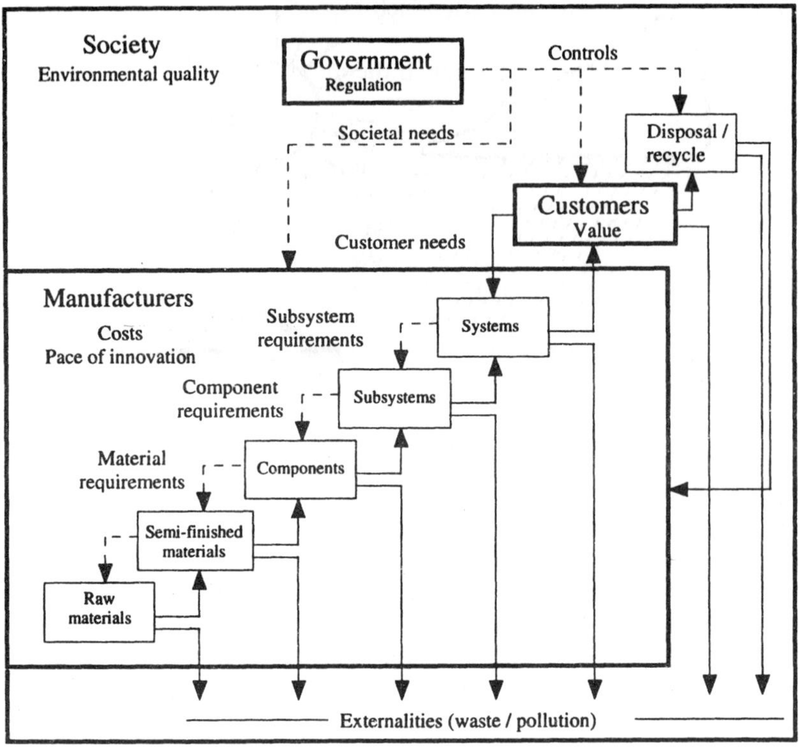

Figure 1.13 Stakeholder involvement and interaction over the entire product life cycle.

1.8 Touchstones for product realization

Although we will devote considerable depth to several aspects of product realization, there are some important touchstones or thought starters which should be considered before beginning work on a new product. The Committee on Engineering Design Theory and Methodology of the National Research Council [12] has compiled an extensive list (© 1991 National Academy Press, used by permission):

- **Customer:** who is the customer? What does he or she really need?
- **Stakeholders:** understand the positions of those who have stakes in the product's success or the status quo.
- **Ease of use:** human factors design needs to be addressed early in the process.
- **Documentation:** essential; match to user's needs; start early.
- **Cultural change:** if the development or production of this product or process requires cultural change, its introduction will not be easy or swift.
- **Patent/copyright:** plan for this early to avoid pitfalls and to get high quality coverage.

- **Legal/regulatory:** consider early. Such obstacles have delayed or damaged many projects.
- **Environmental impact:** determine if the manufacture or use of any product may adversely affect the environment.
- **Manufacturability:** has the manufacturing engineer been on the team?
- **Aesthetics:** these hard-to-define characteristics are also critical.
- **Dynamics:** how does the product or process behave in non-steady state conditions?
- **Testability:** how will the product be tested?
- **Prototypes:** consider how the final product may differ from the prototype if prototype and production processes are not identical.
- **Universality:** universal solutions almost never work.
- **Simplicity:** strive for beautiful, simple designs. They often work well.
- **Appearance:** if the design doesn't look right, watch out!
- **Interfaces:** many otherwise sound designs fail because of unanticipated problems at interfaces.
- **Maturity:** where is the product on its 'S-curve?' Is it time to jump to a new approach?
- **Partitioning:** consider partitioning to provide additional degrees of freedom.
- **Models:** do the mathematical models used in design apply over the anticipated range of use?
- **Scale-up:** do not undertake this lightly. Proceed by small increments.
- **Transportation:** what happens to the product in transportation?

1.9 Robust design

Taguchi's four-step structure (Figure 1.14) for 'robust design' is perhaps the most operationally powerful paradigm for product realization [13]. The first step is discovering customer needs. The flow described earlier of how product order builds up is triggered only after customer needs have been assessed. Once this is done, system design is initiated which involves setting target specifications by balancing customer needs against what is achievable in terms of the cost and capability to manufacture. It is also useful to have an assessment at this point of what competitors may be doing in regard to their product specifications. The third step is parameter design which generally involves experimentation aimed at discovering the right combination of design and manufacturing processing attributes that minimize deviations or variance from the desired target specifications.

Variance is caused by specific, well understood changes in influential variables and also by fluctuations in processing parameters during manufacture and in environmental conditions once the product is in use. A **robust** design is relatively insensitive to uncontrollable or difficult-to-control variations. The criticality of tolerance design – the setting of allowable ranges for product dimensions – is simplified when care has been taken to make the product robust in the parameter

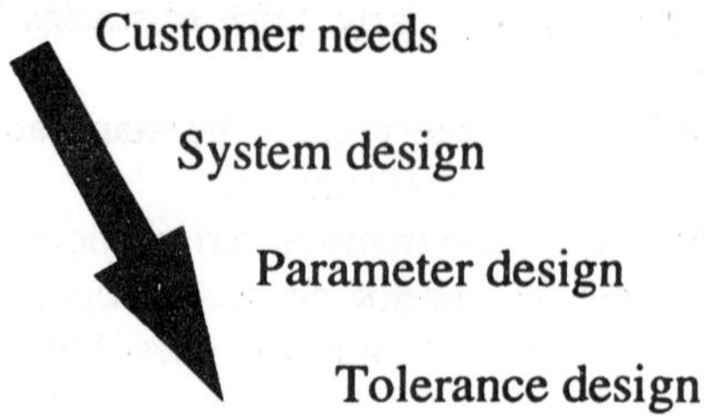

Figure 1.14 Taguchi's paradigm for robust design.

design stage as that generates greater latitude for variation from design specification. When tolerances are set very tightly, manufacturing costs are generally high.

1.10 Innovation

1.10.1 QUANTUM LEAPS

Innovation can be divided into two forms:

1. continuous improvement;
2. discontinuous improvement or 'quantum leaps.'

Both are essential for having a long-term supply of goods of ever-increasing quality. With continuous improvement, the difference between the steps in the evolution of the product is **quantitative** in nature. For example, the formability of a steel is improved by 8% by the addition of a minor constituent and a modification of the heat treatment; the speed of a computer is doubled by increasing the circuit density on the chip, improving packaging, and increasing the clock speed. The alloy is still steel and the chip is still made of silicon. Quantum leaps, on the other hand, represent significant **qualitative** discontinuities in technology such as the replacements of vacuum tube diodes by semiconductors, electronic signals by light pulses, wood-bodied cars by steel, aircraft piston engines by jets, horses by cars, smoke signals by telegraph, telegraph by telephone, and radio by television.

1.10.2 THE S-CURVE

In time, all manufacturing enterprises must consider replacing their existing technologies with new technologies. This is a vexing decision for two reasons. First, the facilities are in place for manufacturing the product using the existing technology and second, the new technology is seldom sufficiently developed to offer a low risk. Richard Foster [14] has described the S-curve relating the performance generated by a new technology versus the effort as measured by the cumulative investment put into the product over time (Figure 1.15). The change to a new technology involves switching to a new S-curve (Figure 1.16). Do we continue to ride technology A or is it now time to switch to B? This is a central question in the management of technology.

Moreover, it may not be clear which new technology is most appropriate to carry the product in the future. Will gallium arsenide semiconductors replace silicon in future integrated circuits, will it be diamond, or will silicon continue to dominate?

Foster [14] has listed several important examples of discontinuous changes in technology that affected major businesses (with permission of McKinsey & Company):

1. Nylon versus polyester tire cord.
2. Sailing ships versus steam.
3. Germanium versus silicon.
4. Mechanical versus digital watches.
5. Electro-mechanical versus electronic cash registers.
6. Propeller versus jet power for aircraft.

Some companies waited too long to change from the old to the new technology discontinuities. The Fall River Ship and Engine Building Company built the *Thomas W. Lawson* in 1902 as a sailing ship hoping that more sails would make it competitive with steam powered ships. The *Lawson* had seven masts and as much sail as possible and could generate 22 knots in favorable winds. However, the *Lawson* was difficult to handle in unfavorable winds and capsized at anchor in a gale in 1907. DuPont stayed too long with nylon while Celanese moved aggressively to polyester tire cord; Texas Instruments and Motorola adopted silicon early but Hughes and Sylvania held with germanium; NCR lingered with electro-mechanical cash registers while Burroughs and others moved to electronics.

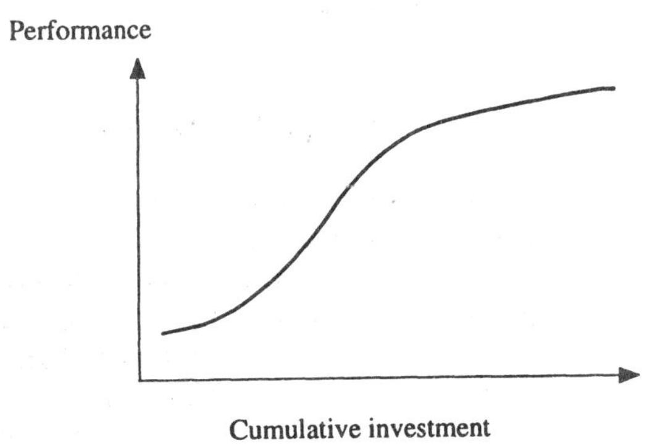

Figure 1.15 S-curve of product performance versus cumulative investment. (Source: R.N. Foster (1986) *Innovation, The Attacker's Advantage*, Summit Books, New York, © 1986 McKinsey & Company. Adapted with permission.)

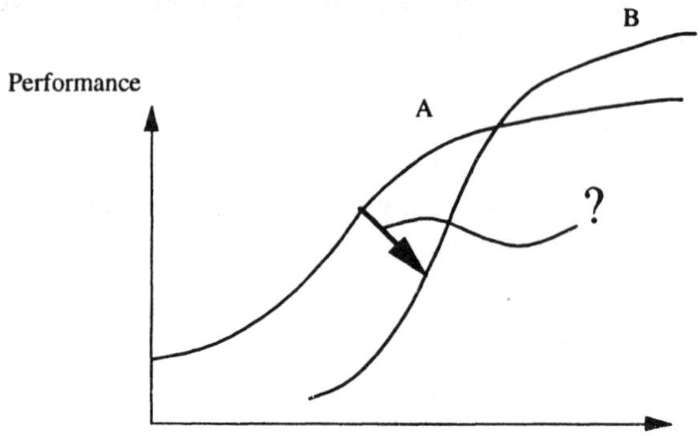

Figure 1.16 The decision to switch to the S-curve for a new technology. (Source: R.N. Foster (1986) *Innovation, The Attacker's Advantage*, Summit Books, New York, © 1986 McKinsey & Company. Adapted with permission.)

The automobile, if the vertical axis of the S-curve is measured only in terms of physical performance such as speed and roominess, seemingly reached its final level of maturity in the 1970s. However, this conclusion would be incorrect because major investments were being made in response to societal concerns for safety and emissions. As a result, the overall performance of automobiles was significantly increased during this period.

When the 1979 energy crisis caused many U.S. buyers to demand small, fuel-efficient cars, the Japanese producers were able to make significant penetration into the U.S. small car market. Because quality improvement was recognized by the Japanese automotive manufacturers as an opportunity during this same period, buyers soon discovered that Japanese automobiles were not only fuel efficient, they also had very high quality. As a result, the Japanese manufacturers were able to enter other automotive market segments with relative ease.

1.11 Product development lead time

The lead time required to bring a new product innovation to market is especially critical whether the nature of the improvement is of the continuous or discontinuous variety. The lead time between the customer ordering a product and receiving it is also very important and should be measured from the actual moment that the customer requests the product to the time the customer receives it. Internal processing times by the enterprise may be competitive but the distribution system may be glacial. There are a variety of important decisions that need to be made prior to the appearance of the new product in the market (Figure 1.17). The time t_r when the decision is made to target the first resources to exploring the concept and feasibility of the new product is the official start of the development of a new product. However, a more critical

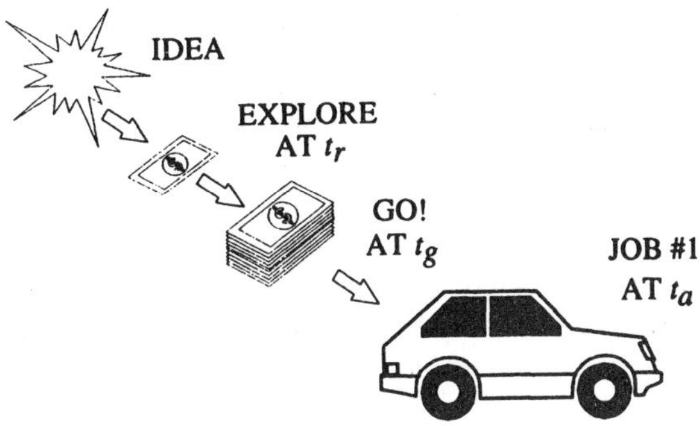

Figure 1.17 Critical design points in the product development process after getting a good idea.

moment is the 'go-time' point, t_g, when the decision is made to take the proposed new product to production which is set for the time t_a.

Thus there are several lead times (L_i) of importance to product development:

$$L_1 = t_g - t_r,$$

$$L_2 = t_a - t_g,$$

$$L_3 = t_c - t_a.$$

Of these, L_3 which measures the difference between your strongest competitor's time, t_c, and your time, t_a, to introduce a product is the most critical. You want L_3 to be long. The next most critical lead time is L_2 which you want to be short. When L_2 is short, you are able to more accurately target the product properly for the customer as tastes can change significantly over time. Also by having L_2 short there is a better chance of having L_3 long. Moreover, many costs are fixed and occur at a constant rate over time and these costs are reduced with L_2 short. Certain development costs may be increased in order to reduce lead time and these need to be offset by the benefits that accrue of having the go decision made as late as possible for a fixed production date, t_a. Having flexibility in manufacturing tools and facilities also helps in reducing L_2 and total development costs.

Ideally you would like to have all laboratory, prototype and pilot development work completed before committing funds for production tooling at time t_g as actual expenditures increase dramatically after the go-time and thus you do not want to be in the position of betting these investments against new concepts that are not well validated or proven out before t_g.

1.11.1 FACTORING PRODUCT DEVELOPMENT LEAD TIME

The combined product development lead times, $L = L_1 + L_2$, can be factored into four major components:

$$L=\tau_s-\tau_p+\tau_w+\tau_c,$$

where τ_s is the lead time required if all tasks, using today's best design tools, were done sequentially and perfectly with all tasks understood and not requiring negotiation, τ_p is the total lead time saved by doing tasks in parallel, τ_w is the increase in lead time caused by wasted efforts, and τ_c is the increase in lead time resulting from conflicts and problems in communication.

A good tool to use in discovering ways to reduce lead time is as-is/to-be analysis. This process begins with the as-is process flow examination of product development (described in Chapter 6) as currently practiced by the enterprise. Surprisingly, many in the enterprise will think that they understand the process today only to find after a careful as-is study that the process is quite different from what they imagined with many opportunities for reducing both lead times and costs by eliminating (1) operations that add little or no value and (2) duplicative operations. As-is analysis is also a way of highlighting interfaces and interactions between units that are suboptimal.

1.12 Customer/company needs loop

The theoretical development of the structured methodology advanced here follows the customer/company needs loop shown in Figure 1.18 [15]. Once identified in terms of the 'voice of the customer,' customer needs must be expressed in terms of changes required in system-level product attributes. The value improvement for the new product is then determined from the changes in its system-level attributes relative to the baseline product. The changes in value and cost are then used in forecasting the price of the new product and demand is forecast from value and price. With this information, the profit expected to be generated by the new product is forecast. Profit from the new product is used to generate the working capital needed to develop improved products in the future to meet the growing expectations of customers. The loop shown in Figure 1.18 is fundamental to product management and can be considered as time-invariant. It provides the high level structure for guiding the work of the enterprise.

References

1. Feldman, E.J. (1985) *Concorde and Dissent*, Cambridge University Press, Cambridge, U.K.
2. Dertouzos, M.L., Lester, R.K., and Solow, R.M. (1989) *Made In America*, MIT Press, Cambridge, MA, p. 44.
3. Hayes, R.H. and Wheelwright, S.C. (1984) *Restoring Our Competitive Edge: Competing Through Manufacturing*, Wiley, New York.
4. Cyert, R.M. (1988) *Leadership in Developing an International Economy*, W.R. Sweatt Lectures, University of Minnesota.

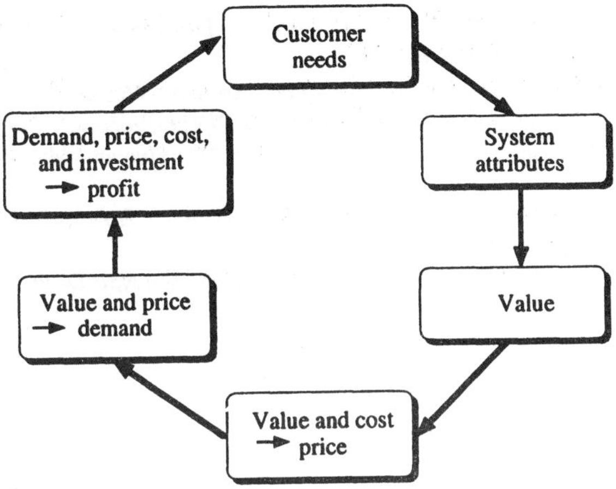

Figure 1.18 Time-invariant customer/company needs loop.

5. Bennis, W. (1989) *Why Leaders Can't Lead*, Jossey Bass, San Francisco, CA.
6. Dertouzos, M.L., Lester, R.K., and Solow, R.M. (1989) *Made In America*, MIT Press, Cambridge, MA, p. 118.
7. Womack, J.P., Jones, D.T., and Roos, D. (1990) *The Machine That Changed The World*, Rawson Associates, Macmillan, New York, pp. 48–69.
8. Kuhn, T.S. (1970) *The Structure of Scientific Revolutions*, 2nd edn, University of Chicago Press, Chicago, IL.
9. Drucker, P.F. (1989) *Innovation and Entrepreneurship*, Harper and Row, Perennial Library Edition, New York, Chapters 1 and 2.
10. Hayes, R.H., Wheelwright, S.C. and Clark, K.B. (1988) *Dynamic Manufacturing*, Free Press, New York.
11. Drucker, P.F. (1989) *The New Realities*, Harper and Row, New York, p. 87.
12. Hoover, C.W. and Jones, J.B. (eds) (1991) *Improving Engineering Design*, National Academy Press, Washington, DC, p. 31.
13. Taguchi, G. and Wu, Y. (1980) *Introduction to Off-line Quality Control*, Central Japan Quality Association, Nagoya, Japan.
14. Foster, R.N. (1986) *Innovation, The Attacker's Advantage*, Summit Books, New York.
15. Cook, H.E. (1996) A unified decision support tool for product management, in *Life Cyle Modeling for Innovative Products and Processes* (eds H. Jansen and F.L. Krause), Chapman & Hall, London, pp.146–157.

Further reading

An incomplete list of books is shown below which provide important insight and perspective to the overall product realization process:

Baxter, M. (1995) *Product Design*, Chapman & Hall, London.

Clark, K.B. and Fujimoto, T. (1991) *Product Development Performance*, Harvard Business School Press, Boston, MA.

Cooper, R.G. (1993) *Winning at New Products*, Addison-Wesley, Boston, MA.

Ertas, A. and Jones, J.C. (1993) *The Engineering Design Process*, Wiley, New York.

Moore, W.L. and Pessemier, E.A. (1993) *Product Planning and Management*, McGraw-Hill, New York.

Pardee, W.J. (1996) *To Satisfy & Delight Your Customer: How to Manage for Customer Value*, Dorset House Publishing, New York.

Schonberger, R.J. (1996) *World Class Manufacturing: The Next Decade*, Free Press, New York.

2 Motivation and consumer behavior

2.1 Product appeal

The value of a product will almost always contain a 'soft' component whose appeal is subjective. Consequently, an understanding of the motivational and behavioral aspects of customers is vital in developing and marketing a successful product. In fact, if the product is a work of art, we might expect its appeal to lie almost entirely on the softer side. On the other hand, if the product is a road grader, we might expect that the customer, a construction company, will be able to compute the value of a grader rather precisely based upon its performance specifications.

Now let's re-examine these positions. Even with the road grader, the manufacturer should also expect that some degree of softness will enter the buyer's decision process. Perhaps the experience of the construction company with the manufacturer's products in the past has been favorable which will effectively add additional value to the product after the importance of the more tangible performance specifications have been weighed. Or perhaps the soft value arises from the fact that the manufacturer has sponsored a favorite television program of the owner of the construction company. Both of these soft attributes can be grouped under brand loyalty. By contrast, the value of a work of art can be as commercial as that of a road grader when viewed only for its 'investment quality' based upon its recent auction price and its price history or the price history of similar artwork, the 'name value' of an established artist being the most significant factor in generating its market appeal. To a lesser extent, but nevertheless importantly, the 'name values' of manufacturers also contribute to the value of road graders and many other products.

Then how does one go about combining a product's soft appeal with its technical performance specifications to arrive at an overall assessment of the value of a product? Dupuit [1] showed how this could be done using the concept of the customer's 'willingness to pay.' For example, during the development of new or improved products, persons representative of the market segment of interest are given a chance to evaluate prototypes and afterwards are asked to state their willingness to pay for such a product.

2.2 Human needs

A somewhat tautological premise guiding product development is that the things which appeal to customers are those things which satisfy their needs. Maslow argued that we are motivated to satisfy a hierarchy of needs [2]. The most basic of these are physiological – hunger, thirst, sleep, sex, maternal behavior and communication – which must be first reasonably satisfied before higher needs are considered. They can be ordered in terms of the arrow of satisfaction (Figure 2.1).

When the physiological needs are satisfied, safety becomes the next concern. Love of family and friends and the feeling of belonging to a group is the next need. Self esteem and the desire of the esteem of others follows love. We want to confidently go about our daily business and have the admiration of others as we deal with the burdens and vicissitudes of life. At the top of the pyramid is the need to accomplish through our daily experience those things which we envision are best suited to our strengths. We desire self-fulfillment. Importantly, the society which we live in needs to give us sufficient freedom and opportunity, including the freedoms to learn and to act, if we are to satisfy all of the needs in the hierarchy.

2.3 Aggregate behavior

Consumers' spending patterns give us insight into the importance of many of the needs described in Maslow's theory. As seen in Figure 2.2, housing is the dominant expenditure followed by transportation and food [3]. Food and housing reside in levels one and two, respectively, of Maslow's hierarchy. What perhaps may be unexpected in the chart is that the percentage of income spent is roughly constant across income levels, an empirical fact, well known to behavioral economists. As might be expected though, the lower income group spends somewhat more, proportionally, for the necessities of food and housing. Because of the relatively large percentage of income left over once the basic necessities have been met, significant financial resources are available for those included in the income groups in Figure 2.2 to pursue the higher needs enumerated by Maslow.

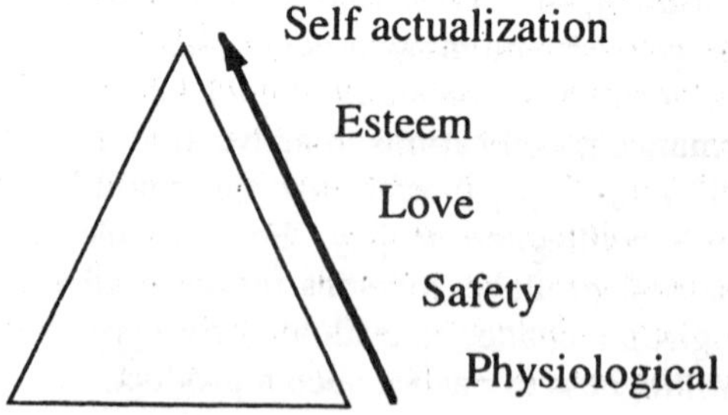

Figure 2.1 Maslow's hierarchy of needs.

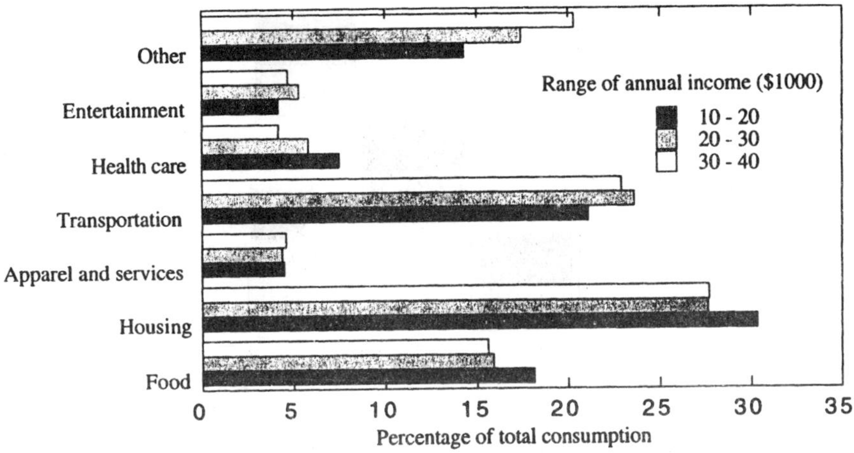

Figure 2.2 How consumers spend their money. (Source: U.S. Department of Labor, Bureau of Statistics, Second Quarter 1988, Report 757, as summarized by Salvatore [3].)

Because a fixed percentage of income is spent across income levels, more expensive items or more items or both are purchased in each category as income is increased. Differences in aggregate consumer purchase patterns as a function of their segmentation by income are reflected in the type of brands purchased (e.g. Escort, Lumina, Taurus, Accord, Buick Roadmaster, Cadillac Seville, Lincoln Town Car, Mercedes 300SL and Lexus) within a product class (e.g. automobiles).

An understanding of consumer behavior by market segment is important because the overall market is not homogeneous but it is impossible in a large market to track the interests of specific individuals. Thus division into segments is a compromise in which individuals are grouped according to income, education level, age, gender, geography, job, lifestyle, etc. Within each segment, consumer behavior and preferences are assumed to be similar – the variance in behavior being small enough that products can be developed that will be worthy of consideration by a major fraction of the segment. If a product is developed in hopes of appealing to several different consumer segments, it runs a risk of appealing to none. Market segmentation by product value, price, and lifestyle (spirited and conservative) is shown in Figure 2.3. There are many other lifestyle segments including the types of recreation and hobbies a person has; the type and frequency of vacations taken; the frequency of eating outside the home; the manner of dress; location of home in city, suburb, or country; residing in single-family dwelling or apartment; heavy user of a certain type of product; ownership of a certain kind of pet animal; number and age of children; reader of fine books; and nature of occupation.

In the fall semester of 1991 students who were at UIUC and at General Motors were surveyed separately as to their perceptions of the positioning for

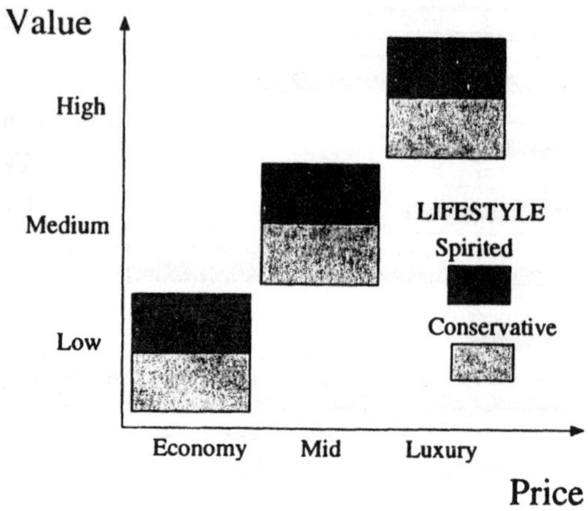

Figure 2.3 Market segmentation by value, price, and lifestyle (spirited versus conservative).

a variety of vehicle brands. The results are shown in Figures 2.4 and 2.5. From conservative to spirited and from affordable to luxury, the positions stated are very similar for both groups. Because Lincoln and Cadillac have a strong image for luxury that appeals to the conservative lifestyle segment, they may experience problems in successfully generating new models that appeal to those in the spirited lifestyle segment.

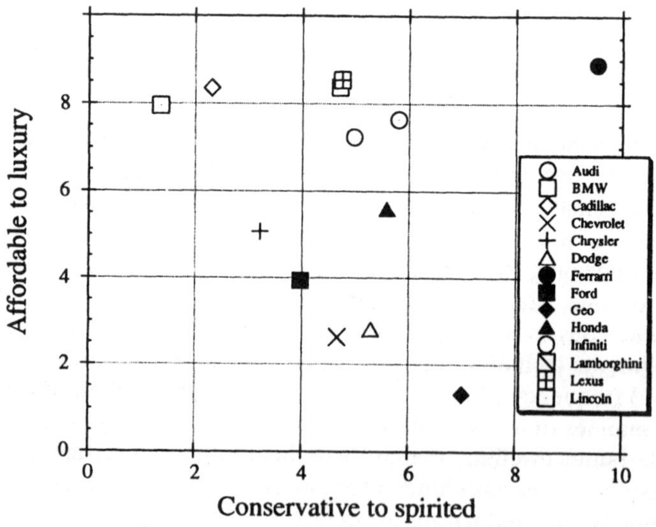

Figure 2.4 Positioning of vehicle brand names by UIUC students.

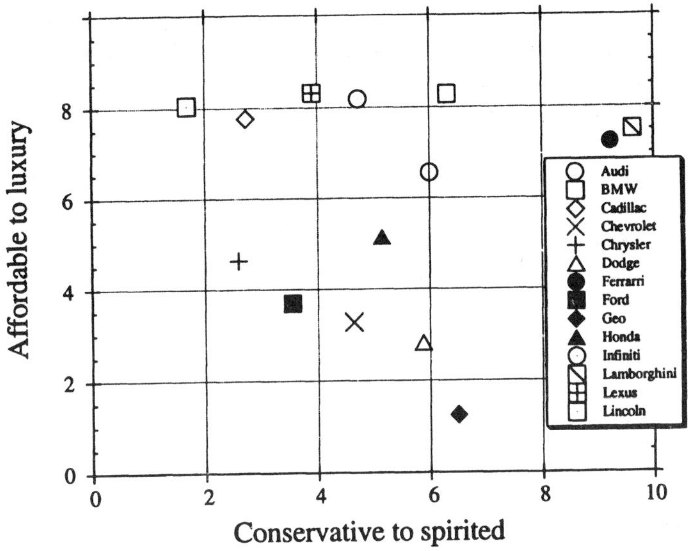

Figure 2.5 Positioning of vehicle brand names by GM students.

2.4 Flow charting buyer behavior

It is useful to apply the flow-chart methodology of system analysis in considering buyer behavior which is described by a box or boxes having arrows denoting input, output and controls. Each box represents a task which, using the set of tools assigned to the task, transforms the input received to the required output. A single transformation box is shown in Figure 2.6. When a sequence of discrete tasks are used to make the transformation, arrows denoting feedback and feed forward between the tasks would also appear.

For example, when the low-fuel-level light appears as you are driving a car, it acts as an input stimulus or cue which generates an output response to buy

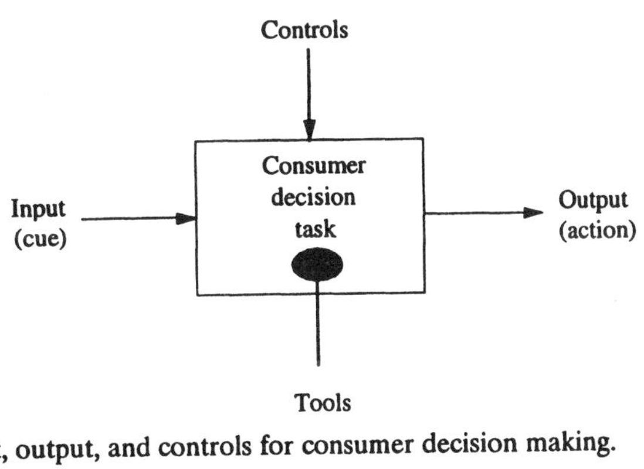

Figure 2.6 Input, output, and controls for consumer decision making.

fuel. The controls will be availability of stations, various prices of fuel, and type of credit card which you carry. The tools used to make the transformation from a nearly empty tank to a filled tank will be the vehicle itself to help locate a station, and you, the driver, for pumping the gas.

The actions leading to customer purchases generally fall into the category of learned responses rather than reflexive actions such as the leg jerk when the knee is tapped or the conditioned response of students gathering their notes and books when the class bell rings. The consumer also often operates in a very open environment in that the product purchased is conspicuous, examples being a new house, a new car, or a new suit. Learning takes place by experience with the product and feedback from peer groups after a purchase is made. Do I like the product after buying it? Do my peers accept or reject my action? In Figure 2.7 the dotted line is used to indicate that the output action is conspicuous to the peer group. The wish not to be rejected by your peer group is a reflection of the natural desire for acceptance and companionship [4] which is found in level three of Maslow's hierarchy.

More importantly, the anxiety of peer response can be reduced by using a feed forward condition as shown in Figure 2.8 in which the potential customer studies what his or her peer group is driving or wearing and purchases accordingly.

Although risk is a subject often considered by the manufacturer when investing in a new product, it is clear from the above discussion that the consumer also takes a social risk in buying a new product. In their consideration of consumer risks, Louden and Della Bitta have identified six types [5]:

1. **Financial risk:** the product may require much heavier sums for maintenance or wear out much sooner than expected.
2. **Performance risk:** the product may fall well short of performance expectations.
3. **Physical risk:** the product may be unsafe.
4. **Psychological risk:** the image conveyed by the product may not be that intended.
5. **Social risk:** the peer group may not approve of the purchase.
6. **Time-loss risk:** the product may require your constant attention to maintain performance.

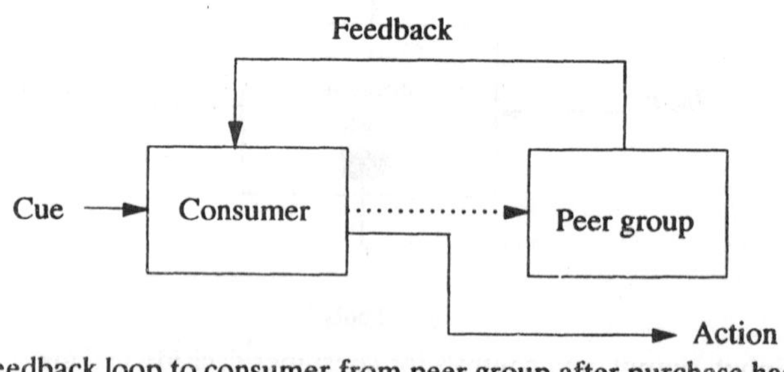

Figure 2.7 Feedback loop to consumer from peer group after purchase has been made.

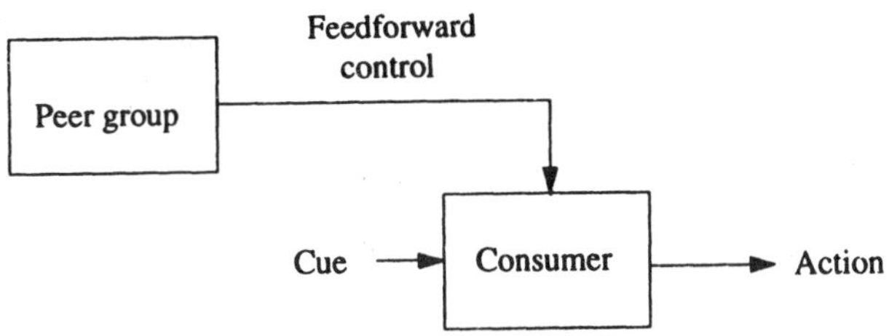

Figure 2.8 Feed forward to consumer from peer group before purchase is made.

The current state of mind of the consumer is also a very important influence in his or her buying behavior. If recent experience engenders pessimism and insecurity, for example, during a recession when lay-offs are occurring, the consumer will not be in the mood to buy, whereas consumer optimism and security promote spending [6]. Also aspirations tend to grow when achievements are being made and decline when they are not.

One area of study has been the relationship between personality – aggressive, passive, secure, insecure, domineering, charming, warm – and consumer behavior. Roughly 5–10% of buyer behavior correlates to personality traits. Although small, it is certainly not insignificant [7].

2.5 Three stages of problem solving

Howard and Sheth [8] have defined three phases of the consumers' cognitive processes (Figure 2.9) over the lifetime of a product which describe the process used by the consumer to manage risk:

- **Extended problem solving (EPS):** should I buy this new class of products? (Example: instant coffee versus ground coffee.)
- **Limited problem solving (LPS):** which brand in this class of products should I buy? (Nescafe, Maxwell House, etc.)
- **Routine response behavior (RRB):** I like brand 'X' and will buy it habitually or routinely now. ('Honey, we're out of Taster's Choice. Would you pick some up on your way from work?')

The input or stimulus or cue for the EPS stage of consumer behavior is the appearance of a new class of product. A control perceived by the consumer in the use of instant coffee when it was introduced in the 1940s and 1950s was 'Will I be considered lazy by family and friends if I serve instant coffee?' Sellers over time have countered this concern by pointing out that such time-saving products give you more time 'to be with those you love.'

The stimulus or cue for LPS is which brand to favor once you have decided on the product class. The cue for RRB is being out or nearly out of the preferred brand. The concept of a brand, its identification and the development of

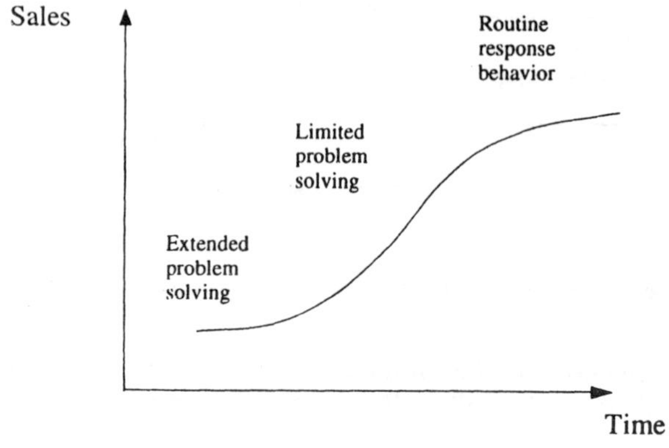

Figure 2.9 Sales history over the three stages of problem solving. (Source: adapted from J.A. Howard and J.N. Sheth (1969) *The Theory of Buyer Behavior*, Wiley, New York.)

loyalty to the brand in the RRB stage is important in influencing consumer purchases. The brand will have a name and often a logo: Dodge – ram; Apple Computer – apple; NBC – peacock; U.S. Postal Service – eagle; Ford – blue oval; Pontiac – red arrow.

Competing brands within a segment are generally positioned close to each other on charts of key attributes. For example, there are two key segments for luxury cars. Amero Lux vehicles are designed for the famed 'boulevard ride' and Euro Lux are more performance oriented in terms of speed and handling. Amero Lux vehicles are best suited for long journeys on well designed interstates and Euro Lux vehicles are more suited for high-speed autobahns as well as the alpine curves and tighter streets of Europe. The manner in which representative vehicles of the two segments are positioned on a horsepower versus weight chart is shown in Figure 2.10. The Euro Lux vehicles follow a relationship that yields a higher horsepower-to-weight ratio versus the Amero Lux vehicles.

Performance attribute positioning represents part of the physical positioning of the product. Paint gloss, fit, and finish are also part of the physical packaging for an automobile as well as the under-the-hood look of the engine compartment for the more studious consumer. The physical package can improve on or detract from the overall acceptance of the product. The product's psychological positioning is determined by the manner in which the brand is advertised and distributed. The dealerships for several new vehicle brands – Saturn, Lexus, and Infinity – have gone to special lengths to generate a less stressful atmosphere and to show more true concern for the customer than ordinary dealerships. The product's visual package relates to the style of the product. The three types of positioning should be in harmony for a given product and combine to make an overall statement by the brand to the consumer.

When there are few solid performance attributes with which to differentiate competing brands, the responsibility falls heavily on psychological packaging

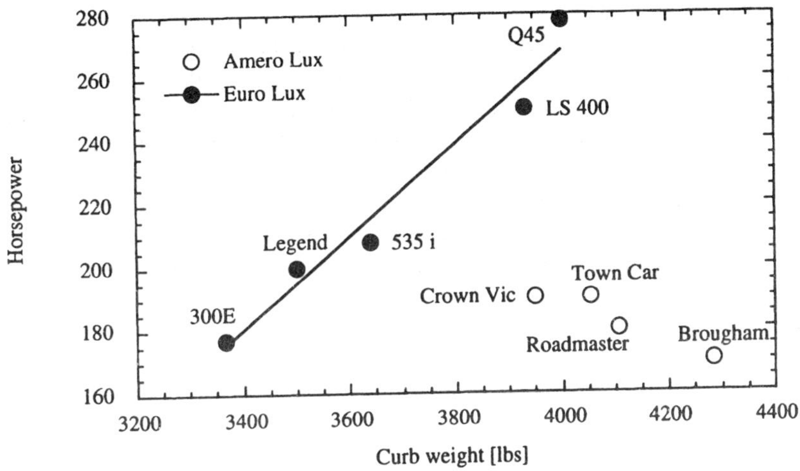

Figure 2.10 Positioning of Amero Lux and Euro Lux vehicles by horsepower and curb weight.

through advertising to create the statement or image. For example, Marlboro cigarette advertisements appeal directly to rugged individualism (cowboy) and indirectly to virility [9] in its users; Pepsi, 'uh-huh,' is the drink of the young and spirited; and Coke brings friendship all over the world. These attempts at differentiation are needed to promote brand loyalty because the products in each class are close in tangible attributes.

Therefore, the difference in Figure 2.11 between the sales rate of brand 'X' and brand 'A' beyond that accounted for by differences in tangible attributes can be ascribed to the soft 'residual' preference for 'A' over 'X.' A specific example of this would be the consumer who will either (1) never buy a foreign car versus domestic or (2) never buy a domestic car versus foreign even though in terms of hard attributes many foreign and domestic cars are very similar.

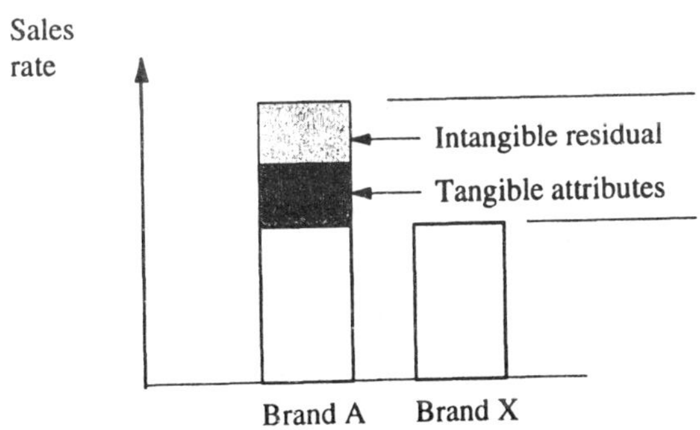

Figure 2.11 Hypothetical contribution of tangible and intangible (residual) attributes to sales rate differences.

The soft attribute may be the result of poor experience in the past with a vow never to buy that brand again. In such a situation the consumer is into a form of routine response behavior against the former brand. In the early 90s, a well-known automobile executive took a direct approach to the issue of relative quality of American automobiles versus import: 'We built a lot of crap!' referring to vehicles built in the late 70s and early 80s but claimed that the quality of the cars currently built by his company to be as good as those from Japanese competitors, even better with airbags and lower prices. Another U.S. manufacturer took a less strident approach by asking if you had driven one of their products 'lately' implying that significant improvements had been made since you did. Patience is needed by both because customers 'lost' to foreign manufacturers will have to move away from their favorable routine problem solving with their new brand and engage once again in the limited-problem-solving phase in reconsidering the brand to buy. Conversion back to a prior brand could be a lengthy process.

2.6 Behavioral economics

2.6.1 ELEMENTARY UTILITY THEORY

When faced with a decision between alternatives, a person is expected to choose the alternative which will give him or her the greatest utility. In their seminal work, von Neumann and Morgenstern [10] introduced the concept of a lottery to arrive at a dimensionless measure of utility, $U(g)$, for the attribute, g, under consideration involving uncertainty. The attribute ranked highest is usually given a utility equal to one, $U(g_{max})=1$, and the baseline or minimum attribute given a utility equal to zero, $U(g_{min})=0$. The distinction between utility and value is not profound [11] but their methods of measurement are different. Some prefer to use the term utility when the choice involves risk and value when it does not.

A person's expected utility for a single attribute (often expressed in textbook problems as an amount of money) is written as:

$$\begin{aligned} U(g) &= h(g_{max}; g)U(g_{max}) + [1 - h(g_{max}; g)]U(g_{min}) \\ &= h(g_{max}; g) \end{aligned} \tag{2.1}$$

where $h(g_{max}; g)$ is the probability determined by the person being questioned that makes this individual indifferent to having g as a certainty versus being given one chance in a lottery of receiving the ideal attribute with the probability $h(g_{max}; g)$ and the least desirable attribute with the probability $[1 - h(g_{max}; g)]$. In writing Equation 2.1, we make the assumption that utility increases in a monotonic fashion with g.

An example of a hypothetical utility curve is shown in Figure 2.12 for g_{max} = 4 and g_{min} = 0. For a given g, the utility $U(g)$ for a given individual is equal to the

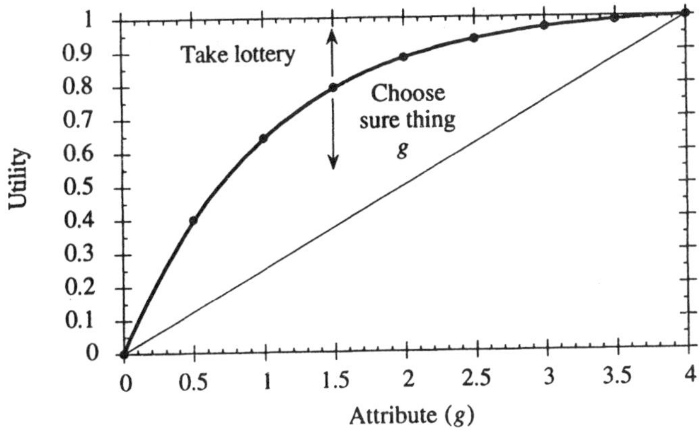

Figure 2.12 Determining the utility function of a single attribute from the indifference point between choosing a sure thing and taking the lottery.

indifference probability $h(g_{max}; g)$. Assume that the attribute in Figure 2.12 represents money in thousands of dollars an individual would be paid and we wish to determine the utility of \$1500. We could use the survey shown in Figure 2.13 for this. It asks the respondent to make five separate choices between a sure thing of \$1500 on the left and a lottery on the right of winning \$4000 or nothing for the stated probabilities. As the respondent moves down the list, the odds of winning increase.

\$1500 for sure ☒	⇐ select one ⇒	30% chance of winning \$4000 70% chance of winning \$0 ☐
\$1500 for sure ☒	⇐ select one ⇒	50% chance of winning \$4000 50% chance of winning \$0 ☐
\$1500 for sure ☒	⇐ select one ⇒	70% chance of winning \$4000 30% chance of winning \$0 ☐
\$1500 for sure ☐	⇐ select one ⇒	90% chance of winning \$4000 10% chance of winning \$0 ☒
\$1500 for sure ☐	⇐ select one ⇒	99% chance of winning \$4000 1% chance of winning \$0 ☒

Figure 2.13 Survey for determining the utility of \$1500 for a single individual.

In Figure 2.13, the respondent has checked a preference for the sure thing versus the lottery up to and including the 70% chance of winning the $4000 but switched to the lottery at the 90% chance level of winning. This indicates that the individual would be neutral between the choice of a sure thing of $1500 versus an 80% chance of winning $4000. This neutral or indifference probability, according to Equation 2.1, represents the respondent's utility for $1500 on a scale where the utility of $4000 is unity. Other sure thing amounts between $0 and $4000 can be evaluated in the same manner to map out the respondent's entire utility curve as shown in Figure 2.12.

The expected value of the lottery is equal to the award multiplied by the chance of winning the award. A person is deemed risk averse if he or she chooses a sure thing of value *g* when it is less than the expected value of the lottery. A person that chooses the lottery when *g*, the sure thing, is greater than the lottery's expected value is deemed a risk taker. Not surprisingly, most persons given the opportunity for a desirable sure thing are risk averse. The fact that the utility curve in Figure 2.12 is above the straight line indicates that the respondent was risk averse.

Now let's make the utility curve in Fig 2.12 represent something less artificial by assuming that (1) the attribute *g* is the mach number for the top speed of future military fighter aircraft and (2) that the individual questioned in the lottery was an air force general who, wise in the way of air combat, was the decision maker for the specifications of the next generation of fighter aircraft. The maximum top speed that the general felt would be effective for such aircraft was mach 4.

When offered the lottery of a sure thing at *g* equal to mach 1, the general said he was neutral to a development program yielding a 65% possibility of a mach 4 aircraft and a 35% probability of a program failure yielding no aircraft. The utility curve was developed from eight such lotteries over the range of 0 to 4 mach.

The resulting utility in Figure 2.12 is seen to be almost as much (90%) for an aircraft that has only half the effective top speed as it is for the top performing aircraft. This is an example of diminishing returns in regard to top speed. Since the cost of developing and manufacturing an aircraft at mach 2 is likely to be several times less than for the mach 4 aircraft, the best procurement decision would likely be to develop the mach 2 aircraft.

Now the outcomes of real air combat situations depend not only on the performance of the aircraft but on the skills of the pilots which, for example, can be developed from extensive 'top gun' mock air combat training. Well-trained pilots usually win such battles against less skilled pilots flying aircraft having significantly higher performance. The point being made here is not military tactics but to stress that the full system needs to be evaluated in assessing the value of any product.

2.6.2 PROSPECT THEORY

In their studies of persons making decisions under uncertainty, Kahneman and Tversky have found behavior to be more complex than that assumed by the axioms of utility theory [12]. One of the tenets of utility theory is that the utility curve is always concave downwards as shown in Figure 2.12. Another is that the utility for a loss of $-g$ is simply minus one times the utility for a gain of g.

Consider the following prospect and determine which alternative, A or B, you would choose:

- **A:** a loss of \$50 000 for certain;
- **B:** a lottery having 50% chance of losing \$100 000 and 50% chance of losing \$0.

Now consider the mirror image of this prospect:

- **A:** a gain of \$50 000 for certain versus,
- **B:** A lottery having 50% chance of winning \$100 000 and 50% chance of winning \$0.

If you chose A in one of the above prospects and B in the other, then, for you, $|U(-g)| \neq |U(g)|$. By considering a variety of prospects of this nature, Kahneman and Tversky arrived at the following important points:

- The absolute value of a loss of $-g$ is greater than the positive value of a gain of $+g$.
- The value of a prospect is a function of both the gain, g, and the probability, h, of the gain.
- The 'framing', i.e. the structure or description of the prospect, significantly affects the resulting value associated with it.
- The value of a prospect is influenced by the nature of the baseline condition.

A consequence of the above findings which has been amply demonstrated [13] is that people state that they are willing to pay an amount x for a specific good that they do not already have, but, if they already have the good, they state that the price that they are willing to accept for the good is considerably more than x. This is an expression of the so-called 'endowment effect.' It is related to the fact that when someone wants to buy something from you, you need to be compensated for the loss of that good. But when you buy something, you are simply paying for a gain which is less in absolute value to the equivalent loss.

A prospect, in general, is a lottery having one of two possible payoffs. One payoff g has a probability h of occurring and the other payoff g' has a probability h' of occurring. Both g and g' are measured from a common baseline. Two equations were found for the value of a prospect depending upon the nature of the probabilities. One is given by:

$$V(g, h; g', h') = \pi(h)V(g) + \pi(h')V(g') \tag{2.2}$$

which applies for a prospect for which $h + h' < 1$ or $g \geq 0 \geq g'$. The other is given by:

$$V(g, h; g', h') = \pi(h)[V(g) - V(g')] + V(g') \tag{2.3}$$

which applies when $h + h' = 1$ and the prospect is strictly positive, $g > g' > 0$, or when it is strictly negative, $g < g' < 0$. Neither equation for the value of a prospect is identical to the classical utility expression, Equation 2.1. The function $\pi(h)$ weights the probability, h, with respect to the desirability of the prospect under consideration. Value is measured from a baseline reference point and is a function of this reference point. A hypothetical value curve for prospect gains and losses relative to a baseline reference point (undefined) is shown in Figure 2.14.

The weighting functions have been found to have the following properties:

$$\pi(rh) > r\pi(h)$$
$$\pi(h) > h \text{ for low probabilities}$$
$$\pi(h) + \pi(1 - h) < 1$$

and are not well defined at the upper and lower limits on h (1 and 0, respectively).

The consequences of the findings of prospect theory are of considerable importance to product management. The value curve is concave in the positive regime and convex in the negative regime because the value of a gain or loss is in relation to its proportional impact and not to its absolute impact which is an expression of the psychophysical principle known as Weber's Law.

It also follows that products which do not meet customer expectations are more costly to the manufacturer in terms of the losses in customer satisfaction than they might seem when evaluated in strictly economic terms. An unexpected repair that cost the manufacturer a certain amount in warranty to fix may well have generated an even larger loss of perceived value to the customer. Moreover, when the customer has to pay for an unexpected repair the loss is doubly painful. Another consequence is that negative information about a product can easily offset what, on the face of it, may seem to be an equal amount of positive information about it. A related phenomenon is that effective political campaigns are based more on 'dirty tricks' than a positive

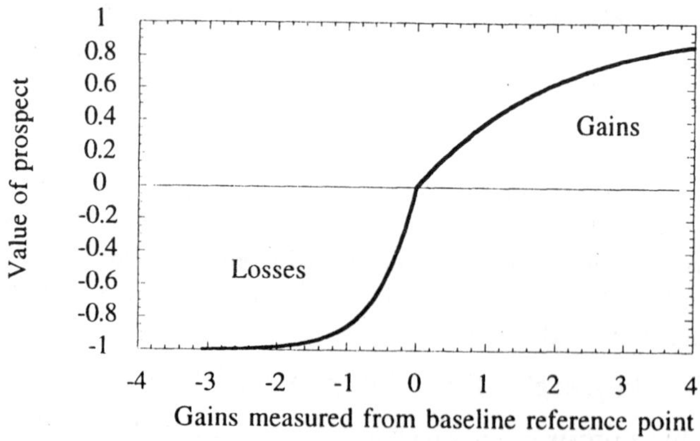

Figure 2.14 Schematic representation of the value of a prospect relative to a baseline point.

approach to the issues [14]. Also in making decisions between alternatives, persons often disregard the things that the alternatives have in common and focus on the things that differentiate them [15].

2.6.3 CONTINGENT VALUATION

Another way of evaluating utility or value is simply to follow Dupuis' insight and ask people what they are willing to pay. This approach, which is very direct, is known as contingent valuation [16]. Instead of determining the utility of $1500 using Figure 2.13, let's use the contingent valuation approach based upon the survey in Figure 2.15 to determine the value of a lottery ticket having an award of $4000 with the odds of winning set at 80%.

The respondent is asked to choose between not buying the lottery ticket by checking the box on the left and buying the lottery ticket at the price shown by checking the box on the right. The respondent in Figure 2.15 would purchase the lottery ticket at the prices offered up to and including $1600 but would switch to not purchasing the ticket above this price. The contingent valuation survey in Figure 2.15 discovered the price switchpoint given an 80% chance of winning

Lottery ticket: $4,000 award, 80% chance of winning

Not buy		Buy
☐	Price = $400 ⇐ select one ⇒	☒
☐	Price = $800 ⇐ select one ⇒	☒
☐	Price = $1200 ⇐ select one ⇒	☒
☐	Price = $1600 ⇐ select one ⇒	☒
☒	Price = $2000 ⇐ select one ⇒	☐
☒	Price = $2400 ⇐ select one ⇒	☐
☒	Price = $2800 ⇐ select one ⇒	☐
☒	Price = $3200 ⇐ select one ⇒	☐

Figure 2.15 Contingent valuation survey for the value of a lottery ticket.

$4000. This is the inverse of the utility problem of discovering the switchpoint given the amount of the sure thing. Not surprisingly, the $1800 switchpoint found in Figure 2.15 is near the sure thing of $1500 found in Figure 2.13.

The two surveys, nevertheless, measure different things because their framings are different. The survey in Figure 2.13 yields the respondent's utility of 0.8 for $1500 on a scale of $4000 equal to unity, whereas the survey in Figure 2.15 yields the maximum amount the respondent is willing to pay for a lottery ticket having an award of $4000 with 80% odds of winning. The contingent valuation approach is measuring something more tangible and direct, the price of a lottery ticket with the award and odds clearly stated. The first survey yielded something more abstract, the utility of a certain amount of money on a scale of $4000=1 utile. The choice of $4000 was for illustrative purposes only and was arbitrary. When developing the utility for an attribute in a real problem setting, the upper limit for the attribute should be set at the level considered by the decision maker to be ideal as was done for the problem regarding the speed of future combat aircraft.

2.6.4 PRODUCT VALUE

Product value as discussed above is a vital concept for the stakeholders in the product realization process to grasp and we need to arrive at a clear, quantifiable definition for value before going further. The simplest approach might seem to define value as the price that a customer paid for a product. But the customer might have also purchased the product if the price had been higher. If value is a fundamental property of a product, then it can not be set equal to its price which can be arbitrary. We need a definition for value that is operationally sound in that we can determine it in an unambiguous manner and yet allows for the fact that consumers view it subjectively.

The measure of product value which we will use can be understood by considering the simple transaction between buyer and seller shown in Figure 2.16 which is producing smiles for both.

The reason for the smiles on the seller and buyer is that both are receiving a net gain or net value from the transaction. The buyer receives a net value equal

Figure 2.16 The three stakeholders – seller, buyer, and the rest of society – in a market transaction.

to the value, V, gained through the use of the product less the price, P, paid to the seller, and the seller receives a net value equal to the price minus the cost, C, needed to manufacture the product. If price exceeded value, the buyer would not purchase the product. If price is less than V, then there is a finite probability that the buyer will purchase the product and, as the difference between V and price increases, the probability of purchase increases.

The gains for the seller and buyer can be expressed as free value. For the seller, free value is $P - C$ and for the buyer it is $V - P$. If the buyer and seller bargain with equal strength in arriving at a mutually agreed balanced selling price $P = P_B$, in the sense that the buyer and seller receive an equal amount of free value, we can write:

$$P_B - C = V - P_{B'} \qquad (2.4)$$

and it follows from this pricing strategy for the simple market transaction shown above that the balance price lies halfway between cost and value:

$$P_B = (V + C)/2. \qquad (2.5)$$

There is also an additional stakeholder to the transaction, which is the rest of society (Figure 2.16). Society is concerned with protecting its own welfare vis-à-vis undesirable externalities of market transactions. The aggregate effect of the transactions can be a loss of environmental quality, G.

2.7 Consumer behavior assumptions of microeconomic theory

Economists have traditionally taken a more rational view of consumer behavior than psychologists. The key behavior of consumers from the viewpoint of classical microeconomic theory is that demand for a product generally decreases as price increases. There can, of course, be exceptions to this behavior such as when a hurried individual customer is unfamiliar with a brand in a product class and chooses the brand with the higher price on the belief that it will likely have superior quality. Urgency compresses the three stages of problem solving. With time available for sampling competing products, it is a very good assumption that the brand with the lower price will be favored in a given market segment, other things being equal.

Demand for a product as a function of price is expected to be convex downwards as shown in Figure 2.17. (As a result of historical precedence, economists draw the supply and demand curves differently with the price axis vertical and the demand axis horizontal.) The intercept on the price axis at zero demand is known as the reservation price [17], P_{Res} (= 10). In most markets the actual price will lie somewhere between the limits as defined by the manufacturer's cost, C, and the price, P_{Res}, as neither limiting condition is favorable to generating a profit. At a price, P, below cost, C, the supplier is losing money and at $P = P_{Res}$ there is no demand.

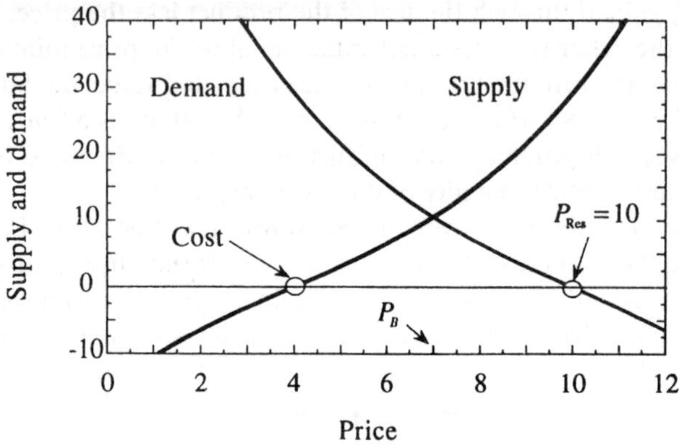

Figure 2.17 Schematic representation of supply and demand as a function of price showing the variable cost and reservation price.

2.8 Aggregate product value for a market

Our procedure for establishing the aggregate value of a product in a real market for a monopoly is shown by the graph in Figure 2.18. The real value curve is shown as convex downwards and a tangent has been drawn from the curve from the point on it where $P = 6$. The price where this line intercepts the zero demand line is defined here as the value of the product at price $P = 6$. For a curving demand line, value will not be fixed as a function of price and as such it behaves like a marginal quantity. The fact that a product may not have a fixed value should not be troubling as it is an indicator that the product has multiple values. Water, for example, is used for many different purposes, to drink, to wash with, and to irrigate crops and lawns. Its demand curve should be convex downwards because, as its price increases, we would expect its marginal uses to decline and thus its value, as determined from the intercept, to increase with price. The ultimate value of water, of course, is immeasurable as it is required to sustain life. For lesser products which are not required to sustain life, there will be no incentive, by definition, for customers to buy the product if price exceeds its value as they can either do without or have substitutes available.

The intercept in Figure 2.18 does not define how a given individual values the product. It defines an average (marginal) value at the price of 6 for persons buying the product. In the strictest sense, value as defined in this manner is a phenomenological constant. Nevertheless, it has all of the attributes that the value of the product should have:

1. It has the same units as price.
2. As price approaches value, demand goes to zero.
3. If value increases, demand for the product increases; if value decreases demand for the product decreases.

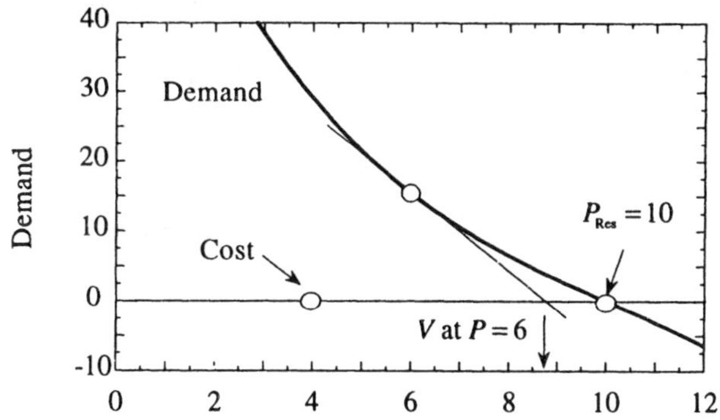

Figure 2.18 Demand relationship with price for a monopoly. The intercept of the tangent at $P = 6$ with the line for zero price determines the marginal value of the product at the demand level for $P = 6$.

Seldom is the demand curve for a product known. However, the slope $-K$ of demand, D, with price is often known at price, P, which is sufficient for determining the value of a product offered by a monopoly from the relation:

$$V = P + \frac{D}{K} \quad \text{(monopoly)} \tag{2.6}$$

With this approach, we do not need to ask customers individually to evaluate or judge the product to establish its value; instead the market response is used to obtain a measure of value in the form of a phenomenological constant.

This definition of value, although precise, may be uncomfortable for some because it does not use a personal or individual viewpoint of product value or because it does not equate value to price. It is true that few if any buyers would likely identify the price intercept at zero demand as their value of the product if asked. This is not a flaw and it is not important whatsoever to our effort which is to develop a phenomenological theory for supporting good product decision making.

Value as used here is related to but is not the same as consumer surplus, introduced by Alfred Marshall [18]. A linear demand curve and a linear supply curve having, for simplicity, slopes of opposite sign but equal absolute value are shown in Figure 2.19. The two curves intersect at a price of 7 which is the equilibrium market price for this example from elementary economic considerations in which supply equals demand. If price were 9 instead of 7, demand would be 1 instead of 3. Persons who would have just purchased when the price dropped below 9 have received price savings equal to the distance $CS = 2$ when $P = 7$. (Note that demand is used throughout as a rate of quantity purchased per unit time and not simply a quantity purchased.) The total consumer surplus for all buyers at a price of 7 is equal to the area (1/2)3×3 = 4.5 under the demand curve from $P = 7$ to 10. The line labeled $PS = 2$ is the surplus at $P = 7$ for producers who would have just produced when the price rose above 5. The total producer surplus is equal to the area $(1/2)3\times3 = 4.5$ under the supply curve from $P = 4$ to 7.

The interpretation of the demand curve here is different. For a given segment of the buying public (persons of similar age, gender, income, etc.), the product is assumed to have a single value for a linear demand curve. As price drops below value, the probability that persons in this segment purchase the product increases. The net value per year to all of the buyers in Figure 2.19 at $P = 7$ is therefore 9 which is twice the consumer surplus. The variable cost to the producers is assumed constant over the demand range of interest, and gross profit to all of the producers before subtracting out fixed costs and investments is also 9 which is twice producer surplus for the example in Figure 2.19. Thus using the value model, net value to the producers and net value to the buyers are equal as already noted in Equation 2.4. Given a variable cost, C, and a market for the good having a value, V, and a monopoly producer (for simplicity), profit to the producer is maximized at a price of $(V + C)/2$. At this price, the producers and the buyers share in net value (Equation 2.5). The same price point would be reached if the buyers and producers bargained with equal strength as discussed earlier.

2.9 Value speculation and market dynamics

The equilibrium price is reached when the demand rate is equal to the rate of products being offered for sale. The dynamics by which a real market approaches equilibrium will be strongly dependent on the specific market as defined by the product, its method of manufacture and distribution, the nature of compe-

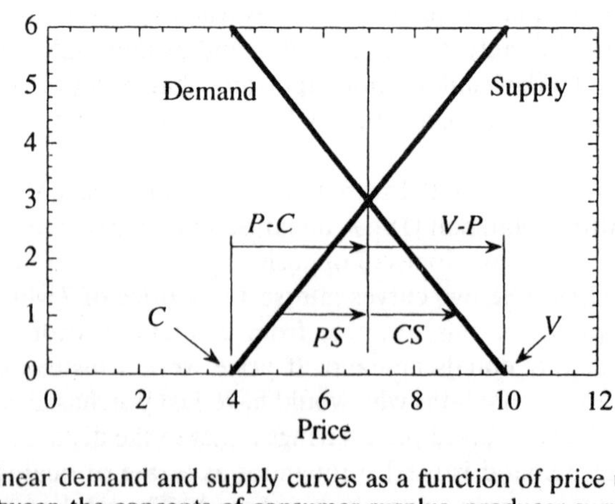

Figure 2.19 Linear demand and supply curves as a function of price used to show the differences between the concepts of consumer surplus, producer surplus, net value to producer, and net value to customer.

tition from like products and substitutions, the nature of the transaction costs involved, and the nature of the information available to buyers and sellers alike about the attributes of the product that generate value. In contrast to real markets, a hypothetical market displays 'perfect competition' when the following conditions are obeyed [19]:

1. There are many buyers and sellers of the product.
2. No single buyer or seller is strong enough to influence the market price.
3. All products from the sellers are exactly alike.
4. Everyone has perfect knowledge of the value and cost of the product.
5. There are no barriers to entry to the market (finance charges, patents, investments, etc.) either as a buyer or seller.

The stock market is often described as the real market which is closest to being an example of the hypothetical, perfectly competitive market. Nevertheless, it is still far from being perfectly competitive due to such factors as sales fees and large institutional investors. But the major reason for it exhibiting behavior strongly different from that expected for perfect competition is tied to the observations of Gennotte and Leland [20] that 'only a small proportion of investors actively gather information on future economic prospects or asset supply. Other investors look to current prices to impute information about future prices.'

Considerable interest in the behavior of the stock market was generated by the crash of October 19, 1987 when its valuation dropped by more than 20% resulting in $700 billion in losses. Gennotte and Leland's model indicates that a hedging strategy known as 'portfolio insurance' likely created excess supply as prices fell and led to the crash much in the same manner as stop-loss orders created excess supply as prices fell in the crash of 1929. It is instructive to consider the dynamics of a model stock market using the approach developed here because the exercise demonstrates how the important concepts of product value, supply, demand, and human behavior influence general market behavior.

Classical market dynamics [21] are generally modeled as starting with an arbitrary price and excess supply or demand. Price then moves up in proportion to the excess demand and downwards in proportion to excess supply. In terms of our model, a stock market in its simplest form is described by two groups of people (two segments) who in themselves are homogeneous. One group which we denote as 'buyers' values the stock higher, and the other group which we denote as 'sellers' values the stock lower. Thus, instead of having the normal intersecting supply and demand curves, we have two demand curves that are separate but parallel to each other as shown in Figure 2.20.

For this example, the buyers value the stock at 14 and the sellers value it at 6. How can this occur? Assume that the sellers need to pay for the college tuition of triplets who just graduated from high school and the buyers just won the state lottery. The terms buyers and sellers are strictly applicable to the two groups only when the price is between 6 and 14. If price rose above 14, both

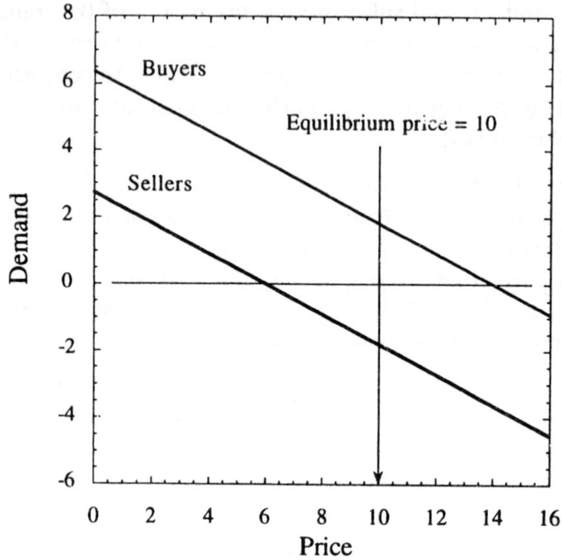

Figure 2.20 Demand curves of so-called buyers and sellers of stock as a function of price. The only differences between buyers and sellers is that buyers value the stock higher than sellers.

groups would desire to sell but there would be no buyers and if it fell below 6, both groups would want to buy but there would be no sellers. The equilibrium price, 10, is halfway between the two value points.

In agreement with the classical approach to market dynamics, we write the time-dependent price equation as:

$$P(t + \delta t) = P(t) + \frac{P(t)\alpha'\delta t}{N_{Sh}}[D_B + D_S], \tag{2.7}$$

where t is the time, δt is the time increment, α' is a dimensionless rate constant, N_{Sh} is the number of shares outstanding, D_B is the demand of 'buyers', and D_S is the demand of 'sellers.' The annual rates at which these persons offer to buy and sell shares are assumed for simplicity to be given by the linear expressions:

$$D_B = K\,[V_B - P(t)] \tag{2.8}$$

and

$$D_S = K\,[V_S - P(t)] \tag{2.9}$$

where V_B and V_S are the values placed on the stock by buyers and sellers, respectively. Substitution of Equations 2.8 and 2.9 into Equation 2.7 yields, after rearrangement and on taking the time step arbitrarily small, the nonlinear, first-order differential equation ($\alpha = \alpha' K$):

$$\frac{dP}{dt}=\frac{2\alpha P}{N_{Sh}}\{[(V_B+V_S)/2]-P\}, \tag{2.10}$$

which on dividing through by $P\{[(V_B+V_S)/2]-P\}$ and multiplying both sides by dt becomes a standard form in a Table of Integrals having the solution:

$$P=\left\{\frac{P(0)\left[\dfrac{(V_B+V_S)}{2}\right]\exp\left\{\dfrac{\alpha t(V_B+V_S)}{N_{Sh}}\right\}}{\left[\dfrac{(V_B+V_S)}{2}\right]-P(0)\left[1-\exp\left\{\dfrac{\alpha t(V_B+V_S)}{N_{Sh}}\right\}\right]}\right\}$$

When $P=(V_B+V_S)/2$, the market price is at equilibrium as $dP/dt=0$. Price approaches its equilibrium point in a diffusive, exponential decay fashion.

The above equation expresses the dynamics of price for the case where both buyers and sellers have a fixed understanding of the value of the stock independent of its price behavior although their values differ. Equilibrium is dynamic in that it is the rate at which buyers want to buy that is matched by the rate at which sellers want to sell. Price is well behaved as long as value is well behaved. If the company is a mining company and it strikes gold, its value would rise and so would its stock price according to the model. However, if either or both of the two values change, the stock price will not overshoot or undershoot as it moves to its new equilibrium point.

When buyers and sellers also set a portion of the stock's value on the rate at which its price changes over time, price overshoots and undershoots will occur if the information that the buyers and sellers have regarding price is not precisely up to date. For example, consider that buyers value the stock according to the equation:

$$V_B=V_{B,0}+\psi[dP/dt] \tag{2.11}$$

and sellers value the stock according to the equation:

$$V_S=V_{S,0}+\psi'[dP/dt] \tag{2.12}$$

where ψ and ψ' are constants that weight the degree to which buyers and sellers, respectively, add value to a stock based upon its rate of movement in price. If the information on the time derivative of price is not current but generated for a prior time equal to $t-\Delta t$, then the quantity that should appear in the two equations above is not dP/dt but the quantity $\{[dP/dt]-\Delta t[d^2P/dt^2]\}$ instead which is equal to the price derivative with respect to time at the earlier time. Substitution of this relationship into Equation 2.10 results in a nonlinear, second-order differential equation given by:

$$\begin{gathered}dP/dt-[2\alpha/N_{Sh}]\,P\{[(V_{B,0}+V_{S,0})/2]+\\ [(\psi+\psi')/2][(dP/dt)-\Delta t(d^2P/dt^2)]-P\}=0.\end{gathered} \tag{2.13}$$

The above equation is equivalent to that for a damped, nonlinear oscillator. As a result of this analogy, we can think of the stock price as having an effective mass proportional to the product of the time delay Δt and the quantity $(\psi + \psi')/2$.

Several examples of price behavior as a function of time are shown in Figures 2.21 through 2.24 as determined from numerical solutions to the differential equation for different choices of the coefficients as evaluated using *Mathematica*®. The notations P′ and P″ in Figure 2.25 refer to the first and second derivatives, respectively, of price with time.

Behavior of the stock price as an overdamped oscillator, Figure 2.21, about its equilibrium point of 10 occurs for the coefficients in the first example, Figure 2.20, whereas the coefficients for the second example lead to a strong overshoot followed by a crash. The demand rate of the so-called buyers as a function of time for the stock in Figure 2.22 which crashed is shown in Figure 2.23. The sell rate of the so-called sellers as a function of time for the above stock is shown in Figure 2.24.

If the equation were integrated beyond the point shown in Figure 2.22, the price of the stock would soon rise and overshoot even more and then again crash. However, it is likely that after the first crash shown here buyer behavior would change sufficiently so that there would be less speculation and the price might find its equilibrium point without much additional drama. Paul Greywall analyzed the crash of November 1987 using Equation 2.13 and found that a reasonable fit to the Dow-Jones average from May to December could be obtained using a delay of two and one-half days [22].

Speculative bubbles of the type described by the above model can occur in virtually all goods traded including land, housing, silver, vacation homes, and office buildings when true value is confounded with value based upon the rate of change of price and this information is not current. Each bubble has its own dynamics and all such markets have some persons that lose and others that gain. But, as we know from prospect theory, the cries of those with losses should drown out the cheers of those that gained by buying in at the bottom of the crash.

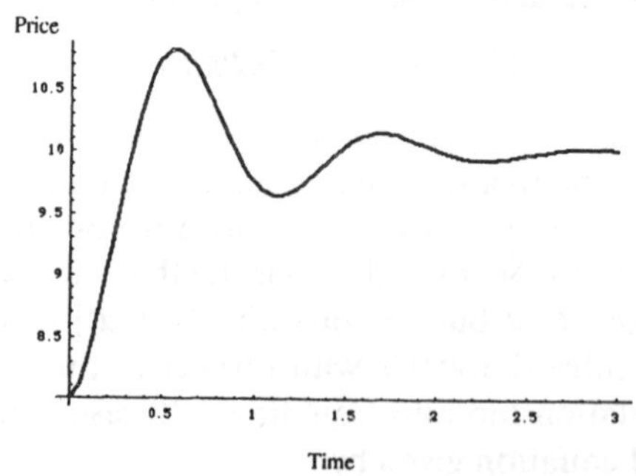

Figure 2.21 Stock price behaving as overdamped oscillator as a function of time according to Equation 2.13.

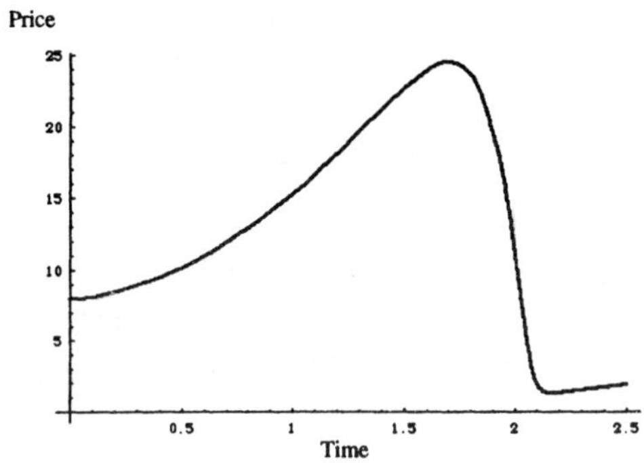

Figure 2.22 Stock price exhibiting a crash as a function of time according to Equation 2.13.

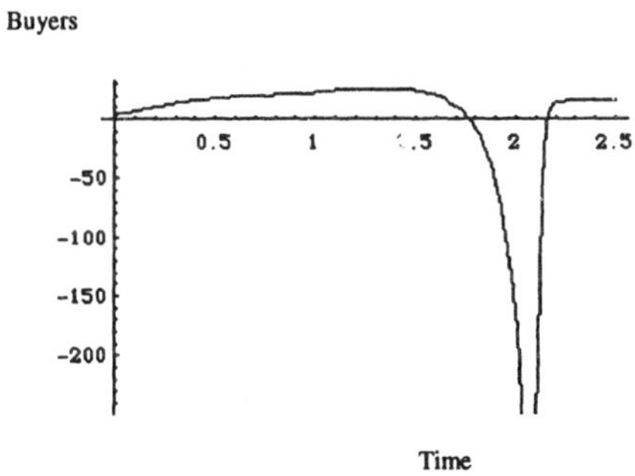

Figure 2.23 Number of potential buyers (positive numbers) for stock over time as described by Figure 2.22 which exhibits a crash.

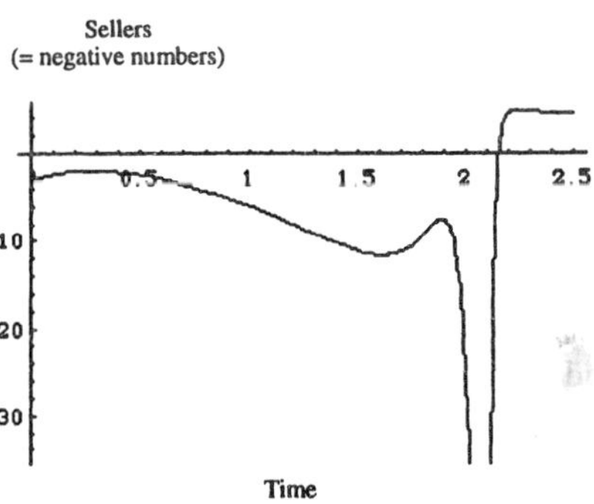

Figure 2.24 Number of potential sellers (negative numbers) for stock over time as described by Figure 2.22 which exhibits a crash.

Fig. 2.21 NDSolve
```
[{P'[x] - P[x] (10 + 0.01 P'[x] - 0.03 P''[x] - P[x]) == 0,
P[0] == 8, P'[0] == 0.5 }, P, {x, 0, 3.0}]
Plot[Evaluate[ P[x] /. %], {x, 0, 3.0}]
```

Fig. 2.22 NDSolve
```
[{P'[x] - P[x] (10 + 1.15 P'[x] - 0.1 P''[x] - P[x])/10 == 0,
P[0] == 8, P'[0] == .5 }, P, {x, 0, 2.5}]
Plot[Evaluate[ P[x] /. %], {x, 0, 2.5}]
```

Fig. 2.23 NDSolve
```
[{P'[x] - P[x] (10 + 1.15 P'[x] - 0.1 P''[x] - P[x])/10 == 0,
P[0] == 8, P'[0] == .5 }, P, {x, 0, 2.5}]
Plot[Evaluate[ 14 - P[x] + 2 P'[x] - 0.15 P''[x] /. %], {x, 0, 2.5}]
```

Fig. 2.24 NDSolve
```
[{P'[x] - P[x] (10 + 1.15 P'[x] - 0.1 P''[x] - P[x])/10 == 0,
P[0] == 8, P'[0] == .5 }, P, {x, 0, 2.5}]
Plot[Evaluate[ 6 - P[x] + .3 P'[x] - 0.05 P''[x] /. %], {x, 0, 2.5}]
```

Figure 2.25 Parameters for non-linear equations in Figures 2.21 through 2.24.

References

1. Dupuit, J. (1884) De la mesure de l'utilité des travails publics. *Annales des Ponts et Chaussées*, 2ieme Série, 8.
2. Maslow, A.H. (1943) A theory of human motivation. *Psychological Review*, **50**, 370–396.
3. Salvatore, D. (1991) *Microeconomics*, HarperCollins, New York, p. 102.
4. Wasson, C.R. (1975) *Consumer Behavior: A Managerial Viewpoint,* Lone Star, Austin, TX, p. 54.
5. Louden, D.L. and Della Bitta, A.J. (1988) *Consumer Behavior: Concepts and Applications*, 3rd edn, McGraw-Hill, New York, p. 610.
6. Katona, G. (1977) Rational behavior and economic behavior, in *Classics in Consumer Behavior*, (ed. L.E. Boone), PPC Books, Tulsa, OK, pp. 24–39.
7. Kassarjian, H.H. (1977) Personality and consumer behavior: a review, in *Classics in Consumer Behavior,* pp. 153–84.
8. Howard, J.A. and Sheth, J.N. (1969) *The Theory of Buyer Behavior*, Wiley, New York.
9. Wasson, C.R. (1975) *Consumer Behavior: A Managerial Viewpoint,* pp. 130–3.
10. von Neumann, J. and Morgenstern, O (1944) *Theory of Games and Economic Behavior*, Princeton University Press, Princeton, NJ.
11. von Winterfeldt, D. and Edwards, W. (1986) *Decision Analysis and Behavioral Research*, Cambridge University Press, Cambridge, MA, p. 213.

12. Kahneman, D. and Tversky, A. (1979) Prospect theory, an analysis of decision under risk. *Econometrica*, **47**, 263–91; Tversky, A. and Kahneman, D. (1981) The framing of decisions and psychology of choice. *Science*, **211**, 1981, 453–8; Loss aversion in riskless choice a reference-dependent model. *The Quarterly Journal of Economics*, November, 1039–61.
13. Kahneman, D., Knetsch, J.L. and Thaler, R.H. (1990) Experimental tests of the endowment effect and the Coase theorem. *Journal of Political Economy*, **98**, 1325–47.
14. Little, J.D.C. (1992) Are there laws of manufacturing?, in *Manufacturing Systems, Foundations of World-Class Practice*, (eds J.A. Heim and W.D. Compton), NAE Press, Washington, DC.
15. Tversky, A. (1972) Elimination by aspects: a theory of choice. *Psychological Review*, **79**, 281–99.
16. Carson, R.T. (1991) Constructed markets, in *Measuring the Demand for Environmental Quality*, (eds J. Braden and C. Kolstad), Elsevier, Amsterdam, The Netherlands.
17. Gabor, A. and Granger, C.W. (1966) Price as an indicator of quality. *Economica*, **33**(129), 43–70.
18. Marshall, A. (1920) *Principles of Economics*, 8th edn, Macmillan, London, p. 124.
19. Salvatore, D. (1991) *Microeconomics*, HarperCollins, New York, pp. 253–5.
20. Gennotte, G. and Leland, H. (1990) Market liquidity, hedging, and crashes. *The American Economic Review*, **80**, 999–1021.
21. Arrow, K.J. (1988) Workshop on the economy as a complex evolving system: summary, in *The Economy as an Evolving Complex System*, **5**, (eds P.W. Anderson, K.J. Arrow, and D. Pines), Santa Fe Institute Studies in the Science of Complexity, Addison-Wesley, New York, pp. 275–81.
22. Greywall, P. (1994) Speculative bubbles heard popping. *Technical Report UILC-ENG 94-4009*, University of Illinois at Urbana-Champaign, Urbana, IL.

3 Enterprise model

3.1 Purpose of model

To this point, we have discussed, in general terms, the process of product realization and certain aspects of consumer behavior. The next step is to join these elements within the framework of an enterprise model for planning, developing, and pricing future products. The model quantifies how the means of the enterprise generates the end; how the right actions achieve the goals of the enterprise. The goals represent the ends, the so-called bottom line. The means represent the independent variables which the enterprise can manage to reach its goals.

Both the dependent and independent variables in the means/ends analysis are called metrics [1]. The most important means are the people that work in the enterprise. Return on investment and market share are dependent, bottom-line metrics. They are a function of another set of metrics which we call fundamental because they provide the coupling between the means and the ends [2] (Figure 3.1).

According to results of comparative studies [3, 4, 5, 6] the fundamental metrics are believed to be quality (Q), costs (C) or productivity, innovation (I), and speed (S) or lead time. However, there is no uniform agreement on their relative importance. Buzzell and Gale [6] have found a direct connection, as described schematically in Figure 3.2, between return on investment, market share, and a measure of quality.

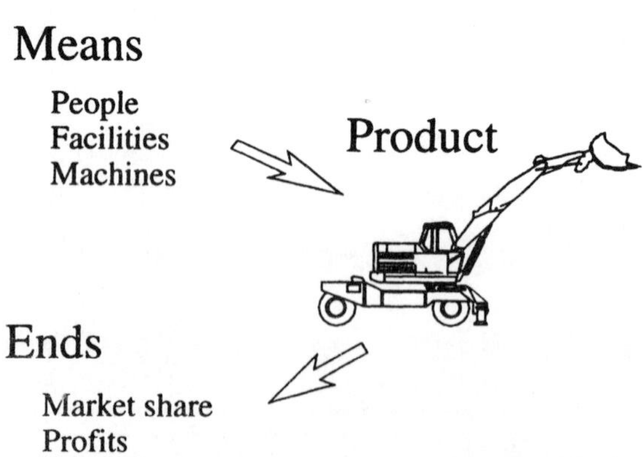

Figure 3.1 The product's fundamental metrics couple the means of the enterprise to its ends.

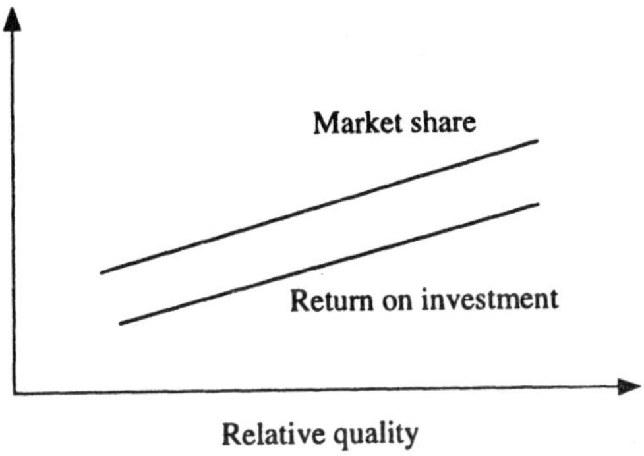

Figure 3.2 Schematic depiction of the connection found by Buzzell and Gale [6] between market share, return on investment, and quality.

Stalk and Hout [7] have argued that speed is the dominant metric. Krupka [8] has also emphasized the importance of time. The MIT study [4] found that firms that exhibit world-class performance emphasize simultaneous improvements in quality, cost, and speed of delivery. Clark and Fujimoto [5] found similar results. The study of Leong, Snyder, and Ward [1] also identified innovation as an important metric. Although the comparative studies are in reasonable qualitative agreement as to what the fundamental metrics are, questions remain (Figure 3.3) as to their relative importance and their inter-relationships.

The results of the comparative studies strongly suggest that functional relationships exist between the bottom-line metrics of return, r; share, m; profits, A; and demand, D and the four fundamental metrics:

$$r = f_1(Q, C, I, S)$$

$$m = f_2(Q, C, I, S)$$

$$A = f_3(Q, C, I, S)$$

$$D = f_4(Q, C, I, S)$$

The enterprise model wc are seeking should yield the explicit forms of the above relationships and thereby resolve the questions of the inter-relationships and relative importance between quality, cost, innovation, and speed as illustrated by the metric puzzle (Figure 3.4).

Figure 3.3 The need to discover and target the fundamental metrics which drive the bottom-line metrics is central to forming a strategic plan.

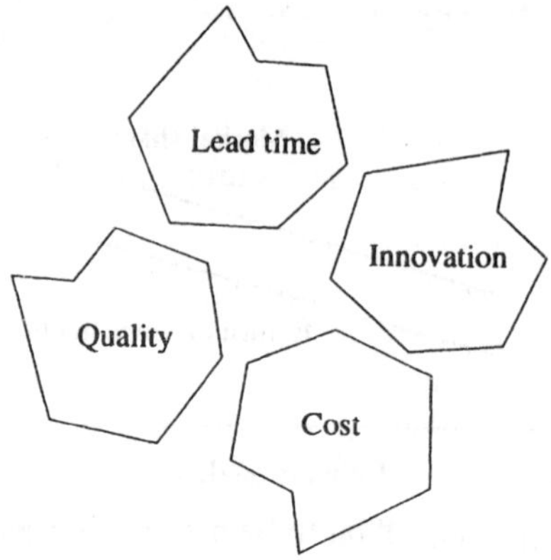

Figure 3.4 Schematic representation of the metric puzzle. (Source: adapted from H.E. Cook, and R.E. DeVor, (1991) On competitive manufacturing enterprises I: the S-model and the theory of quality. *Manufacturing Review*, **4**, 96–105, © 1991 ASME International. Used with permission.)

In addition to sorting out the relative importance of the metrics, the model should also unify and integrate the subjects shown in Figure 3.5 that are important to product realization but are often seen as separate elements of the process. Moreover, the need for continuous improvement for the enterprise to survive in highly competitive global markets requires that we consider the bottom-line metrics in terms of their time rate of change given by $\delta r/\delta t$; $\delta m/\delta t$; $\delta A/\delta t$; and $\delta D/\delta t$.

3.2 Criteria for a model of the enterprise

Models fall between two extremes – those that are so simple that they do not tell you anything that you did not already know and those that are so complex that you do not have enough information to specify the almost unlimited array of unknown parameters and coefficients involved. Little's criteria [9] states that

Product planning
Pricing
Statistical process control
System design   Cost management
Value engineering
Competitive benchmarking
Total quality management
Parameter design
Metrics
Design of experiments

Figure 3.5 Topics important to product realization.

a model should be easy to understand, not easily give bad answers, and treat the major issues of interest. These criteria generate a balance between simplicity and rigor and were used to guide the development of the S-model [10] and its extensions [11, 12]. The major issues of interest that an enterprise model needs to treat are as follows:

- a set of goals;
- some degree of internal order;
- a work plan to achieve its goals;
- a process called management;
- interaction with and acceptance by society.

The goal of the enterprise will be to be profitable, or to maximize return on investment, or to gain market share over the long term, or some combination of these three. The internal order of the enterprise is characterized by its structure, culture, and technology which represent the tools used by the enterprise to transform input to output. The work plan for the enterprise model is structured around the customer/company time-invariant loop (Figure 1.18). Management is defined as the process of directing and adjusting the internal order within the enterprise to best suit changing conditions in the marketplace. The acceptance of the enterprise by society is determined by its net value to society as defined by the total quality of its products (Chapter 4).

3.3 S-model

3.3.1 CARTEL REFERENCE STATE

We examine demand within a product segment by considering an initial reference state where all of the N competitors in the segment are manufacturing identical products. The products from each are equally available to customers and all are offered at the same price but that price can change. In other words, the manufacturers are a strict cartel in the reference state. The demand curve for each product for this condition for small price changes is of the form shown in Figure 3.6. The intercept of the extrapolation gives the value of the product in the reference state. Now we allow the competitors to compete by making small differentiations in their products in terms of value and price. However, the prices and values do not differ so much that the products move out of the original product segment defined by the reference state. An example of the resulting demand is shown in Figure 3.7 for three competitors. The change in demand for each from the reference state in Figure 3.6 is given by the Taylor's expansion:

$$\delta D_i = \Sigma_j K_{ij} [\delta V_j - \delta P_j] \tag{3.1}$$

in which the summation is over all competitors in the segment. It follows from Equation 3.1 and from the symmetry of the problem at the reference state

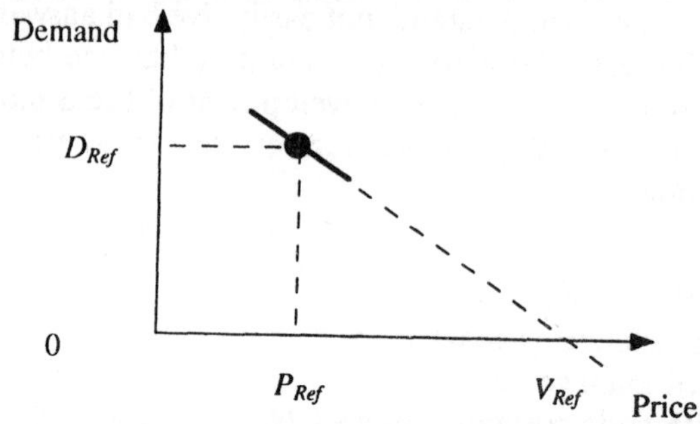

Figure 3.6 Demand when products are identical and at the same price. The subscript '*Ref*' denotes the reference state for the Taylor's expansion for demand in terms of price and value.

where the coefficients are evaluated that $K_{ii}=K_{jj}=K$ and $K_{ij}=K_{ji}=K_{kl}=K\kappa$ for $i\neq j$ and $k\neq l$ because all of the products are identical in the reference state.

When the values of the products are not changed but all competitors change price by δP in unison from the reference state, the change in demand is given by:

$$\delta D_i = \delta D = -\delta PK + \delta PK(N-1)\kappa \tag{3.2}$$

Each competitor in this situation should receive $1/N$ of the amount that would have gone to a monopolist, δD_M. On equating $N\delta D_i$ determined from Equation 3.2 to $\delta D_M = -\delta PK$ (determined from Equation 3.2 for $N = 1$), we find that κ must equal to $1/N$ for this relationship to hold.

The coefficient K is determined from the change in demand in product i when only competitor i changes price:

$$K = \frac{-\delta D_i}{\delta P_i} \tag{3.3}$$

The index i is arbitrary for the competitors in the segment which means that we are assuming that K is competitor independent within a product segment.

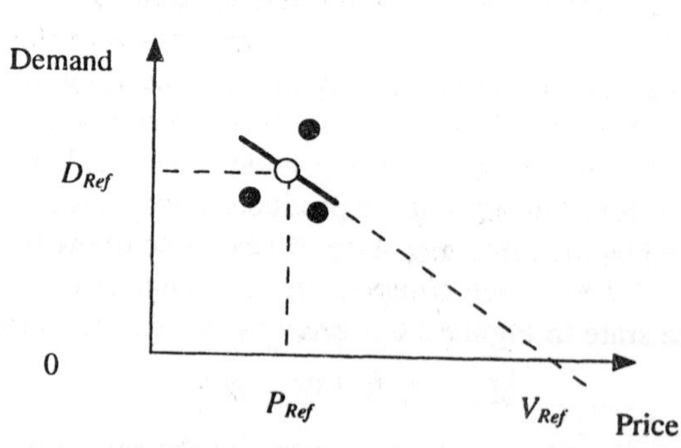

Figure 3.7 Movement from the reference state when each product is differentiated in value and not offered at the same price.

The coefficient K can also be determined from the price elasticity of demand defined as:

$$E_1 \equiv \frac{-\delta D_i / D_i}{\delta P_i / P_i} = \frac{KP_i}{D_i}. \tag{3.4}$$

when only product i changes price. We assume this elasticity to be the same for the other products in the segment.

The change in demand for product i when all products in the segment are increased in price by δP gives rise to another price elasticity defined by:

$$E_2 \equiv \frac{-\delta D_i / D_i}{\delta P_i / P_i} \tag{3.5}$$

which we assume to be independent of the type of product, i. Because the elasticity E_2 is independent of the number of competitors, it is more fundamental than the elasticity E_1 which is related to E_2 by the expression $E_1 = NE_2$. When competitors are added to the segment, E_1 increases but E_2 remains unchanged.

3.3.2 LINEAR DEMAND EQUATION

Price in Equation 3.1 is well defined. The question is value. How can we determine what the change in value is if we change a product attribute? The answer is found by considering what happens to demand. For example, if a single competitor changed an attribute of product i from the reference state in Figure 3.6 but held price constant and the other competitors did not change any attribute nor change price, then the change in value of product i is determined from the relation:

$$\delta V_i = \frac{\delta D_i}{K} \tag{3.6}$$

The change in demand of the other products for $j \neq i$ is:

$$\begin{aligned} \delta D_j &= K_{ji} \delta V_i \\ &\approx \frac{-K \delta V_i}{N} \end{aligned}$$

Because of the symmetry in the expansion coefficients, Equation 3.1 can be written as:

$$\delta D_i = K\left\{\delta V_i - \delta P_i - \frac{1}{N}\sum_{j \neq i}(\delta V_j - \delta P_j)\right\} \tag{3.7}$$

In the vicinity of the reference state, the above differential form can be written in terms of the linear approximation to the demand curve:

$$D_i = K\left\{V_i - P_i - \frac{1}{N}\sum_{j \neq i}(V_j - P_j)\right\} \tag{3.8}$$

The coefficient K should be set equal to its value at the reference state, which we can estimate using the approximate relation:

$$K \approx \frac{E_1 \overline{D}}{\overline{P}} \tag{3.9}$$

where $\overline{P}$ is the average price for the segment and $\overline{D}$ is the average demand per competitor. An important point is that K should be constant for the competitors if their products have comparable availability to customers and are advertised to the same degree.

3.3.3 S-MODEL VALUE DETERMINATION FROM MARKET SHARE AND PRICE

Other useful relationships follow from the above equations. The average value $\overline{V}$ for a segment is related to the average price $\overline{P}$ for the segment by the relation:

$$\overline{V} = \overline{P}\left[\frac{1+E_2}{E_2}\right] \tag{3.10}$$

For a market segment of N competitors, there are N simultaneous equations of the type given by Equation 3.8. Although demand is written as a dependent variable, demand and price represent known quantities for products already in production. When this is the case, the set of simultaneous equations can be solved for the unknown values of the products competing in the segment. The solution, obtained by Gaussian elimination, is given by:

$$V_i = \frac{N[D_i + D_T]}{K[N+1]} + P_i \tag{3.11}$$

where D_T is the total demand for the N products. Often we will only need to consider the value difference between two products i and k which is given by:

$$V_i - V_k = \frac{N}{K[N+1]}\{D_i - D_k\} + P_i - P_k \tag{3.12}$$

The expression for the value of product i can also be written as

$$V_i = \left[\frac{N\overline{P}}{[N+1]E_2}\right][1 + m_i(N)] + P_i$$

where $m_i(N)$ is the market share for product i among the N competitors.

The process of 'value benchmarking' developed in Chapter 5 and applied in Chapter 7 uses Equations 3.11 and 3.12 to estimate the values and value differences of products currently competing in the marketplace. The origins of the value differences can then be examined in terms of the differences in the attributes of the competing products. This process gives insight into how the attributes of a product determine its value to the customer.

The value of a current product is also useful in establishing the baseline for estimating the value of the future product by adding the value increases resulting from product improvements to the baseline value determined from Equation 3.11. Once an estimate of the value of the future product has been made, the demand for this product can then be forecast using Equation 3.8.

3.3.4 LOGIT MODEL

By making some additional assumptions, the requirements of the linear demand model that the demands and prices of the competing products be near the cartel point can be relaxed somewhat. A particularly useful non-linear model is the so-called logit model [13] which yields an expression of the form:

$$\frac{D_i}{D_k} = \exp(U_i - U_k)$$

for the ratios of the demands of any two competing products in a segment where U_i and U_k are the utilities of products i and k. The relationship between the utilities and the linear model prices and values is obtained by taking the natural logarithm of the above expression and expanding it about the same $\bar{D} = D_T / N$ cartel point. On doing this, we find that the demand difference $D_i - D_k$ is equal to the utility difference times $\bar{D}$. Setting this equal to the demand difference $D_i - D_k$ obtained from Equation 3.8, we find:

$$U_i - U_k = \frac{[N+1]K}{D_T}\{V_i - V_k - [P_i - P_k]\}$$

The above expression can be rearranged to the more useful and insightful form given by:

$$U_i - U_k = \frac{[N+1]E_2\{V_i - V_k - [P_i - P_k]\}}{\bar{P}} \tag{3.13}$$

3.4 Profits and pricing

In addition to demand, the profit for each product alternative under consideration needs to be forecast before deciding upon which to make. This requires that their costs be understood and a realistic price established for each. The general expression for annual profit (strictly speaking, cash flow) of product i in year j is given by the relation:

(3.14)

where $F_i(j)$ and $M_i(j)$ are the fixed costs and investment for year j, respectively. The variable cost per unit and fixed costs which appear in the profit equation are obtained from the total costs given by:

$$TC_i = D_iC_i + F_i, \tag{3.15}$$

over the range of D_i about the point of tangency of the line to the total cost curve (Figure 3.8).

Fixed costs, F_i, are defined by the intercept of the straight line approximation at $D_i = 0$. The slope of total costs with sales $C_i \equiv \partial TC_i/\partial D_i$, the so-called marginal costs in economic theory, equals variable cost. The definition of fixed and variable costs in this manner gives an exact description of total costs for small changes in demand about the reference baseline demand where the slope (variable cost) is evaluated.

The question as to which costs are truly variable and which are fixed depends on the time period involved for the change in demand. As the period becomes large, all costs are variable because machines can be added or sold as required to restructure the enterprise to a new level of demand and direct and indirect labor can be adjusted. When the period is very short, however, all costs are fixed (Figure 3.9).

The demand curves shown in Figure 3.10 are for product 1 having two competitors noted as products 2 and 3. The curves are based upon Equation 3.8 with K=150 and the values of the three products equal to 20. The line with the steeper slope is for the case where prices of products 2 and 3 are held constant at 10 as the price of 1 changes. The demand curve for product 1 when the prices of all three products are the same is the cartel line having the less steep slope which intersects the zero price line at 20, which is equal to the value of the products.

The profit for product 1 based upon Equation 3.14 is shown in Figure 3.11 for the case where its variable costs are 4 and there are no fixed costs or investment. Profit is seen to peak at 8.5 when the price of product 1 is being varied and the prices of products 2 and 3 are fixed at 10, whereas it peaks at 12, halfway between variable cost and value, if the products are priced together as a cartel. Fixed costs and investment are assumed to be zero for the calculations.

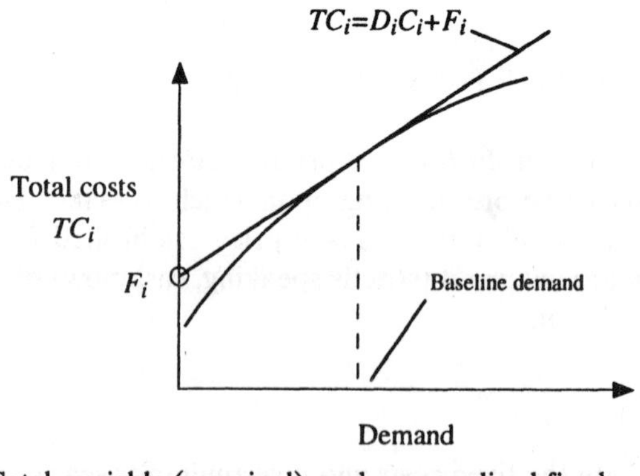

Figure 3.8 Total, variable (marginal), and renormalized fixed costs.

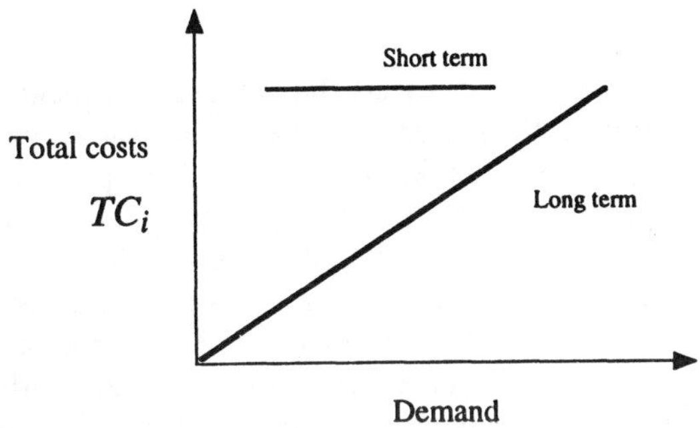

Figure 3.9 Total costs versus demand over short term and over long term.

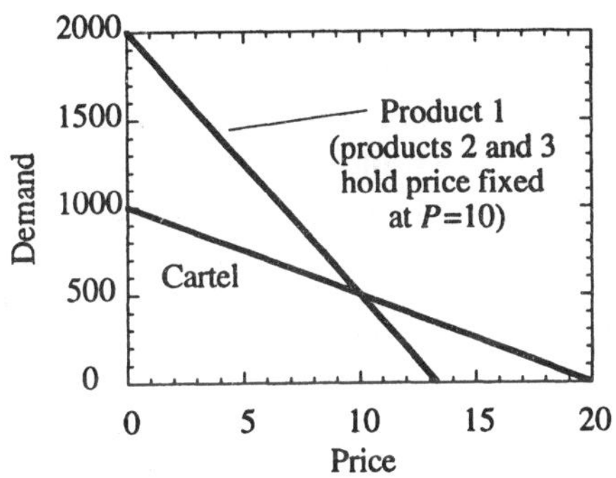

Figure 3.10 Demand curves for product 1 when priced as a cartel with products 2 and 3 and when products 2 and 3 hold price fixed at 10.

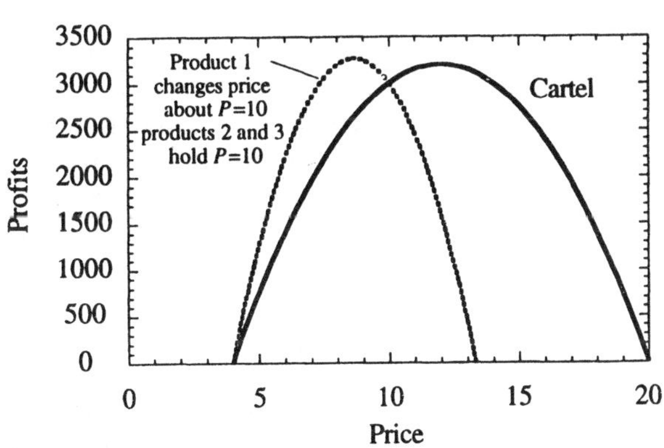

Figure 3.11 Profit curves for product 1 when part of cartel with 2 and 3 and when products 2 and 3 hold price fixed at 10.

3.4.1 PRICE WAR

If the three products are currently priced the same at 10, producer 1 could choose to reduce price to gain the increase in profitability shown in Figure 3.11. However, the other two producers would likely reduce price by the same amount to recoup some of their profits lost by producer 1's action. The profitability of each product returns to the cartel curve but at a lower level. Repetition of this behavior generates what is known as a **price war** and is destructive to profits over the short term. Over the long term, such behavior could increase the profitability of some firms if one or more other firms go bankrupt or choose to leave the market because of insufficient resources to withstand a price war.

If producer 1 had increased price instead of decreasing it, the other two producers may not have followed because their profitability would have improved. Consequently, the three producers may jockey about with price. From this exercise, they should learn how their profits respond to their collective actions and should discover that a form of cooperative behavior in which one becomes a **price leader** and the other two **price followers** can aid profitability by having the leader take the group up the cartel curve toward the maximum in profitability which is at a price of 12 for this example. When the competitors in a segment choose to price the same and are producing products having identical value and cost, their profits, as stated already, are maximized when the price is given by:

$$P = \frac{V + C}{2} \tag{3.16}$$

However, producers should not conspire in pricing as it represents **price fixing** which is patently illegal and carries stiff penalties, as it should. Nevertheless, there is a natural and pervasive market force as a result of the shape of the profitability curve which over time can lead producers to exhibit cooperative behavior when their strengths in product realization are approximately equal and when they have similar resources.

It is hard to argue convincingly that customers are being swindled by the natural (as opposed to the conspired) price leader form of cooperative behavior as the end result should never be worse for the customer, in theory, than simply a sharing of net value between the producers and their customers. This result was seen earlier in the balance price given by Equation 2.4 which also follows on rewriting Equation 3.16 as:

$$P - C = V - P \tag{3.17}$$

to equate the net value received by the seller to that received by the customer.

A greater concern regarding anti-competitive behavior should be that producers may cease to compete through aggressive investments in cost reductions and value-adding product innovations. Such behavior neither serves customers nor society. The best assurances that this does not happen is not by the stick represented by tight monitoring and control by the anti-trust division of the Department of Justice but by the carrot of a market in which strong global competitors can readily enter because the number of competitors increases the likelihood that the market prices will stay well below the point of profit maximization.

3.4.2 VALUE-INDUCED PRICE WAR

Just as a price reduction by one producer can lead to a price war, it is possible to start a price war by increasing the value of a product relative to the other products in the segment but not increasing price or increasing it by too little. The reason is that an increase in value is equivalent to a decrease in price in regard to its impact on the demand of competing products. This result is shown in Figure 3.12.

Without a change in price, product 1 has a large profit increase, arrow 1A, due to an increase in its value over products 2 and 3. The profit for product 2 (and 3) is reduced as a result, arrow 2A. In the short term, producer 2 (and 3) can react in the same manner as if producer 1 had made a price reduction, i.e. by reducing its price to recoup some of the lost profit. Producer 1 may want to avoid this reaction if it sees that the other two competitors are viable and will in due time increase the value of their products to match or better the value of product 1. The chances of producers 2 and 3 reducing their prices and initiating a price war can be diminished if producer 1 increases the price of its product as shown by the arrow to point 1B. This moves the profit for producer 2 to the point 2B. Another reason to increase price with increasing value is that adding value without a price increase may create an increase in demand which cannot be met with existing capacity constraints.

3.4.3 ENDING PRICE WARS AND EQUILIBRIUM MODELS

When a price war is initiated for whatever reason, a combination of price points can be reached where the partial derivative of profits of each producer i with

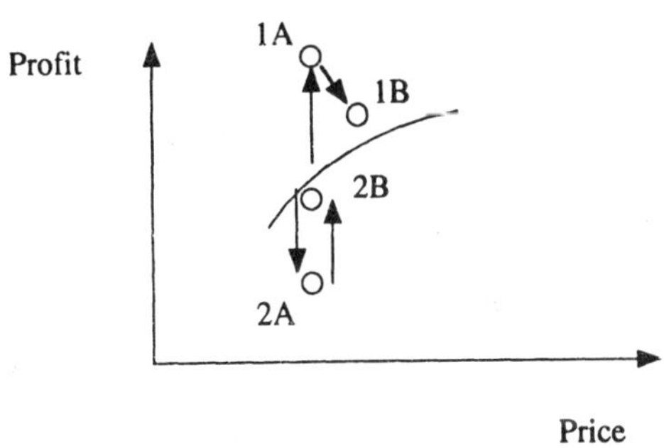

Figure 3.12 Change in profits of products 1 and 2 resulting from increase in value of product 1. For condition A, product 1 changes value but not price. For condition B, product 1 also increases value and price.

respect to its price is zero. For the example of three producers, these prices noted as P_1*, P_2* and P_3* are given by the matrix form:

$$\begin{bmatrix} P_1 * \\ P_2 * \\ P_3 * \end{bmatrix} = \begin{bmatrix} 2 & -\kappa & -\kappa \\ -\kappa & 2 & -\kappa \\ -\kappa & -\kappa & 2 \end{bmatrix}^{-1} \begin{bmatrix} 1 & 0 & 0 & 1 & -\kappa & -\kappa \\ 0 & 1 & 0 & -\kappa & 1 & -\kappa \\ 0 & 0 & 1 & -\kappa & -\kappa & 1 \end{bmatrix} \begin{bmatrix} C_1 \\ C_2 \\ C_3 \\ V_1 \\ V_2 \\ V_3 \end{bmatrix} \tag{3.18}$$

where $\kappa=1/3$. This price combination, if reached during a price war, should impart considerable resistance against any further reductions as there are no gains for any producer acting alone in reducing price further. If the three producers have the same costs (= 4) and values (= 20), we find that $P_1*=P_2*=P_3*=8$. The profit for a price change by a single producer about this point is shown in Figure 3.13.

Although the matrix solution given by Equation 3.18 can be written by inspection for an arbitrary number of competitors, the general solution for these prices for N competitors can also be written in the more insightful closed form given by [14]:

$$P_i* = \frac{(N^2+2N)C_i + (N^2+N+1)V_i + (N^2-N)(\overline{C}_{Cmpt} - \overline{V}_{Cmpt})}{2N^2+3N+1} \tag{3.19}$$

where $\underline{C}_{Cmpt}$ and $\underline{V}_{Cmpt}$ are the variable cost and value, respectively, averaged over competitors other than i. When the costs and values of the products are the same, we see from Equation 3.19 that the prices approach variable cost in the limit as N, the number of competitors in the segment, becomes large.

According to the Cournot/Bertrand model of pricing, each producer prices its product to maximize its profit under the assumption that competitors will not change their price. When each keeps repeating this strategy, then the same

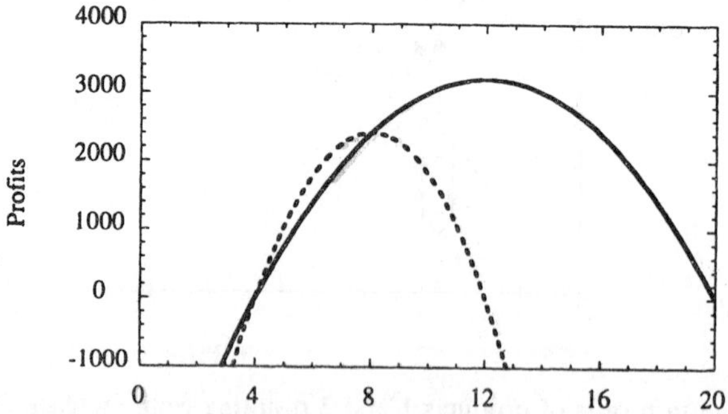

Figure 3.13 The price point $P* = 8$ when a competitor cannot improve profitability (even temporarily) by unilaterally reducing price.

price points given by Equation 3.19 would eventually be reached and price would then remain constant. In this context, the solutions given by Equation 3.19 also represent a generalization of the Cournot/Bertrand equilibrium price model for the case where the values and costs of the products are different. However, profitability at the Cournot/Bertrand equilibrium price can be negative when there are many competitors in the market and fixed costs are high. Such a price position is certainly not stable and it is probably not wise to think of these price points as representing a true equilibrium situation no matter what the fixed costs are. The view offered here is that the prices given by Equation 3.19 are best interpreted as the points where price wars should end if they have not ended before due to some sort of cooperative behavior emerging between the competitors to allow prices to rise. Also, over time, prices will change due to changes in variable cost and value as products are improved. In time, price wars can again appear. This interpretation of pricing is dynamic not static.

The plots of profits in Figures 3.11 and 3.13 show a relatively wide range of prices where profits are positive. However, these curves were constructed, as already stated, by ignoring fixed costs and investment which subtract uniformly from the profit curves and thereby significantly narrow the range of prices where the enterprise may be able to operate profitably. As fixed costs increase, the range of operating prices narrows but it remains centered about $(V + C)/2$.

3.5 Objective functions for making product decisions

3.5.1 A RULE OF THUMB FOR PRICING NEW PRODUCTS IN A COMPETITIVE MARKET

A dynamic viewpoint to pricing is taken here because we are interested in highly competitive markets where value and cost changes are occurring frequently in all of the products in the segment. As a result, a rule of thumb for making price changes from a given baseline will be all that we seek. We are not too concerned about what the price of products currently in production should be. Current prices are reflective of whatever strategy, sound or unsound, was used in the past to set them. The emphasis is on how to arrive at the price for a new product that is reasonable based upon the price of the prior product and the value and variable cost changes determined for the new product, the rule of thumb being:

$$P = P_{i,0} + \alpha_V\, \delta V_i + \alpha_C\, \delta C_i \tag{3.20}$$

The term $P_{i,0}$ is the price of the baseline product. The terms δV_i and δC_i are the change in value and the change in variable cost, respectively. The coefficients α_V and α_C are dimensionless quantities that are expected to lie between zero and one and whose sum is approximately one. The numerical choices for these coefficients represents a **pricing scenario**. For a monopoly, they are both equal to 1/2 if the baseline price has been set to maximize profit and the objective of the new product is also to maximize profit. It can be seen, on taking the partial derivatives of Equation 3.19 with respect to value and cost, that these coefficients are also approximately equal to one-half over a wide range of N in the vicinity of P_1*, P_2* and P_3*. On the other hand, if it is desired to keep demand constant for the new product, which could occur, for example, if there were no additional capacity available, then the pricing scenario would be $\alpha_V = 1$ and $\alpha_C = 0$.

3.5.2 PROFIT AS AN OBJECTIVE FUNCTION

The equation for the change in the forecast annual profits, A_i, is given by:

$$\begin{aligned}\delta A_i &= (P_{i,0} - C_{i,0})\delta D_i + (\delta P_i - \delta C_i)D_{i,0} \\ &-\delta F_i - (\delta M_i / Y_{RS})\end{aligned} \tag{3.21}$$

in which δD_i is the change in annual demand. The terms $C_{i,0}$ and $D_{i,0}$ are baseline variable cost and demand, respectively. The price change $\delta P_i (= P_i - P_{i,0})$ is computed in terms of the value and variable cost changes using Equation 3.20. The term δC_i is the change in variable cost, δF_i is the change in annual fixed costs, and δM_i is the change in total investment where:

$$M_i = \sum_{j=1,\, Y_{RS}} M_i(j);$$

and Y_{RS} is the number of years that the model is to be in production. The total investment in Equation 3.21 is prorated, for simplicity, equally over the years, Y_{RS}, of production. Traditional accounting practice would replace the term $\delta M_i / Y_{RS}$ in Equation 3.21 with the change in the annual depreciation of the total assets used to manufacture the product.

For incremental changes in the value or cost of product i, the enterprise should assume that competitors may not only react but are, in fact, planning their own incremental product actions whether or not others do anything. Decisions regarding such proposed changes should use Equation 3.7 for the change in demand because the actions of any single competitor (unless the number of competitors is very large) can greatly impact the profits of the others as we have already shown. If the results for the change in demand in Equation 3.7 are substituted into Equation 3.21, the expression below for the change in profits follows after rearrangement:

$$\begin{aligned}\delta A_i = K\Bigg\{&\delta P_i\Big[V_{i,0} + C_{i,0} - 2P_{i,0} - \kappa\sum_{j\neq 1}(V_{j,0} - P_{j,0}\Big] \\ &- \delta C_i\Big[V_{i,0} - P_{i,0} - \kappa\sum_{j\neq 1}(V_{j,0} - P_{j,0})\Big] \\ &+ (P_{i,0} - C_{i,0})\Big[\delta V_i - \kappa\sum_{j\neq 1}(\delta V_j - \delta P_j)\Big]\Bigg\} \\ &- \delta F_i - (\delta M_i / Y_{RS})\end{aligned} \tag{3.22}$$

This expression can be used as an objective function for weighing the merits of incremental product or process changes and various pricing scenarios if the externalities associated with environmental quality, as described in Chapter 4, have been internalized in the profitability equation. One basic point to note is that the price regime, where a loss in profitability from a variable cost increase can be offset to some degree by a corresponding increase in price, is restricted. When price satisfies the inequality given by:

$$P_{i,0} > \left[V_{i,0} + C_{i,0} - \kappa \sum_{j \neq 1} (V_j - P_j) \right] / 2, \tag{3.23}$$

a price increase will exacerbate the loss of profitability from a cost increase.

3.5.3 PRESENT WORTH

For major product proposals involving complex investment and product change actions over a period of time, the forecast profits $A_i(q,j)$ over the years j = 0, 1, 2, 3, ... given by:

$$A_i(q,j) = D_i(q,j)[P_i(q,j) - C_i(q,j)] - F_i(q,j) - M_i(q,j) \tag{3.24}$$

should first be computed for each product scenario q under consideration. The merit of each scenario can be defined in terms of its present worth, $\Pi_i(q)$, given by:

$$\Pi_i(q) = \sum_{j=0}^{Y_{RS}} \frac{A_i(q,j)}{(1+I_R)^j} \tag{3.25}$$

where I_R is the interest rate. The index j refers to the end of the year. At the end of year zero, the starting point for the summation, there is generally no demand as the product is not ready but there is often a large investment.

The so-called internal rate of return, I_{RR}, is the interest rate which when substituted into Equation 3.25 for I_R gives zero for present worth. It can be computed by trial and error substitutions for I_R in Equation 3.25. The outcomes for each product scenario can also be evaluated for different possible product actions of competitors by postulating what they are likely to do in terms of value increments and pricing and when they are likely to make them. The introduction of the possible actions of competitors allows an estimate to be made of the robustness of a product scenario under investigation. Other sources of variation that should be evaluated are ranges of material costs, interest rates, labor rates, transportation rates, tax policy, and so forth.

Another property of interest to new product programs is the time to break even, Y_{BE}, for a given scenario. This is given by the solution to:

$$\sum_{j=0}^{Y_{BE}} A_i(q,j) = 0 \tag{3.26}$$

3.6 Market segmentation by price and value

Assume that you are a manufacturer of an automobile which competes in the entry level of the product segment for family sedans represented schematically in sketch A in Figure 3.14. This automobile has been successful and you now want to add another to your product line as shown by sketch B. The question you are wrestling with is, 'How large should the price increment δP be?' If it is too small, the two vehicles will compete against each other but, if it is too large, few people will be able to afford it.

For the new car to command a higher price, it has to have attributes which differentiate it favorably from the entry level car. In other words, price differentiation between two products also requires that their value also be differentiated as shown in sketch C of Figure 3.14. The differentiation in value must be large enough for persons to perceive the two vehicles as being distinct from each other. One of the more direct methods for differentiating the vehicle in the higher segment, of course, would be to make it larger than the entry level vehicle by an appropriate amount.

Products within a segment are also differentiated as already discussed in developing the equations for demand in this chapter. For automobiles, such attributes would include reliability, ride, handling, interior noise, comfort, and appearance. Size differences can also be used to differentiate automobiles within a segment as amply demonstrated by the Chrysler's LH vehicles in the upper middle product segment and its Neon in the small product segment.

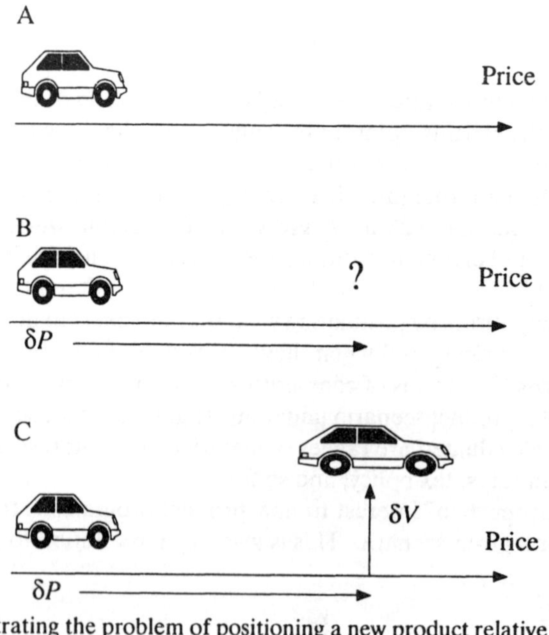

Figure 3.14 Illustrating the problem of positioning a new product relative to an existing product in terms of value and price.

As a result, we can view product segmentation of family sedans in terms of a two-dimensional value versus price plot as shown in Figure 3.15 where a dotted line has been drawn joining the centers of the four product segments. The centers represent the reference points for the Taylor's expansions given by Equation 3.1 for each product segment and the dotted line expresses the value/price relationship between product segments given by Equation 3.10.

An example of a value versus price segmentation plot for family automobiles is shown in Figure 3.16 [14]. The values and prices have been normalized by dividing through by a single price, P_0. Only the top six vehicles in terms of sales were analyzed for each segment and values were computed using Equation 3.11.

Within each product segment, the value/price relationship is given by Equation 3.11. This relationship has been drawn for convenience as a straight line which, if extended, runs below the product segment above it. This implies that there is a value advantage to vehicles in the higher product segment which can be attributed not so much to certain tangible attributes but to image. If a manufacturer has vehicles in each of the product segments shown in Figure 3.16, care must be given not to be at the high end in a given product segment and at the low end in the segment above it. This situation is undesirable because the image difference could lead to a dramatic loss of sales of the vehicle in the lower product segment to the vehicle in the higher segment. Such competition and loss of sales between brands of the same manufacturer is called 'cannibalization.'

Thus manufacturers should simultaneously consider the positioning of brands within and between product segments. If price is too high with respect to value, the manufacturer will not compete well within the product segment and, if the price positioning between brands from a single manufacturer in adja-

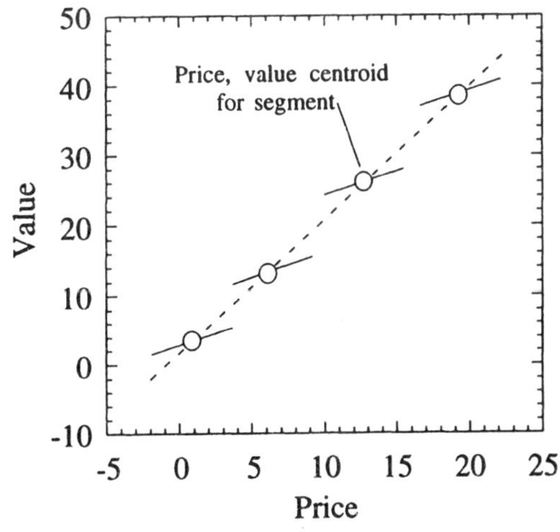

Figure 3.15 Schematic representation of value/price positioning between segments. The solid lines through each of the segment centroids represents the distribution of competing brands within a segment. (Source: reprinted with permission from SAE Paper No. 970765 © 1997 Society of Automotive Engineers, Inc.)

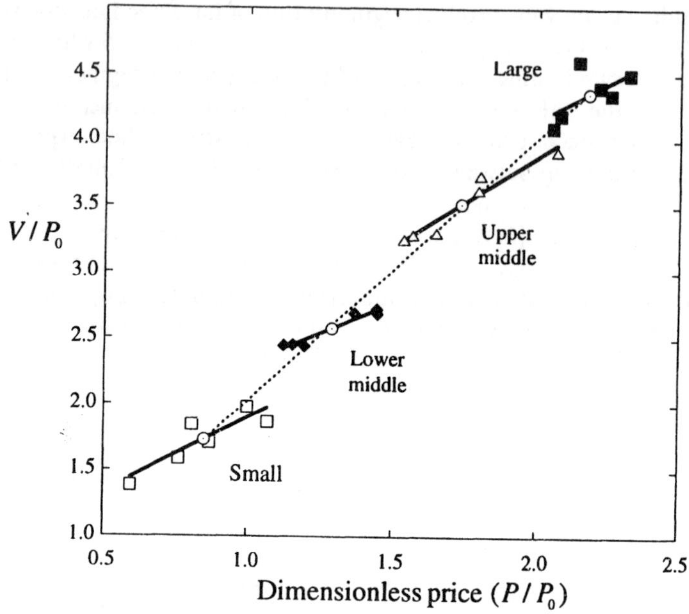

Figure 3.16 Value versus price plots for top selling brands in U.S. family sedan segments for the 1993 model year. (Source: reprinted with permission from SAE Paper No. 970765 © 1997 Society of Automotive Engineers, Inc.)

cent segments is too close, the problem of cannibalization arises. Finally, if the positioning is too far apart, sales can be lost to competitors which fill the gap.

3.6.1 PSYCHOMETRIC PRICING OF A PRODUCT LINE

Kent Monroe [15], using psychometric arguments, has suggested that manufacturers should price their brands in a product line by a percentage amount. As the price gets larger the difference in price between adjacent brands from the same manufacturer should increase proportionally. The report by Louden and Della Bitta, discussed in Chapter 2, that a price reduction of at least 15% is needed to attract people to a sale [16], is supportive of psychophysical arguments in relation to price. The findings of prospect theory are also supportive of psychometric arguments in relation to price and value.

If the entry level brand for a product is at price $P(1)$ and if the psychometric price detection point for the second product in the line is located at $[1+\theta]P(1)$, then it follows that for an arbitrary product segment level j:

$$P(j)=[1+\theta]P(j-1) \tag{3.27}$$

In terms of the entry level price, we find that:

$$P(j)=[1+\theta]^{(j-1)}P(1). \tag{3.28}$$

On taking logarithms, we find that the price points for the product line will appear at a constant distance apart when graphed along the scale j = 1, 2, 3, ...:

$$\mathrm{Log}(P(j)) = \mathrm{Log}\left(\frac{P(1)}{1+\theta}\right) + j\mathrm{Log}(1+\theta) \qquad (3.29)$$

the slope of log price being $\mathrm{Log}(1+\theta)$.

If we interpret the 15% rule of Louden and Della Bitta such that the lower priced product at j–1 should be no closer in price than $P(j-1)=0.85P(j)$ to the product priced above it, then $\theta \approx 0.18$. As the separation between the two prices becomes larger, it is straightforward to show that an opportunity should arise for inserting a new product between the two prices when $P(j)>(1+\theta)^2P(j-1)$. Thus a range for θ between 0.18 and 0.39 is expected if the price spacing follows psychophysical behavior and the 15% rule is valid for this application.

The four-door family sedan market was examined to see if the pricing by seven major manufacturers followed the psychometric model or simply a linear model for pricing increments [14]. Analysis of the data, although sparse in terms of a wide range of prices needed for an unambiguous test, favored the psychometric model. The resulting θ values computed for the vehicles, shown in Figure 3.17 as a normal probability plot, lie within the range expected of 0.18 to 0.39 except for one manufacturer. The results are seen to cluster around 0.3, midway between the two limits.

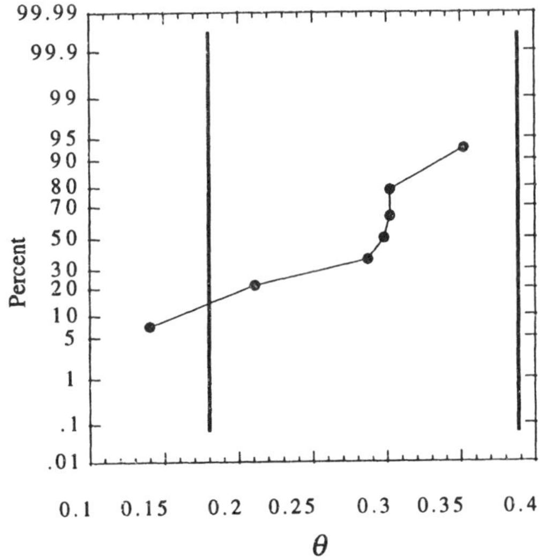

Figure 3.17 Determination of θ from slopes of plots of log(price) versus segment index j using Equation 3.29 for four-door family sedans in 1993 model year. (Source: reprinted with permission from SAE Paper No. 970765 © 1997 Society of Automotive Engineers, Inc.)

References

1. Leong, G.K., Snyder, D.L., and Ward, P.T. (1990) Research in the process and content of manufacturing strategy. *Omega,* **18**, 109–22.
2. Cook, H.E. (1992) Organizing manufacturing enterprises for customer satisfaction, in *Manufacturing Systems: Foundations of World Class Practice*, (eds J.A. Heim and W. Dale Compton), National Academy Press, pp. 116–27.
3. Leong, G.K., Snyder, D.L. and Ward, P.T. (1990) Research in the process and content of manufacturing strategy. *Omega,* **18**.
4. Dertouzos, M.L., Lester, R.K., and Solow, R.M. (1989) *Made In America*, MIT Press, Cambridge, MA.
5. Clark, K. and Fujimoto, T. (1991) *Product Development Performance,* Harvard Business Press, Boston, MA.
6. Buzzell, R.D. and Gale, B. (1987) *The PIMS Principles*, Free Press, New York.
7. Stalk, Jr., G. and Hout, T.M. (1990) *Competing Against Time*, Free Press, New York.
8. Krupka, D.C. (1992) Time as a primary system metric, in *Manufacturing Systems: Foundations of World Class Practice,* pp. 166–72.
9. Little, J.D.C. (1970) Models and managers: the concept of a decision calculus. *Management Science*, **16**, B-466–85.
10. Cook, H.E. and DeVor, R.E. (1991) On competitive manufacturing enterprises I. The S-model and the theory of quality. *Manufacturing Review*, **4**, 96–105.
11. Cook, H.E. (1992) New avenues to total quality management. *Manufacturing Review*, **5**, 284–92.
12. Cook, H.E. and Kolli, R.P. (1994) Using value benchmarking to plan and price new products and processes. *Manufacturing Review*, **7**, 134–47.
13. Ben-Akiva, M. and Lerman, S.R. (1985) *Discrete Choice Analysis: Theory and Applications to Travel Demand*, MIT Press, Cambridge, MA, pp. 103–4.
14. Monroe, E.M., Silver, R.L., and Cook, H.E. (1997) Value versus price segmentation of family automobiles. *SAE Paper 970765*, Society of Automotive Engineers, Warrendale, PA.
15. Monroe, K.B. (1990) *Pricing, Making Profitable Decisions*, 2nd edn, McGraw-Hill, New York, pp. 309–12.
16. Loudon, D.L. and Della Bitta, A.J. (1988) *Consumer Behavior: Concepts and Applications,* 3rd edn., McGraw-Hill, New York, p. 405.

4 Definition of quality

4.1 Attributes of quality

Product quality has many interpretations. Garvin [1] defined eight dimensions of quality:

1. performance
2. features
3. reliability
4. conformance
5. durability
6. serviceability
7. aesthetics
8. perceived quality

Let's consider these eight dimensions as applied to an automobile. Performance relates to its acceleration, fuel economy, emissions levels, ride, vibration level, interior noise, door closing efforts, steering efforts, stopping distance, time for the air conditioner to cool down the interior after sitting in a parking lot, etc., etc. Features refer to various options and gadgets including things like map pockets, radio, cellular phone, automatic transmissions, V8 engine, tinted glass, reclining seats, and air bags. Reliability is a measure of the degree of confidence that can be expressed in the automobile and its options performing without failure. Conformance refers to the degree to which the actual performance of the automobile meets its stated performance specifications. Durability is a measure of the amount of time in service before the automobile needs repair because something broke. Serviceability measures the ease of making routine and non-routine maintenance. Aesthetics refer to the softer specifications such as exterior and interior styling, fit and finish of the panels, the sound of the door closing, the feel of the switches and brake pedal, and the arrangement of components under the hood. Perceived quality is the overall satisfaction that the customer has in the automobile as a result of experiencing the other seven dimensions.

4.2 The quality loss function

A different approach to quality embracing the loss function concept [2] was used by Taguchi and Wu [3] to define, instead of the quality itself, the loss of quality to society per unit sold. Taguchi's premise is that no product is perfect so it is appropriate to measure the loss of quality in terms of the deviation of a specific performance level from perfection. Quality loss according to Taguchi is the sum of the cost to manufacture the product and the 'cost of inferior quality' caused by the product being off its unattainable, ideal specification. The summation of these two terms when displayed as a function of the product attribute of interest will generally lead to a minimum in the loss of quality as shown in Figure 4.1 for a so-called 'smaller is better' condition for the attribute. The manufacturing enterprise should target the attribute which gives the minimum point for the loss because it generates the minimum loss to society even though it may be far from the ideal specification which will generally be prohibitively expensive to manufacture.

Manufacturing variations, however, cause the actual attributes to deviate about the target specification and, as a result, the average loss of quality per unit of product produced is increased over what it would be at the minimum. The amount of loss measured from the target g_T can be computed from the statistical distribution of the attributes $g(1)$, $g(2)$, ..., $g(n_P)$ where n_P is the total number of products manufactured. This calculation requires that the second derivative of the loss function, $\partial^2 L/\partial g^2$, in the vicinity of the target be known. A rough estimate of $\partial^2 L/\partial g^2$ can be made using the relationship:

$$\frac{\partial^2 L}{2\partial g^2} \approx \frac{C_W}{\Delta^2} \tag{4.1}$$

where C_W is the average warranty cost in dollars per unit generated by an attribute deviation from the target equal to Δ. In the limit of small deviations $\delta g_j (= g_j - \mu)$ about the population mean of the attributes g for the manufactured products:

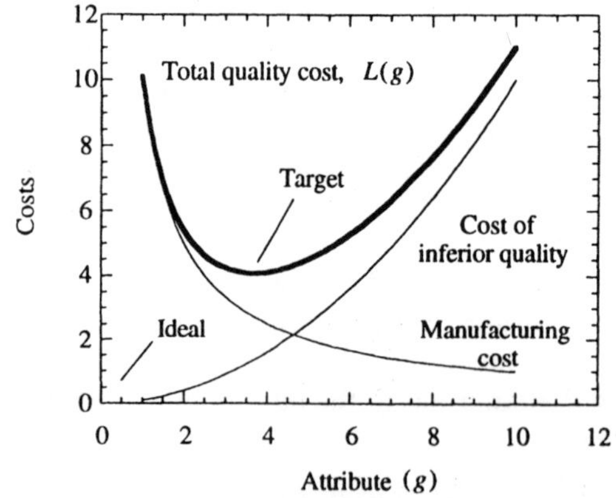

Figure 4.1 The components of quality according to Taguchi and Wu. (Source: adapted from G. Taguchi and T. Wu (1980) *Introduction to Off-Line Quality Control*, Central Japan Quality Association, Nagoya, Japan.)

$$\mu = \frac{1}{n_P} \sum_j g_j, \tag{4.2}$$

it follows that the loss for a single part is given by:

$$L(g_j) = L(\bar{g}) + \left[\frac{\partial L}{\partial g}\right]\delta g_j + \left[\frac{\partial^2 L}{2\partial g^2}\right]\delta g_j^2 \tag{4.3}$$

Summing this expression for the total loss for the n_p parts produced and taking the average yields:

$$\bar{L}(\mu, \sigma) = L(\mu) + \frac{\partial^2 L}{2\partial g^2}\sigma^2 \tag{4.4}$$

where the variance σ^2 has been used in averaging the summation over $(\delta g_j)^2$:

$$\sigma^2 = \frac{1}{n_P}\sum_{j=1}^{n_P}(g_j - \mu)^2 \tag{4.5}$$

The quantity $L(\mu)$ is given by:

$$L(\mu) = L(g_T) + \frac{\partial^2 L}{2\partial g^2}(\mu - g_T)^2. \tag{4.6}$$

Defining the loss function as measured instead from its minimum at the target specification in the manner of Taguchi, we have:

$$L = \bar{L}(\mu, \sigma) - L(g_T),$$

which from Equations 4.4 and 4.6 can be written as:

$$L = \frac{\partial^2 L}{2\partial g^2}\left[(\mu - g_T)^2 + \sigma^2\right]. \tag{4.7}$$

There are three things suggested by Equation 4.7 to improve quality:

1. Reduce variance σ^2.
2. Move the mean μ of the population toward the target g_T.
3. Reduce the curvature, $\partial^2 L/\partial g^2$, which is sometime referred to as improving latitude, of the loss function.

A very important fourth possibility also exists which is to discover ways of reducing manufacturing costs, thus being able to lower $L(g_T)$ itself.

Taguchi's model has proven to be a very useful guide to improving quality. His design of experiment methodology has a large following with product and process engineers because the formalism, discussed in Chapter 8, is a practical and powerful tool for reducing variance once the target specifications have been set.

4.3 Total quality

Because the setting of system-level targets is a vital step in the product realization process, we need to explore how to combine and extend the breadth of quality as seen by Garvin with the important and powerful operational formalism for quality introduced by Taguchi. These two approaches to quality are reconciled in the S-model which defines quality as the net value of a product to society [4]. This definition is more encompassing than Garvin's eight dimensions. It also is a more complete and general definition of quality itself than the loss-of-quality description given by Taguchi for parameter and tolerance design. Importantly, its mathematical formalism can use all the tools and most of the concepts already developed for design of experiments, Taguchi methods, and elementary principles of microeconomic theory.

The annual net value of the product to society, Q_i, is a sum of the net value received by its customers $D_i(\Psi_i - P_i)$, the net value received by the enterprise $D_i(P_i - C_i) - W_i$, and the net value received by the remainder of society G_i, termed environmental quality:

$$Q_i = D_i(\Psi_i - C_i) - W_i + G_i \tag{4.8}$$

where W_i is the annual cost of the investment, M_i, plus fixed costs. The quantity Ψ_i is the average of product value actually received by the customer over the time period that the product was used until resale or scrap. For Q_i to be positive, the costs have to be less than $[\Psi_i + (G_i - W_i)/D_i]$. If perceived value is taken as approximately equal to Ψ_i in Equation 4.8, it follows that the net value to society generated for products is given by:

$$Q_i \approx D_i(V_i - C_i) - W_i + G_i \tag{4.9}$$

When price is set equal to $(V_i + C_i)/2$, a comparison of Equation 4.9 to Equation 3.24 shows that total quality and profit depend similarly on $(V_i - C_i)^2$ if Equation 3.8 is substituted for D_i. High quality and profit occur when value is high and costs are low. Quality is a bottom-line metric just like profit and market share in that it is a dependent variable defined in terms of value and costs.

The expression for Q_i given by Equation 4.8 and its less general form Equation 4.9 represents an absolute aggregate measure of quality (defined for convenience on an annual basis) and can be used to evaluate or forecast the merits of rather diverse products or actions. For example, the total aggregate quality for an inexpensive but reliable watch that millions use may be much greater than the total aggregate quality of an expensive watch used by very few people. Value per unit, of course, could be much higher for the expensive timepiece.

The inclusion in the definition of quality of the total cost to manufacture the product supports the efficient use of resources. Society suffers a loss when resources are not effectively used – the quality of life is less than it could be. Two identical products may have the same value to the customer but the one manufactured most efficiently will have the higher net value to society provided the G_i terms are the same.

4.4 Environmental quality

In considering new products for potential development, it is quite evident that forecasts of profits and market share are necessary but not sufficient conditions for deciding on the merits of the investment as the total societal impact of the product must also be considered. Environmental quality is controlled by the integrity of the enterprise and the integrity of the state through its enforcement of laws, standards, and regulations that are designed to safeguard the environment, persons in the workplace, unsuspecting users, and the rest of society from adverse impact resulting from the manufacture, sale, and use of the product. These form part of the externalities of the market transaction between buyer and seller in microeconomic theory [5].

The costs needed to meet these standards are part of the total manufacturing costs – variable, fixed, and investment – for labor, stock, tools, and facilities. The cost–benefit analyses needed to establish standards are complex and controversial [6].

For certain products, the requirements for the term G_i have been set by legislation, e.g. radiation emissions by video terminals, exhaust emissions by automobiles, and smokestack emissions by power plants. In some instances, legislation has generated significant debates of benefits versus cost. Fuel economy legislation for automobiles has been a major item of contention. An alternative to direct legislation on vehicle fuel economy would be to levy a fee on gasoline to reflect the total cost of externalities to society for the use of this fuel. In spite of these issues, there is, nevertheless, considerable merit in having legislation establish control over environmental quality as this results in a common requirement for all manufacturers and limits unscrupulous manufacturers from gaining a short-term profit at the expense of society over the long term. Moreover, even in the absence of regulation, manufacturers should not ignore the G_i term because it could lead to tragic results and costly actions later as has happened with toxic waste dumps and unsafe products. Thus, the market transaction, as described earlier in Chapter 2, has and needs a third party to represent the needs of the rest of society. Without this third party, the buyer and seller can be myopic to the longer term effects on society of product manufacture and use.

Let's consider an example problem posed by an environmental quality factor, attributed to an automobile as follows: a federal agency has concluded that the environmental factor 'e' relating to vehicle emissions which it has been studying extensively over the past eight years results in a net loss to society of $300e$ dollars per year over the lifetime of each vehicle in service.

To illustrate how the S-model evaluates the trade-offs relative to environmental quality as contained in the expression for total quality, let's consider further how the environmental factor problem outlined above may impact an automotive manufacturer in more detail. We assume for this example that manufacturer A is a monopoly for simplicity and has a product which historically had a slope of annual demand of $-K = -50$ unit/\$ and fixed costs of zero and that the value and cost of the vehicle were \$20 000 and \$12 000, respectively, before the ruling by the agency on the environmental factor, ***e***. After the ruling, the manufacturer studied how it could best reduce the level of the environmental factor and concluded that the added variable costs $C(e)$ per vehicle needed to reduce it from its current, uncontrolled level of $e_0=9$ was given by:

$$C(e) = 50\{(e_0 - e) + 0.05(e_0 - e)^3\}. \tag{4.10}$$

Before the discovery that the environmental factor ***e*** reduced environmental quality, the total quality of the manufacturer's product was calculated to be \$800 000 000 on an annual basis. On discovering the true environmental cost of quality, the agency imposed a levy of $300e$ dollars for each vehicle produced. The effect of all this on the total quality of manufacturer A's product is described in Figure 4.2. The figure should not be interpreted as total quality being reduced from the baseline level but as the baseline level being in error, unwittingly. The true level of quality at $e_0=9$ was \$3.6 10^8 given by the uncontrolled level for the lower curve. Total quality improves by having the manufacturer act to reduce the environmental factor to approximately the level of 3.2 which is the point for the maximum. This is the cost–benefit balance point. The curves of variable cost, value minus the environmental fee, and price are shown as a function of the environmental factor ***e*** in Figure 4.3.

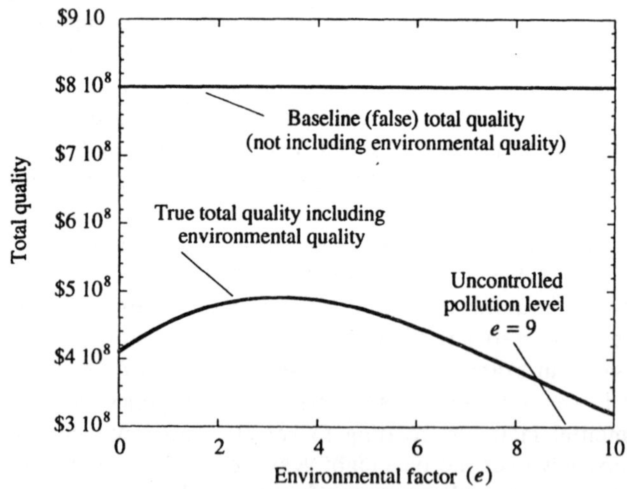

Figure 4.2 Hypothetical example of the influence of an environmental factor on total quality.

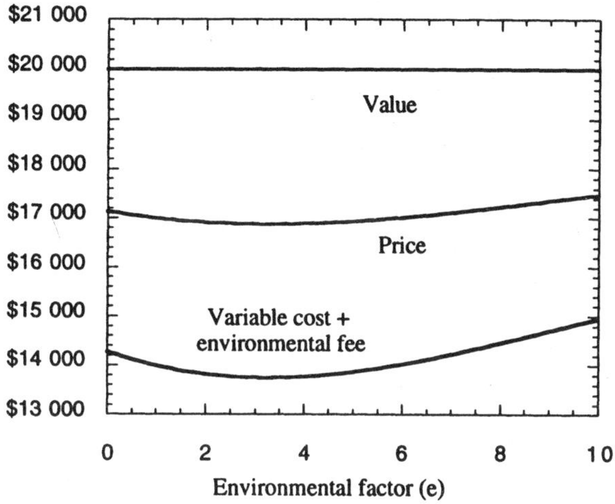

Figure 4.3 Product variable cost, price, and value less environmental fee as a function of the hypothetical environmental factor.

4.5 The driving force for product realization

The term ΔQ_i given by Equations 4.8 and 4.9 represents the forecast aggregate free value that should be given to society on using the product over the period of a year. As such, it is society's driving force for the realization of product *i*. The desire of society is to increase free value according to this viewpoint. The details differ between individuals because of differences in how persons assess free value given an array of alternatives, how they discount future implications of free value, and how they weigh risks in choosing between the uncertain alternatives. Free value, which was the driving force in the stock market model studied in Chapter 2, is very similar to the concept of utility maximization from microeconomic theory; however, we do not propose that individuals truly act to maximize free value but operate in Simon's satisficing mode in the sense that we act on incomplete information because time and other concerns are pressuring us. Moreover, there is a random force in our cognitive decision making that tends to move us one way at one moment and another the next.

The rate at which we generate quality or aggregate free value in developing and improving new products is likely proportional, to a first approximation, to the difference in quality between the current state and the projected new state, $\Delta Q_i(\infty)$. When divided by the characteristic action time τ_E demonstrated by the enterprise in responding to a goal, it follows that the time dependence for quality improvement for such a model is given by

$$\Delta Q_i(t) = \Delta Q_i(0) + \{\Delta Q_i(\infty) - \Delta Q_i(0)\}\{1 - \exp(-t/\tau_E)\} \qquad (4.11)$$

At first glance the equation suggests that quality goals should always be set very high; however, the above phenomenological model does not apply when free value is large.

Seemingly impossible goals may either be demoralizing or ignored. The model suggests continuous improvement – that new goals should be developed as current goals are approached because this is when things slow down and that ways should be found to reduce the characteristic time, τ_E, which is proportional to the inertia of the manufacturing enterprise in getting a job accomplished.

The driving forces for product realization and product decay work against each other, one creating local order and the other, which increases entropy, eventually destroying it. The outcome, nevertheless, should be improvement of the product species, the silver lining referred to earlier. These ideas embrace those of a general systems model [7] where we can imagine two highly coupled sets of differential equations, one set governing the living world in which its individual members – people, animals, flowers, and even bacteria – act in ways to improve free value and the other set governing both the living and the inorganic world which attempts to maximize entropy (minimize free energy). The equations for entropy maximization may have the upper hand in the long run but paradoxically entropy production can drive the formation of life because by having living things using the energy of the sun in their search for free value, the overall rate of entropy production is greater than it would be without life. Overall disorder and entropy production is occurring faster with life than it would without life!

References

1. Garvin, D.A. (1984, Fall) What does 'product quality' really mean? *Sloan Management Review*, **26**(1), 25–43.
2. Deutsch, R. (1969) *System Analysis Techniques*, Prentice-Hall, Englewood Cliffs, NJ, pp. 57–65.
3. Taguchi, G. and Wu, T. (1980) *Introduction to Off-Line Quality Control*, Central Japan Quality Association, Nagoya, Japan.
4. Cook, H.E. and DeVor, R.E. (1991) On competitive manufacturing enterprises I: The S-model and theory of quality, *Manufacturing Review*, **4**, 96–105.
5. Salvatore, D. (1991) *Microeconomics*, HarperCollins, New York, pp. 590–614.
6. Tester, J.W., Wood D.O., and Ferrari N.A. (eds) (1991) *Energy and the Environment in the 21st Century*, MIT Press, Cambridge, MA.
7. von Bertalanffy, L. (1968) *General System Theory*, George Braziller, New York.

5 Forecasting the value of future products

5.1 Importance of value

Understanding how value is generated is vital to the development of successful products because value is the only fundamental metric which makes a positive contribution to all of the bottom-line metrics. By contrast, all costs – variable, fixed, and investment – subtract from total quality and profits. Moreover, as already noted, the net impact of manufactured products on environmental quality will also almost always be negative. Thus, not only is product value important to the bottom line, it must be of sufficient size to considerably outweigh and offset the combination of manufacturing costs and environmental losses associated with the manufacture, use, and disposal of the product.

As discussed in Chapter 2, customers are expected to associate value with tangible performance specifications as well as with the softer more subjective attributes of the product. Examples of totally subjective attributes versus performance-oriented attributes are shown in Figure 5.1 for an automobile. A 'value curve' is assumed to exist for each performance attribute. Any product attribute that directly influences customer needs is termed a system attribute for the product. If value as a function of the system attributes have been estimated, the value of a proposed future product can be forecast from the changes to be made in each of the system attributes in going from their current baseline conditions to those for the future product.

The value of a product's soft attributes generally cannot be expressed in terms of a well-defined, continuous function of a specific property of the product. Instead, potential customers need to be asked to sample product alternatives with different soft attributes and then state what they would be willing to pay for each relative to a baseline condition. For example, customers could be asked to compare a rendering of a proposed new design of a coffee pot with the rendering of the current baseline pot and asked, with all other things being equal, how much would they be willing to pay for the alternative design versus the baseline. As value is ultimately subjective for all attributes, the same technique also can be used to determine the changes in value (willingness to pay a premium for a value increase or to accept a price decrease for a reduction in value) for changes in the more tangible performance specifications.

Objective partially

Subjective totally

Brand name
Country of Manufacturer
Color
Shape

Interior noise
Seat vibration
0 to 60 mph time
Fuel economy
Interior roominess
Reliability
Resale price

Figure 5.1 Example of totally subjective and partially objective attributes for a vehicle. (Source: reprinted with permission from SAE Paper 960001 © 1996 Society of Automotive Engineers, Inc.)

Because soft attributes are highly subjective by definition, their value may change rather easily over time as consumer tastes change. Changes in the fashion of wearing apparel is a good example. Certain automotive brands, for example, have been dropped because the experience with the brand was not good. Even if the product were totally redesigned to remove the weak or offending features, the legacy of the poor performance of earlier versions can be difficult to overcome. It may be easier to establish a new brand name in a positive manner than to restore the image of a fallen brand.

5.2 Knowledge of your customer

Understanding how the value of a future product can be improved begins with an understanding of the needs of customers and how those needs are reflected in the system attributes of the product. The importance of gaining detailed knowledge of how customers use the product cannot be over-stressed as they may put the product to use in an effective manner that neither you nor they envisioned at the point of sale and, if this is happening, you had better be the first to know about it. Understanding the true nature of the overall task faced by your customer is also crucial. For example, an improvement in the brute speed of a computer will not be of much help if exchange of data across a problematic network is the bottleneck for the task at hand. Three important questions are:

1. How does your product generate value for your customer as measured by the product's ability to help solve his or her overall task?
2. How does the product reduce his or her total costs and time to do the job?
3. By gaining special knowledge of your customers' tasks, can you assure yourself that what the customer states as a want is all he or she truly needs?

This last question is especially important when the customer is venturing into unfamiliar areas of technology. You may also find that the market is sufficiently segmented to develop variations in your product. What you considered as one consumer segment may be better represented as several.

In getting closer to the customer, you are attempting to define more accurately the continuous relationship between value and the system attributes for your product. Situations can arise, however, when the potential customer may not be the best judge of the merits of a ground-breaking, new product.

5.2.1 CHRYSLER MINIVAN

The Chrysler minivan is a case in point. Initial market research did not show strong acceptance of the proposed product. But there were strong believers in the concept within the company at the middle management level, primarily engineers and planners that worked on the concept for some time and were very familiar with the special attributes of the vehicle. They saw it filling a niche between the family sedan and the large size van that was not fully captured by the station wagon. When top management also became convinced, Chrysler proceeded toward the introduction of this vehicle at high risk by typical standards. The company, however, had little choice at the time as its fortunes were such that it needed a major product innovation to survive. It was a greater risk, approaching 100%, to stand pat with existing products versus going ahead with a bold, new product that did not 'research well.'

An interesting anecdote is that the Canadian UAW conducted its own market survey fearing that the company had targeted a mediocre product for their Windsor, Ontario plant when it was announced that the minivan was to be built there. Not surprisingly, their market research found the same lukewarm reception. Nevertheless, champions prevailed, the result being a remarkably innovative product whose financial success saved the company because it generated wide appeal in the marketplace in contrast to what marketing research projected.

The lesson from this case is that products which are significant deviations from the norm are hard for potential customers to judge until they have had sufficient experience using the product. Were owners of vacuum tube radios in 1950 asking for Walkmans with transistorized, solid state circuitry? The accuracy of the minivan survey could obviously have been improved by having a hundred or so prototypes in the hands of customers for a couple of months for in-depth evaluations. At roughly \$300 000 a copy for a prototype vehicle, this would have been an expensive proposition for the struggling automaker in the early 1980s. It would probably also have delayed the introduction date by at least six months to build, distribute, test, analyze the results and act upon the conclusions, a delay which in itself could have been fatal to the fortunes of the company. Seizing an opportunity almost always requires taking risk.

Another note of general interest is the design approach taken by Chrysler versus that taken by competitors. The Chrysler product was designed by 'car guys' (as opposed to 'truck guys') and thought of internally as a 'tall car' having easy egress and ingress because of front wheel drive and ample headroom. It was a replacement for and an improvement over the station wagon in their minds because of greater command (vision) of the road and interior space. Competitors' minivans were designed by 'truck guys' and were more akin to a shrunk, full-size, rear drive van, having retained significant towing capacity with improved fuel economy. As a result, the products appealed to somewhat different consumer segments. Chrysler afterwards dropped station wagons from its model line-up, whereas competitors have still retained a presence in station wagons. In keeping with its car origins, it is not surprising that Chrysler's minivan was the first to offer four doors, which is the number of doors preferred by most car buyers.

5.3 Conjoint analysis case study

Consider the following hypothetical market research problem. Farsighted, Inc. is developing a new line of binoculars. The various attributes under consideration are three designs or styles, three magnifications, three prices, and a feature that eases use of the binoculars for persons wearing eye glasses in that they can leave their glasses on and not have restricted or uncomfortable viewing. This results in a total of 54 different combinations: $3\times3\times3\times2$. If the value of each attribute is independent of the others, however, there are only eight unknowns to evaluate, a baseline condition and the seven deviations from it: two styles, two magnifications, two prices and the single feature for eye glass wearers. The styles under consideration are shown as 1, 2, and 3 in Figure 5.2.

The strength of customers' preferences for these attributes can be obtained by a versatile survey technique known as conjoint analysis [1]. In a typical application, several product attributes are jointly considered (thus the name) in a survey. In designing the survey, the number of attributes should be kept below seven because people have trouble keeping track of more than six in developing the rankings [2]. In the classical approach to conjoint analysis, which we will now review in part, customers are asked to force rank products as described on a series of cards, each having different attributes defined according to the structure of a specific experimental design chosen for the survey. In addition to stated performance specifications, renderings of various styles for the product are shown on the cards so that their aesthetic appeals can also be ranked.

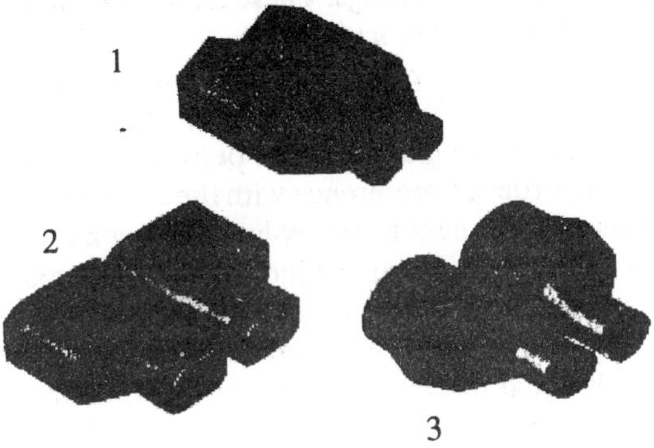

Figure 5.2 Three binocular styles.

The prices and magnifications chosen for evaluation are \$85, \$110, and \$150 and 3×, 7×, and 10×, respectively. The baseline condition is style 1, \$110 price, 7× magnification, and no provision for eye glass wearers. The experimental design shown in Figure 5.3 is in the form of a spreadsheet array whose elements $X_{ij}(q)$ in column ij and row q are zeros and ones. Only the minimum number of trials, eight, equal to the number of unknowns are used for simplicity. Each column represents one of the types and levels of the variables or factors being studied. The column index ij is used as a shorthand notation for the variable listed above it, i denoting the type of factor and j the level of that factor. The baseline level for each factor is defined by $j=0$. If the set point ij is present for a given trial q, then $X_{ij}(q)=1$ in the row for trial q; if not present then $X_{ij}(q)=0$.

The first column shown for data entry (column B) is a list of numbers q = 1, 2, 3, ..., 8 that label each trial. The next four columns are for the baseline factors. The sixth column, with heading 0, is all ones, representing the combined effect of the baseline attributes. The next seven columns represent the seven factors

	B	C	D	E	F	G	H	I	J	K	L	M	N	O	P
4			Baseline factors						Off baseline factors						
5															
6		Style	Mag	Price	Eye		Style	Style	Mag	Mag	Price	Price	Eye		
7		1	7×	110	No		2	3	3×	10×	85	150	Yes		TW
8	Trial	10	20	30	40	0	11	12	21	22	31	32	41	Rank	9-Rank
9	1	1	1	1	1	1	0	0	0	0	0	0	0	5	4
10	2	1	0	0	0	1	0	0	0	1	1	0	1	2	7
11	3	1	0	1	1	1	0	0	1	0	1	0	0	7	2
12	4	0	0	0	0	1	1	0	0	1	1	0	1	1	8
13	5	0	1	0	1	1	1	0	0	0	0	1	0	6	3
14	6	0	0	0	0	1	0	1	1	0	0	1	1	8	1
15	7	0	0	1	1	1	0	1	0	1	0	1	0	4	5
16	8	0	1	0	0	1	1	0	0	0	1	0	1	3	6

Figure 5.3 The experimental design used for the binoculars conjoint analysis experiment.

that differ from the baseline settings. These factors are all zero for the first trial as it represents the baseline binoculars.

The series of eight cards generated according to the above experimental design are shown in Figure 5.4. Listed on Card #1 are the set point attributes noted with 1s for Trial 1 in Figure 5.3. The person being interviewed is asked to arrange the cards in order of preference with the most preferred being first. The rankings are shown in the next to last column of Figure 5.3. Card number one was placed fifth, card two placed second, etc. by the person surveyed. These rankings were then subtracted from nine to arrive at the ordinal trial worths (TW) shown in the last column of Figure 5.3. The most preferred trial has the highest trial worth equal to eight with this transformation. The analysis of the experimental results discussed below is based upon the assumption, for simplicity, that attributes do not interact.

The set point variables, $X_{ij}(q)$, are coupled to the measured ordinal trial worths, TW(q) for q ranging from 1 to 8, through the linear model for the experiments given by the matrix equation (Appendix B):

$$[\mathbf{X}][\lambda]=[\mathbf{TW}] \tag{5.1}$$

The quantity $[\lambda]$ is a column matrix of ordinal 'part worths' given by the transpose of the row of coefficients λ_0, λ_{11}, λ_{12}, λ_{21}, λ_{22}, λ_{31}, λ_{32}, λ_{41}, which need to be solved in Equation 5.1. The coefficient λ_{ij} represents the added strength of the contribution made by changing the set point from $i0$ to ij. The coefficient λ_0 at the top of the $[\lambda]$ column is equal to the trial worth for the baseline trial, TW(1). In Equation 5.1, λ_0 multiplies each of the filled column of 1s in the **[X]** matrix and thereby adds the baseline contribution to each of the trials q. The solution for the unknown $[\lambda]$ array is given by:

$$[\lambda]=[\mathbf{X}]^{-1}[\mathbf{TW}] \tag{5.2}$$

Each off-baseline factor, if present for a specific trial, either adds or subtracts a fixed amount to or from the baseline contribution of four for the above example.

Once a preliminary experimental design has been chosen, the next step is to verify that the **[X]** matrix chosen (for the above example, the enclosed 8×8 portion of the array with the 0s and 1s in Figure 5.3) can be inverted before proceeding with the actual experiment otherwise a solution cannot be found. The inverse of the above **[X]** matrix, represented as $[\mathbf{X}]^{-1}$, exists and is shown in Figure 5.5. The spreadsheet equation

=MINVERSE(G9:N16)

was used to generate the inverse. Note that the rows are labeled according to the ij set points and the columns according to the trial numbers.

The column of λ-coefficients in cells Q19 through Q26 in Figure 5.6 were computed from the rankings using the spreadsheet equation:

=MMULT(MINVERSE (G9:N16),P9:P16) (5.3)

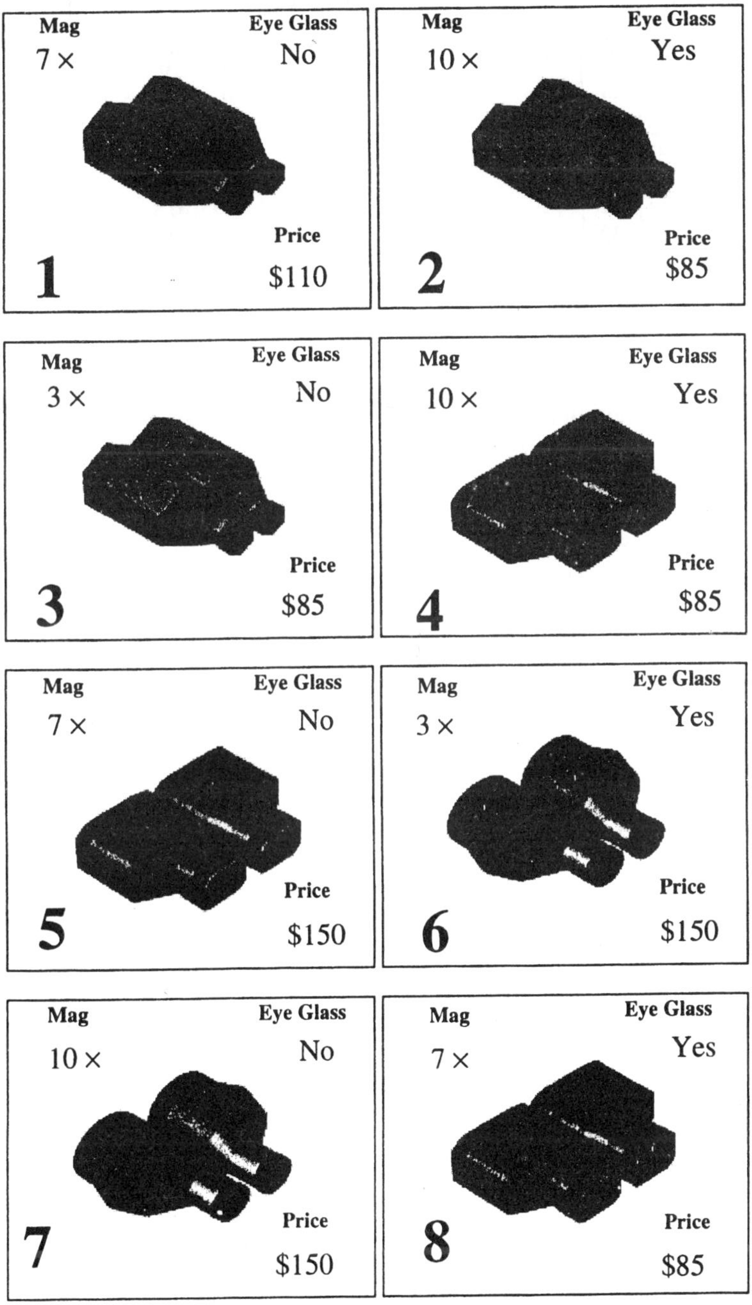

Figure 5.4 Eight cards for ranking in conjoint analysis example.

	B	C	D	E	F	G	H	I	J
17					Trial #				
18	ij	1	2	3	4	5	6	7	8
19	0	1	0	0	0	0	0	0	0
20	11	0	-1	0	1	0	0	0	0
21	12	0	-1	0	0	-1	0	1	1
22	21	0	-0.5	0.5	1	0	0.5	-0.5	-1
23	22	0	0	0	1	0	0	0	-1
24	31	-1	0.5	0.5	-1	0	-0.5	0.5	1
25	32	-1	1	0	-1	1	0	0	0
26	41	0	0.5	-0.5	0	0	0.5	-0.5	0

Figure 5.5 Inverse of [X] matrix from Table 5.1.

The baseline coefficient λ_0 in Figure 5.6 is four and equal to the ordinal worth (nine minus the rank number) of the baseline trial (trial 1). The preferred factors or set points for this problem are those having the most positive coefficients. The 10× magnification (set point 22) is seen to be the most preferred off-baseline set point evaluated, having a coefficient equal to 2.0. This is interpreted as adding two points of worth to the baseline level of four on changing the magnification from 7× to 10× when no other changes are made. The next most preferred were styles 2 and 3 tied at 1.0. The $85 price and the provision for eye glass wearers were preferred at the 0.5 level. The least preferred factor, not surprisingly, was the lowest magnification, 3×, adding a worth of –2.5 to the baseline.

	L	M
18	ij	λ
19	0	4
20	11	1
21	12	1
22	21	-2.5
23	22	2
24	31	0.5
25	32	-2
26	41	0.5

Figure 5.6 λ coefficients (ordinal part worths).

5.4 Direct value method

When the cost differences between the attributes under consideration are minimal, only a relative (ordinal) ranking of customer preferences as obtained above is needed to determine which alternative to pursue. However, if forecasts are to be made for market share and profits, or if price is to be set as some function of value, or if costs are affected, then a cardinal measure of value is needed. The ordinal results obtained from conjoint analysis are usually converted into cardinal utilities through the use of special computer programs designed for this purpose. Louviere and Woodworth [3] have pointed out, however, that the classical conjoint procedure for forecasting consumer choices is 'theoretically inadequate' for several reasons, a major handicap being the fact that the computer programs used to convert the ordinal rankings to cardinal part worths (utilities) employ empirical means for this rather than a choice model. Louviere and Woodworth used a logit model [4] to determine the utilities of attributes for a variety of problems including consumer preferences for soft drinks, vacation travel destinations, and pet foods.

In what follows, we will merge concepts from conjoint analysis, contingent valuation, choice theory, and prospect theory to arrive at a straightforward survey approach for determining cardinal values of product attributes. The major departure from the classical conjoint approach involves how price is used in the survey. Price was just another attribute in the conjoint problem just considered and was ranked along with the aesthetic and functional attributes under consideration. In the modified approach, which we term the direct value (DV) method, price is not used as one of the attributes but is incorporated instead into the survey by asking potential customers what they are willing to pay for the attributes in question. The DV method is in keeping with price being a bottom-line metric directly connected to the value of the attributes. Asking respondents their stated willingness to pay, as noted in Chapter 2, is referred to as contingent valuation [5]. This form of value assessment has been used extensively to assess the value of environmental factors and externalities but has seen limited use, surprisingly, in assessing the value of product attributes.

Consider the demand expression (Equation 3.8) for the situation where two products are being compared ($N = 2$): a baseline product of value V_{0i} and a modification of the baseline product which has value of V_i. When the price of the modified product is changed from its baseline level of P_{0i} to the neutral price of P_N defined such that the demand of the modified product and the baseline product would be equal if both were simultaneously offered, it follows that the value of the modified product is given by:

$$V_i = V_{0i} + P_N - P_{0i}$$

Potential customers are asked to state their preference of purchasing a product with certain added attributes versus the baseline product priced at P_{0i}. The product with the additional attributes carries a price P_H that is estimated to be somewhat higher than customers on average would be willing to pay. This process is repeated (either with the same group or with a different group) with the modified product having the same attributes but with a price P_L estimated to be somewhat low for the additional value added. When just two points are used, the neutral price is given by:

$$P_N = P_H - \{(P_H - P_L)(50 - f_H)/(f_L - f_H)\},$$

where f_H is the fraction choosing the modified product at the higher price and f_L is the fraction at the lower price. A typical solution is shown graphically in Figure 5.7. Although the two points are useful for illustration purposes, more points should be used in an actual survey to increase the confidence that at least two will bracket the 50% response line. A presurvey is also very helpful in establishing the price range for the final survey.

According to the logit model (Equation 3.13) a plot of $\mathrm{Ln}(f/[1-f])$ versus price should be linear with the neutral price being the price where this function equals zero. When considering aggregate data obtained from a simulated choice experiment, the $\mathrm{Ln}(f/[1-f])$ relationship should, in principle, yield a larger range over which a linear plot can be generated thereby yielding a more accurate assessment of the neutral price.

5.4.1 VALUE ANALYSIS OF BINOCULARS USING THE DV METHOD

The binoculars problem is revisited in Figures 5.8a through 5.8d which illustrate a survey based upon the DV method. The respondent is asked to place himself or herself in the position of wanting to purchase a pair of binoculars. Three type of attributes are available: style, magnification, and a provision so that persons need not remove their glasses to conveniently use the binoculars. The baseline binoculars on the left always have the same attributes and same price. For this problem, style 2 was chosen for baseline. Apart from price, each of the alterna-

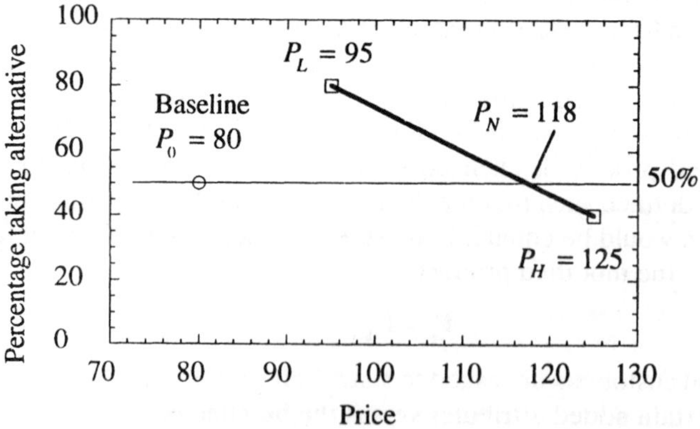

Figure 5.7 Graphical determination of neutral price using only two points.

tives considered on the right have only one attribute that is different from the baseline. For each price comparison, the respondent is asked to select either the baseline binoculars or the alternative. The use of a constant baseline in Figures 5.8a through 5.8d is consistent with the importance of the baseline on value found from prospect theory.

The experimental design for the DV method example in Figures 5.8a through 5.8d is known as 'one-at-a-time' (Appendix B). It can be argued that the experiment is more efficient if attributes are considered jointly using an orthogonal array (Appendix B), but persons surveyed have less cognitive stress and thus may choose to respond more frequently and respond more accurately

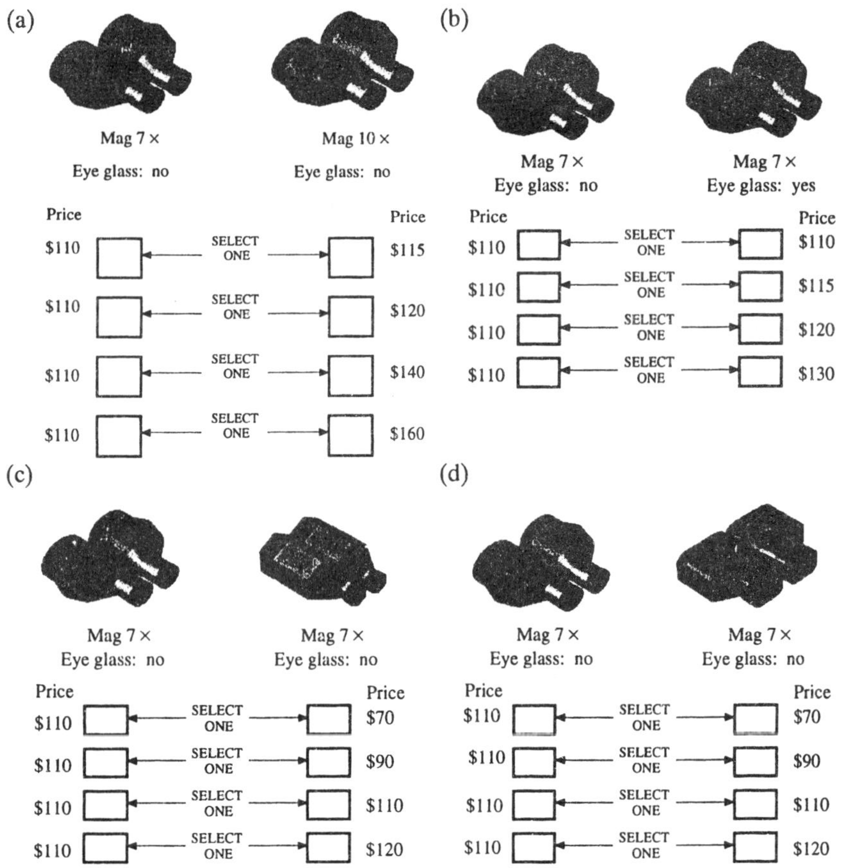

Figure 5.8 (a) A direct value survey to determine the value of 10× versus 7× magnification. (b) A direct value survey to determine the value of a provision for eye glass wearers. (c) A direct value survey to determine the value of style 1 versus style 3. (d) A direct value survey to determine the value of style 2 versus style 3.

when considering one thing at a time. Moreover, the needed sensitivity can always be obtained by surveying more potential customers when using the less efficient design. If specific attributes are expected to be offered as a group, then they should be listed as a single entity in the DV survey as there is little reason to consider them individually. Moreover, if the values of attributes have been identified individually and the value of selected attributes as a group were then desired, they could be grouped together and the value of the group determined using a second survey if there were concern that interactions between the attributes would make the value of the group not equal to the sum of the individual values. An example of using the DV method when considering several attributes jointly without interactions is given in Case Study 6.

5.4.2 VALUE OF MUSTANG OPTIONS

McConville and Cook [6] evaluated three different methods of incorporating price in a DV survey of several popular options for Mustang automobiles. The respondents to the survey were recent Mustang buyers selected randomly from buyers from across the nation. The options considered were V8 engine, convertible, ABS (anti-lock brake system), automatic transmission, air conditioning, and leather interior.

In the first survey, the respondents saw only a single price for each alternative (option). The range of prices needed to determine the neutral price were obtained by mailing out three forms of the survey with each having a different price for a given alternative. In the second survey, the respondents were asked to write in the maximum price they would be willing to pay for the alternative. In the third survey, three price comparisons between the baseline and the alternative were listed in sequence.

Using the plots of the survey percentages versus price, the percentages of persons forecast by each of the three surveys to purchase the options at the published option prices were determined and these results were compared with the actual percentages of options purchased. Extrapolations were needed to reach the neutral prices for the convertible and ABS options.

The three-price sequence for the fraction of persons f selecting the option gave the best agreement with the percentages of persons that actually selected the options. The effectiveness of the single-price method was discounted because the percentage of persons selecting the options in the simulated choice experiments, except for leather interiors, showed very small price sensitivity. Also the single-price method requires three times the number of surveys to be distributed to reach the same confidence level as a survey embracing three price levels. When using the sequential price method, it is advisable to have half of the surveys made with the price ascending and the other half made with the price descending to cancel out possible anchoring effects.

Plots of $\text{Ln}(f/[1-f])$ versus P_{Opt}/V_{Opt} for V8, convertible, ABS, air conditioning, and leather interiors are shown in Figure 5.9 where P_{Opt} is the price of the option and V_{Opt} is its value. The Ln($f/[1-f]$) function comes from the logit model, Equation 3.13, written for a paired choice comparison:

$$\mathrm{Ln}\left(\frac{f}{1-f}\right)=\frac{3E_1}{2\bar{P}}\{V_2-V_1-[P_2-P_1]\}$$

When using this expression for options, we write $E_1 = E_{Opt}$, $V_2 - V_1 = V_{Opt}$ $P_2 - P_1 = P_{Opt}$, and $\bar{P} = V_{Opt}$ yielding the expression used for the plot in Figure 5.9:

$$\mathrm{Ln}\left(\frac{f}{1-f}\right)=\frac{3E_{Opt}}{2}\left[1-\frac{P_{Opt}}{V_{Opt}}\right] \qquad (5.4)$$

The option elasticity in Equation 5.4 is determined at the point given by f = 1/2 and $P_{Opt} = V_{Opt}$. The function Ln($f/[1-f]$), which was determined from the survey results f for each option, was plotted versus the simulated option price and the value, V_{Opt}, of the option was determined graphically from the price where Ln($f/[1-f]$) crossed the zero line. The value of the V8 versus the V6, for example, was found to be $2700 and the value of the convertible was found to be $3300.

The slope of such plots can be used to determine the option's price elasticity once the option value has been determined. The line with a slope of four drawn through the results is for an option elasticity, E_{Opt}, equal to 8/3 which is the self-consistent elasticity for automotive options proposed by Donndelinger and Cook [7] based on the linear demand model. The clustering of the points about the line is seen to be satisfactory enough to use the option elasticity of 8/3 to estimate the value of options for f ranging from 0.25 to 0.75 for the options studied here.

The five options in Figure 5.9 are seen to be 'goods' in that most persons would take them if there were no price increase and many would pay to have

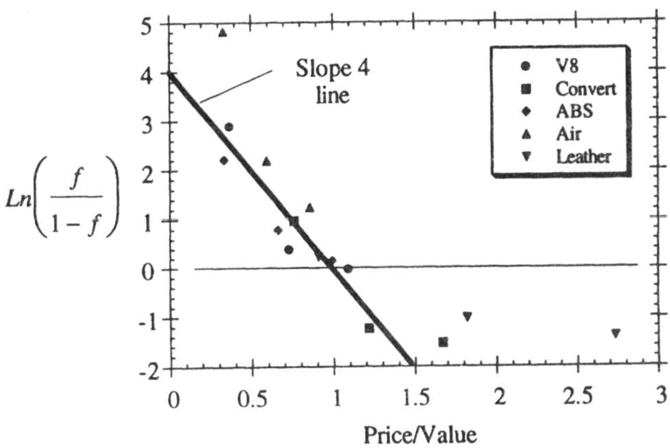

them. The automatic transmission, not shown, was different in that a substantial fraction of persons – roughly 50% – preferred the manual transmission even if there were no additional charge for the automatic. This result is not unexpected for this market segment in which vehicle performance is a major attribute. Because the automatic transmission is 'bad' for a large fraction of persons interested in this type of vehicle, this feature should never be converted to standard equipment but offered only as an option for this market segment.

5.4.3 GUIDELINES FOR CONSTRUCTING A DV SURVEY

The validity of the results of a survey increase with its response rate. Some firms require the rate to be over 40% before decisions can be made based upon the results. It is also of great importance that the persons who receive the survey are in the market segment of interest. Guidelines for generating a high response rate to a mail survey which accurately discovers the preferences of respondents have been developed by Dillman [8]. Many of his recommendations are summarized in the following guidelines:

1. Use a personalized cover letter having a real stamp on the envelope and the reply letter instead of metered postage.
2. Make a brief but convincing case in the cover letter of what the survey is about and the importance of participating in the survey. Avoid using the term survey, however. Guarantee anonymity.
3. Present the survey in the form of a booklet (one or more double size pages folded in half) instead of separate pages stapled together. It should look well-organized and easy to complete.
4. Do not pose questions on the front and back pages of the booklet. Use the front page as a cover page having the title of the survey, a meaningful graphic illustration, and information about the sponsor of the study along with the address of the sponsor.
5. Questions that the respondent will likely deem as being most important should be placed first. Demographic questions should be placed at the end.
6. Questions of a similar content should be placed together and the questions should flow with a sense of continuity.
7. Questions should be clearly posed without indicating a bias.
8. Ask only one question at a time.
9. Pretest the survey to assure that the questions are understood. To gain the best understanding of likely response rate and to fine tune the wording of questions, final pretesting should be made using respondents from the intended segment of interest.
10. Provide an incentive or token of appreciation.
11. Mail out a follow-up card for reminder one week after mailing.
12. Mail the surveys at a time of year perceived as most convenient for the respondent to participate.

Silver [9] followed many of these guidelines in her DV survey of combine harvester owners and realized a 46% response rate. She also sent out a pre-survey which was used to fine tune the final survey and adjust the price points (three) used in determining the neutral prices for the seven attributes considered. This resulted in the neutral prices for each attribute being bracketed by the price points used in the final survey. Silver enclosed a flexible ruler labeled 'Let's Combine for the Future' as a token of appreciation and mailed her survey out in January when farmers are least busy.

When determining the value of products used by commercial firms, special measures must be taken with mail surveys to avoid a low response rate (15% and below) which makes the survey results directional at best. The decision maker that should receive the survey in each firm needs to be identified and contacted to see if he or she is willing to participate and an honorarium should be offered as an expression of gratitude for the person's services. Once the survey has been mailed, a phone call should be made to the decision maker to reconfirm that he or she has received the survey and will respond in a timely manner.

5.5 Functional value

Functional or performance value, as stated earlier, is assumed to be a continuous function of the performance attributes of the product. Although there are a variety of performance attributes for a product at the system, subsystem, and component levels, functional value is defined only in terms of the product system attributes which, as already stated, are those attributes the customer associates directly with his or her needs. For example, the time required for a vehicle to accelerate from 40 to 70 mph to pass a slower moving vehicle is a system attribute. Although vehicle weight is an attribute at the system level that affects the ability of the vehicle to pass, the controlling attribute is the horsepower to weight combination. The scale used to measure the attribute should be the one which best fits how customers sense the attribute. For example, noise is perceived on the dB scale and acceleration is likely experienced as the acceleration force and not the time to go from one velocity to another.

5.5.1 GENERIC PRODUCT SYSTEM ATTRIBUTES

The array of products offered for sale is diverse – things to eat, to diet with, to ride in, to fly in, to relax us, to wear formally, to wear informally, to read, to sing, to listen to, to learn, to compute, to appreciate, to medicate, to entertain, etc. The system-level attributes for the various products that fill these needs, nevertheless, are likely to be contained in a generic list (Figure 5.10) that include the five senses, time between events, friendliness, robustness, generalized dimensions, legal issues, safety, environmental quality, and operating costs [10].

This set of system attributes can be used to gauge the value of any product or service. Most products, of course, do not exhibit the full range of attributes. But having a wide range of potential attributes is important as it can lead to unique ways of product differentiation. For example, Rice Krispies, a breakfast cereal, differentiates itself from competitors through sound – Snap! Crackle! and Pop! No other cereal maker attempts to use sound as a product attribute. Although insulation and exhaust mufflers are used to reduce the noise of an automotive engine, developing a particular engine sound can add value. For example, as the speed of an engine increases during acceleration, the sound quality is better if the sound intensity increases linearly with speed rather than having sudden changes. You can not literally digest the transportation vehicles used today, but in 1847 this was a valid and potentially lifesaving option for those hitching up their mules and oxen in St Louis in preparation for the perilous trek across the 'Great American Desert' and continental divide to reach the California gold fields. Always consider the full range of possibilities for favorably differentiating your product from the competition.

Five senses
- visual excitement
- odor
- sound
- feel
- taste

Time between events
(product functioning)
- between spatial locations
- from input to output
- from start-up to functioning
- from shut-down to off
- from off to store

Time between events
(product malfunctioning)
- between failures
- from failed to fixed
- useful life

Friendliness
- comfort
- ergonomics
- mental effort required

Robustness (tolerance for)
- humidity
- heat
- vibration
- dust
- cold
- snow
- operator differences
- shelf life

Generalized dimensions
- volume
- weight
- moments of inertia
- length.
- diameter
- width
- height
- capacity

Legal ethics/adherence
- state and federal laws
- foreign laws
- moral laws

Safety
- risk of injury
- risk of loss of life
- warn of hazardous condition

Environmental
- ability to recycle
- earth impact
- water impact
- noise impact
- electro-magnetic impact
- direct impact on people
- direct impact on biomass

Costs
- to operate
- to repair
- downtime loss of value

Figure 5.10 Generic product system attributes. (Source: H.E. Cook and M.R. Gill (1993) On system design. *Research in Engineering Design*, **4**, 215–26, © 1993 Springer-Verlag New York. Used with permission.)

5.6 Value curves

For a given customer segment, performance value is assumed to be a function of each system attribute g and to be independent of the design and manufacturing strategy used to arrive at g, other things being equal. Because the value of performance attributes is not expected to change as quickly over time as the value of subjective attributes, we can consider the value curve for performance attributes as a state function for a given consumer segment. Manufacturing costs, however, can be a strong function of the design and manufacturing process (a path-dependent function) used to arrive at g. An example of this is shown in Figure 5.11 for two different manufacturing processes. Cost varies with the type of process but value does not. When investment costs, fixed costs, and environmental quality are ignored, total quality (Equation 4.9) is proportional to $(V-C)^2$ for the region where $V>C$.

This quantity was computed for the curves given in Figure 5.11 and is shown in Figure 5.12. Total quality is seen to shift to a higher amount when the lower cost process is used and the target specification, located at the maximum in total quality, is seen to move toward the ideal specification. It is clear from this diagram that a major opportunity for improving the quality and profitability of products is the discovery of ways to reduce costs in a manner that allows the target specification to be moved toward the ideal specification.

5.6.1 COST OF INFERIOR QUALITY (CIQ)

In discussing Taguchi's loss function in Chapter 4, two specification points for an attribute were described, a target specification and an ideal specification. The cost of inferior quality was a minimum at the ideal specification but the quality loss function was a minimum at the target specification. For the S-model

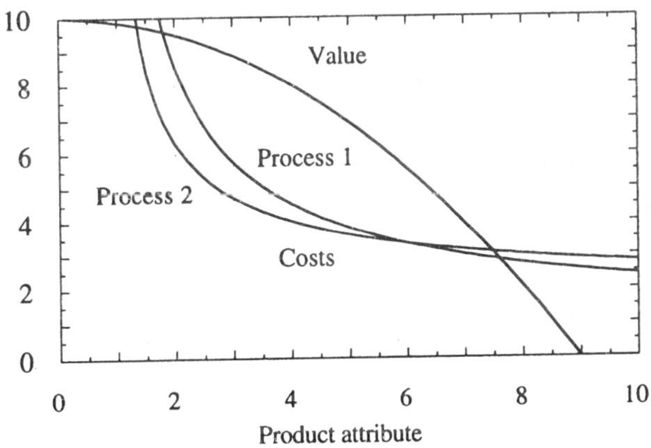

Figure 5.11 Value is a reasonably stable function of the product attribute for a given market segment but costs are not. (Source: reprinted with permission from SAE Paper 960001 © 1996 Society of Automotive Engineers. Inc.)

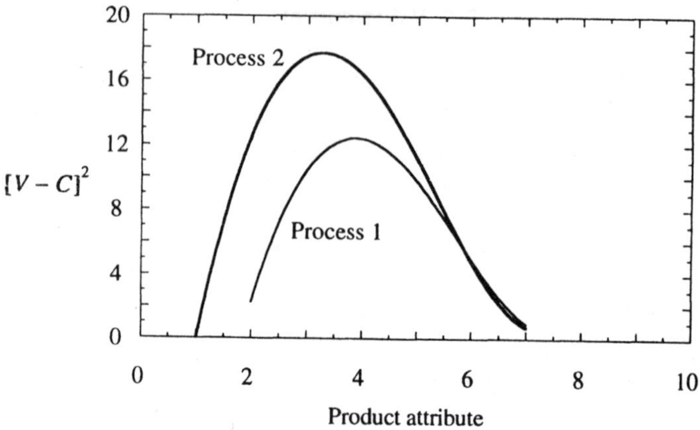

Figure 5.12 Total quality and profit are both proportional to $(V - C)^2$ when fixed costs, investment, and environmental costs are ignored. (Source: reprinted with permission from SAE Paper 960001 © 1996 Society of Automotive Engineers, Inc.)

with price defined by Equation 3.16, total quality is a maximum at the target specification and value is a maximum at the ideal specification. These relationships are summarized in Table 5.1.

Quality is reduced when the product attribute is off the target specification and value is reduced when the product attribute is off the ideal specification. The target specification is not at the position where it would give the highest value to the customer because the costs would be prohibitive. This is in the customer's best interests because the price would likewise be prohibitive if the manufacturer attempted to make the product at the ideal specification which, apart from costs, is often technically impossible.

The target specifications given by the minimum in Taguchi's loss function and the specifications determined by maximizing total quality Q will be identical if Taguchi's cost of inferior quality (CIQ) is defined as:

$$\Omega(g) = V_I - V(g) \tag{5.5}$$

where V_I is the ideal value. The relationship between value and Taguchi's CIQ for a smaller-is-better condition is shown in Figure 5.13. The ideal value is equal to ten and the CIQ is equal to zero at the ideal specification where, for this

Table 5.1 Relationships between Taguchi's loss function and S-model

	Taguchi	*S-model*
Target specification	Loss function at a minimum	Total quality at a maximum
Ideal specification	CIQ at a minimum	Value at a maximum

example, the attribute is zero. Value goes to zero at a critical specification point of nine, the point at which CIQ equals ideal value.

The purpose of Equation 5.5 is to show how the two models are connected; it does not lead to the most effective way of determining value, i.e. by first determining the cost of inferior quality, $\Omega(g)$, and then subtracting it from ideal value, V_I. Value is best determined by more direct means such as the DV method just presented and from the construction of empirical value curves as discussed later. The results can then be substituted into Equation 5.5 to determine the CIQ.

5.6.2 TAGUCHI'S SIB, NIB, AND LIB CONDITIONS

For certain product attributes, smaller is better (SIB) in that value increases as the attribute is reduced. For others nominal is best (NIB) because value is highest between too small and too large, whereas larger is better (LIB) for others. The equations shown in Table 5.2 are based upon Taguchi's relationships for SIB, NIB, and LIB. The subscript *i* for the product has been dropped. The notation g_C refers to the critical specification point, g_0 refers to the baseline specification, and g_I refers to the ideal specification point. For SIB and LIB, value is ideal at 0 and ∞, respectively. The value V_0 for the baseline product is determined from its demand and price using Equation 3.11.

The behaviors of these functions with the product attribute *g* in arbitrary units are shown in Figure 5.14. Each curve represents how the total value of the product should behave as a function of the specific attribute linked to the curve, assuming all of the other attributes remain fixed. The critical points, where value is zero, for each curve are circled. The three curves are assumed to be for three different attributes for a single product but are graphed along a common dimensionless attribute axis for comparison. An important point is that the baseline value, equal to eight in the example, is identical, at the intercepts

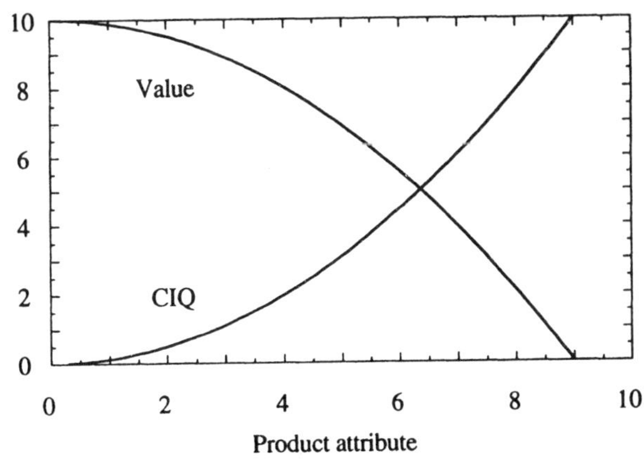

Figure 5.13 Relationship between value and Taguchi's cost of inferior quality.

Table 5.2 Value curves corresponding to Taguchi's SIB, NIB, and LIB conditions

Condition	*Value versus* g *relation* $V = V_0 f_V(g)$	*Ideal value relation*
SIB	$f_V(g) = \dfrac{g_C^2 - g^2}{g_C^2 - g_0^2}$	$V_I = V_0\left[\dfrac{g_C^2}{g_C^2 - g_0^2}\right]$
NIB	$f_V(g) = \dfrac{[g_C - g_I]^2 - [g - g_I]^2}{[g_C - g_I]^2 - [g_0 - g_I]^2}$	$V_I = \dfrac{V_0}{1 - \dfrac{[g_0 - g_I]^2}{[g_C - g_I]^2}}$
LIB	$f_V(g) = \dfrac{\left[1 - \dfrac{g_C^2}{g^2}\right]}{\left[1 - \dfrac{g_C^2}{g_0^2}\right]}$	$V_I = \dfrac{V_0}{1 - \dfrac{g_C^2}{g_0^2}}$

shown for the three curves at their baseline specification points of four, eight, and six, respectively, for SIB, NIB, and LIB. The ideal value points for the three curves, given by their maximum values, are not the same, however.

5.6.3 THE EXPONENTIALLY WEIGHTED PARABOLIC APPROXIMATION FOR VALUE

We have, so far, used one-half of a parabola to interpolate value for SIB conditions between an ideal point at $g = g_I = 0$ and a critical point, $g_C = 9$. For NIB,

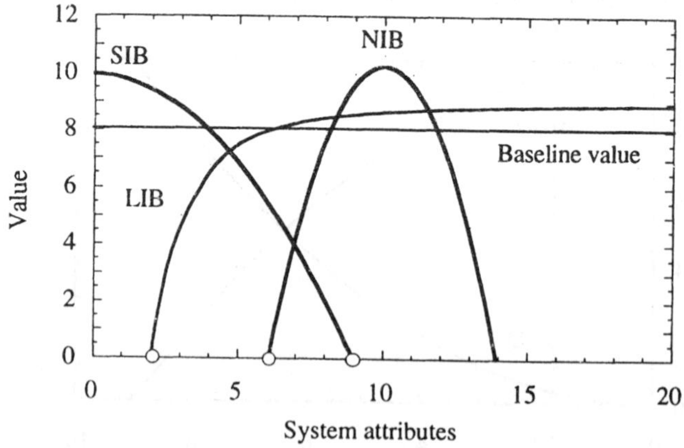

Figure 5.14 Three types of value curves: smaller is better (SIB), larger is better (LIB), and nominal is best (NIB).

the value curve was a full parabola centered about the ideal specification point, g_I. It is also reasonable to use one-half of a parabola for LIB conditions when there is a readily identifiable upper limit where good judgment suggests that customers would be insensitive to improvements in the attribute beyond this point and where there is a lower limit critical point where the product should have no value. The upper limit can be defined as $g = g_I$. It effectively acts like an ideal specification point in that value is never any larger than at this point. There can also be a readily identifiable point for SIB conditions where value is not expected to increase on going below this point.

For interpolating value between its critical point and its ideal point for SIB, NIB, and LIB conditions, the relationship:

$$V(g_i)=V_0 v(g_i) \tag{5.6}$$

can be used for a single attribute where:

$$v(g_i)=\left[\frac{(g_C-g_I)^2-(g-g_I)^2}{(g_C-g_I)^2-(g_0-g_I)^2}\right]^{\gamma} \tag{5.7}$$

This function has a maximum at the point $g = g_I$, and is equal to zero at $g = g_C$. For SIB, only one-half of the parabola is used from g_I to g_C, and for LIB, the half from g_C to g_I is used. For NIB, the full parabola is used. Again, note that baseline value is determined using Equation 3.11. The weighting coefficient γ is an empirical parameter which is expected to be approximately equal to the time fraction that the attribute is of importance when using the product. It is necessary to introduce a weighting factor to have a viable multiattribute expression for value which is assumed to be given by:

$$V(g_1, g_2, g_3, \ldots g_j) = V_0 v(g_1)\, v(g_2)\, v(g_3) \ldots v(g_j), \tag{5.8}$$

because this empirical form renders the system worthless if any attribute reaches a critical point. If the attribute for a subsystem is changed, the impact on value must be obtained by showing how the change affects the system attributes.

Products, as already noted, can also have optional subsystems or attributes (options) such as color screens for display terminals and radios or air conditioners for automobiles. Optional subsystems affecting optional attributes g_i' simply add $\Delta V(g_i')$ to the value of the base product, V_0, as they are installed. If they cease to exist, the total system still has value. If the basic system is rendered worthless, the option could be salvaged and would still have value. The change in the cost of ownership, ΔC_{Own}, versus the baseline product contributes to value. If the attributes of options are noted with primes, the total value forecast for a new product is given by:

$$V = V_0 v(g_1)\, v(g_2)\, v(g_3) \ldots v(g_j) + \Delta V(g_1') + \Delta V(g_2') + \Delta V(g_3') + \ldots + \Delta V(g_k') + \Delta C_{Own} \tag{5.9}$$

5.6.4 GEOMETRIC AVERAGING

The attributes for a specific product are not necessarily constant but depend upon the environment. For example, the interior noise of a car varies with its velocity and acceleration and whether the road is smooth or rough. During the lifetime L of a product, the behavior of each system attribute, g_i, can be divided into the times t_1, t_2, ... during which the attribute, $g_i(t_j)$, to a reasonable approximation, is constant while in the environment noted by the subscripts on the times. The summation of the times over all environments equals L. If the weighting exponent for the attribute is γ, value averaged over all of the environments should be given approximately by the so-called geometrical average given by:

$$v(g_i) \equiv \left\{ \left[\frac{V(g_i(1))}{V(g_{i0}(1))}\right]^{\frac{t_1}{L}} \left[\frac{V(g_i(2))}{V(g_{i0}(2))}\right]^{\frac{t_2}{L}} \left[\frac{V(g_i(3))}{V(g_{i0}(3))}\right]^{\frac{t_3}{L}} \cdots \right\}^{\gamma} \quad (5.10)$$

in which $v(g_{i0}(j))$ is the baseline value in environment j.

5.7 Critical attributes for automobiles

The estimated baseline, critical, and ideal specifications for several important critical attributes for automobiles are listed in Table 5.3 along with the exponent estimated for each [7, 11]. A family of value curves as a function of the interior noise and vibration levels for a vehicle cruising at 70 mph are shown in Figure 5.15. The curves were computed using Equation 5.8 which yielded the product of two individual weighted parabolic functions for noise and vibration with each being calculated using the general expression given by Equation 5.7. The dB scale for vertical vibration used in the computations of the value curves in Figure 5.15 was determined from the relationship given by:

$$g_{vib} = 10\text{Log}\,(A_{vib}/0.1) \quad (5.11)$$

where A_{vib} is the rms vibration amplitude measured in m/sec^2. The critical and ideal levels for the vertical acceleration levels of 14.8 and 3 dB, respectively, were taken from Griffin's findings [11] which showed that a person cannot endure, even for a brief period, a vertical vibration level of roughly 3 m/s^2 rms which yields 14.8 dB when substituted into Equation 5.11.

The critical specification for interior noise at 70 mph in Table 5.3 was taken as 10 dB below the threshold of pain for noise of 120 dB. When the noise level is at the ideal specification chosen of 40 dB, persons are able to communicate using soft whispers [12]. Thus, reductions in noise below this level would add little value. The exponent of 0.5 represents an estimate of the average fraction of time spent cruising at highway speeds.

Table 5.3 Selected critical attributes for automobiles and their estimated critical, ideal, and baseline specifications and weighing factor. (Source: reprinted with permission from SAE Paper 970762 © 1997 Society of Automotive Engineers, Inc.)

Attribute	*Type*	*Units*	*Baseline*	*Critical*	*Ideal*	*Exponent*
Front leg room	LIB	millimeters	N/A	N/A	N/A	N/A
Turning circle	SIB	meters	12	20.54	1.83	0.125
Overall length	SIB	millimeters	4800	7620	762	0.125
Overall width	SIB	millimeters	1780	2743.2	762	0.15
Door-top to ground	LIB	millimeters	1280	1128	1499	0.03
Luggage capacity	LIB	cubic meters	0.45	0	0.85	0.08
Idle noise	SIB	dBA	43	110	40	0.2
Max 1st gear noise	SIB	dBA	71	110	40	0.[illegible]5
70 mph noise	SIB	dBA	67	110	40	0.5
Skidpad lateral acceleration	LIB	Gs	0.72	0.25	1	0.2
Time for 0–60 mph	SIB	seconds	10.5	40	2	0.16
(Times should be transformed into a 10 Log(time) attribute.)						
Vertical acceleration	SIB	meters per second^2 rms	4.8	14.8	0	0.2

(Vertical accelerations should be transformed into a 10 Log(accel./0.1) attribute.)

The importance of many automotive performance attributes will depend upon the lifestyle of the driver. For example, someone with a spirited lifestyle will likely use the handling, acceleration, and top speed attributes of the vehicle more than someone with a conservative lifestyle. Someone that lives in the rocky mountains will favor the snow-handling properties of a vehicle more than someone in south Florida. Interior room will be more important to someone

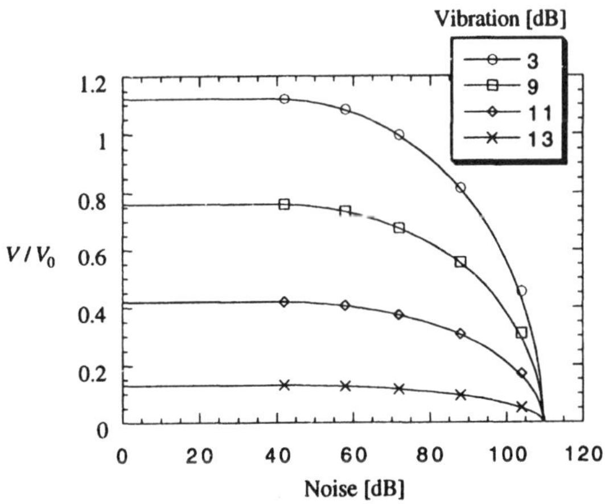

Figure 5.15 Value curves for interior noise for a family of vertical vibration levels in dB.

with a family than a person without a family. These buyers represent different consumer segments and they generate different value curves for the attributes.

It must be emphasized that the individual value curves are established totally independent of any other attribute. This can be done because the final computation of total value using Equation 5.8 sorts out the interplay between attributes. Also it should be understood that the true value function is not a mathematical formula but a subjective, customer-defined relationship as stated earlier. The relation given by Equation 5.7 is simply an interpolation formula used to estimate value when there is minimal customer information regarding its true empirical form. The latter can be mapped out, in principle, by using the DV method to arrive at the neutral prices for a series of attribute levels relative to baseline. The value curves are used for a rapid estimation of how value is expected to change with changes in the level of an attribute.

5.7.1 VALUE OF VEHICLE LENGTH

An instructive example for demonstrating the influence of exponential weighting on the shape of a value curve is overall length. It is tempting, at the outset, to consider this factor as a NIB attribute as a certain amount of length is needed to establish the needed interior room but, on the other hand, too much length would create problems in parking, storing, and turning the vehicle. For example, Ford's new Expedition sport utility vehicle is 15.4 inches shorter than the Chevrolet/GMC Suburban. Ford is expected to stress this point to potential buyers [13].

A reasonable critical point for the overall length of a passenger car can be taken as 300 inches (762 mm) which is 50% longer than the overall length of a large vehicle today. Limousines are an exception to this but they represent a different product segment having professional drivers who can remain with the vehicles when parked and having special garages for storage. The added length to a passenger car does not reduce vehicle value from its baseline point when traveling in a straight line. The added length only becomes a nuisance when maneuvering in traffic, when parking, and when storing. Value curves are shown in Figure 5.16 for five different time fractions – 0.1, 0.2, 0.3, 0.4, and 0.5 – for overall length. The exponent for length is estimated at 0.125 in Table 5.3. The point g_I = 30 inches was chosen for the lower SIB limit. A careful study of driving habits could be made to generate a better estimate for the fractional time. Overall vehicle length also affects the style of the vehicle and changes in this subjective attribute would also need to be evaluated and combined with the value curves for overall length and interior room to arrive at the total change in value.

5.7.2 VALUE OF 0 TO 60 MPH ACCELERATION

An important design trade-off in developing new automobiles is that between fuel economy and acceleration performance. Resolution of this classic design trade-off depends upon the market segment and, of course, the price of fuel.

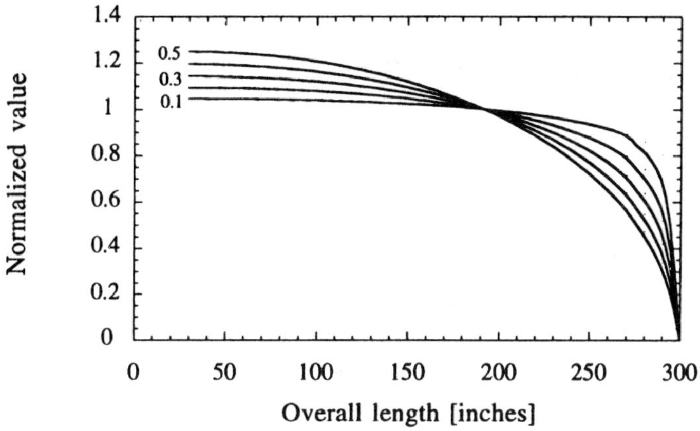

Figure 5.16 Influence of exponential weighting on shape of value curves.

This issue has been recently studied using a drive clinic in which 100 persons participated [14]. Each person drove three vehicles. The vehicles had different acceleration performance but were otherwise identical. The vehicle noted as A had the lowest full throttle acceleration (from 0 to 60 mph in 16.9 seconds) and was designated as the baseline vehicle. Vehicles B and C had times of 12.2 and 10.8 seconds, respectively. A standard test route of 10 miles in length was used having seven specific acceleration maneuvers spaced at convenient sections along the route. The route traversed real, not simulated, urban highways. After evaluating the three vehicles, each person was asked using the DV approach for their willingness to pay for the other two vehicles versus the baseline vehicle. However, they were not told what the acceleration performances were for the vehicles or which, if any, was faster or slower than the baseline vehicle. Most respondents did, however, figure it out for themselves after driving the vehicles. They were also asked their willingness to pay for changes in fuel economy versus a baseline level of 22 mpg.

The resulting value curve constructed for acceleration performance is shown in Figure 5.17. The three interior points shown were obtained from the survey. The critical value was assumed to be 40 seconds (so slow that it is perilous to enter a freeway) and the ideal value was assumed to be 2 seconds. The attribute g_{accel} was taken to be a psychometric variable equal to the logarithm of the average full throttle force experienced by the driver which is proportional to the reciprocal of the time $t_{0,60}$ to accelerate from 0 to 60:

$$g_{accel} = \mathrm{Log}\left(\frac{1}{t_{0,60}}\right) \tag{5.12}$$

The exponent γ should be approximately equal to the time that the acceleration attribute is important to the driver. The best fit shown in Figure 5.17 was for $\gamma = 0.17$. This was fortuitously close to the average time fraction of

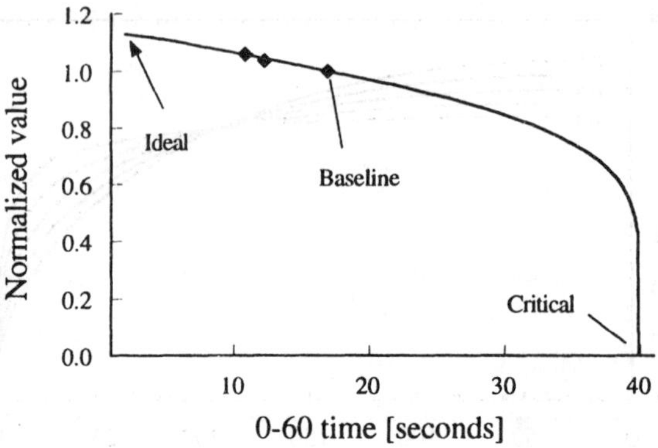

Figure 5.17 Value curve for acceleration time from 0 to 60 mph. (Source: reprinted with permission from SAE Paper 960004 © 1996 Society of Automotive Engineers, Inc.)

0.16 taken up by the acceleration maneuvers during the drive over the test route.

There has been more than one example of a new car not living up to the acceleration performance intimated by its appearance of raw power. This situation has arisen in the not too distant past when the subsystems for a new car were ready for production except for the engine which was far behind schedule. A back-up engine was available but at reduced horsepower and the decision was made to go ahead with the lower performing engine. The marketplace, initially excited, soon lost interest in the vehicle because customers did not like being fooled that performance was there when it wasn't.

5.7.3 DETERMINING THE WEIGHTING EXPONENT γ

As shown by the above study of the value of acceleration, the DV method can be combined with the value curve method to determine the exponential weighting factor, γ. Value needs to be measured at only one additional attribute point, g^*, off the baseline point g_0 to arrive at an estimate for γ. If the DV method is used to determine value at more points, then a best fit can be used as was done for the 0 to 60 acceleration time attribute. Given only one point, γ is computed using the expression:

$$\gamma = \frac{\mathrm{Ln}(V(g^*)/V_0)}{\mathrm{Ln}\left(\left[[g_C - g_I]^2 - [g^* - g_I]^2\right]\right) - \mathrm{Ln}\left(\left[[g_C - g_I]^2 - [g_0 - g_I]^2\right]\right)} \tag{5.13}$$

This approach to determining the weighting factor is preferred because it calibrates the value curve against stated willingness to pay thereby compensating for the effects of errors in estimating baseline value, V_0, which appears in Equation 5.6 as well as errors in estimating the critical and ideal specifications. If time and resources are not available for a customer survey, γ should be determined by having several specialists familiar with the needs of customers to compare a prototype having the attribute g^* against the baseline product having the attribute g_0 and then estimate what customers would be willing to pay. A less reliable estimate for γ, but the easiest to generate, is in terms of the time fraction already discussed.

5.7.4 TIME BETWEEN EVENTS

The times between important product events represent a key set of system-level attributes. The notion of product reliability is linked to the time between repairs. As this time increases, reliability is said to increase. The loss of usage time in making a repair represents a loss of value to the customer in addition to the cost for the repair which may or may not be reimbursed by the seller through a warranty policy.

The cycle time needed to complete a task affects the value of a product used to do the task. If, for example, it takes a wheel loader, as currently manufactured, a time of t_0 to move dirt from a pile and fill a truck and the time to do the same task for the prototype of a future wheel loader is t, then the value increase due to the reduction in cycle time is equal to $V_0 [t_0 - t] / t$ where V_0 is the value of the current wheel loader. This assumes that the task described is the only task for the wheel loader.

5.7.5 VALUE OF INTERIOR ROOM

The value of interior roominess in automobiles has been estimated for leg room, head room, and shoulder room by considering how well the population distributions of adult men and women are accommodated as a function of the pertinent dimensions of the automobiles [15]. The method used is shown schematically in Figure 5.18 which shows (1) a value function for the population for a vehicle having a leg-room dimension equal to LR and (2) the frequency function for the distribution of the buttock-to-heel dimension for the population [15].

Persons whose buttock-to-heel dimension is less than LR can adjust the seat forward to be accommodated without any loss of value. The value function is unity for these people. A fraction of the population whose buttock-to-heel dimension is greater than LR can be partially accommodated by adjusting as much as possible to their ideal position but this generates a loss of value for them, the fall-off from ideal being assumed to follow a parabolic curve. A value coefficient for the vehicle as a function of the vehicle dimension was determined by numerically integrating the product of the value function times the population distribution for a range of vehicle dimensions. The distance over which the value function fell to zero was determined by trial and error adjustment of the fall-off dimension until the computed slope of value with leg room agreed with what people said that they would be willing to pay for additional leg room in a nominal sized vehicle.

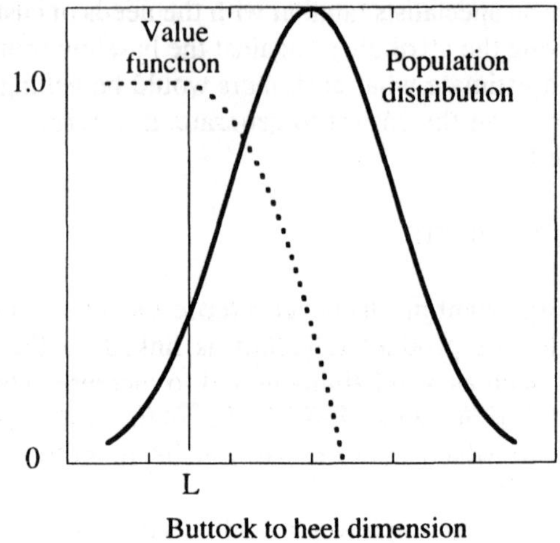

Figure 5.18 Value weighting function and population distribution used to compute the value curve for front leg room. (Source: reprinted with permission from SAE Paper 960002 © 1996 Society of Automotive Engineers, Inc.)

The resulting vehicle value coefficients found by this procedure for front leg room in millimeters is shown in Figure 5.19. A similar process was used to compute the vehicle value coefficients for head room, rear knee room, and shoulder room [15].

5.7.6 COMPARISON WITH MULTIATTRIBUTE UTILITY THEORY

An important common element between the value approach described here and classical multiattribute utility theory [16] is that the high and zero utility

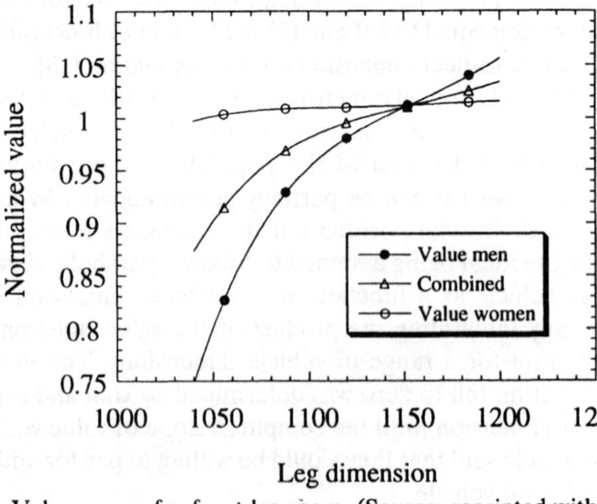

Figure 5.19 Value curves for front leg room. (Source: reprinted with permission from SAE Paper 960002 © 1996 Society of Automotive Engineers, Inc.)

levels for the attributes used in constructing the lottery (Chapter 2) should, in fact, be approximately the same as, if not identical to, the ideal and critical specifications used in the value interpolation formulas given here. For example, there is a close functional behavior between the utility function for the weight of a chair obtained by Thurston [17] using a lottery and the value curve we would compute as a function of its weight. The decision maker in utility theory translates into the person here that decides what the critical attributes and the time factors are. The decision maker can use market research information or his or her own judgment to establish the critical attributes. The approach here, however, to estimating value follows the tenets of prospect theory in that a baseline value is established and then changes in value as a result of changes in the product attributes are determined. In contrast to both utility theory and prospect theory, no lottery is directly involved in assessing the value function itself. The implications of the results from prospect theory on the value curves are discussed at the end of this chapter.

5.8 Perceived versus real value in service

The perceived or expected value that a customer senses in a product at the time of purchase is realized if the product behaves in service as the customer believes or expects it would at the time of purchase. An idealized view of product behavior in service is described by a constant growth rate of value received by the customer, Figure 5.20. When its lifetime is reached, it dies abruptly. The constant growth rate at which value is received by the customer for this idealized case is equal to V/L (Figure 5.21).

5.8.1 VALUE OF RELIABILITY

A more realistic view is that the rate of value received from the product is not a steady function of the time in use and periods exist where the value produced is negative when repair and downtime costs are incurred (Figure 5.22).

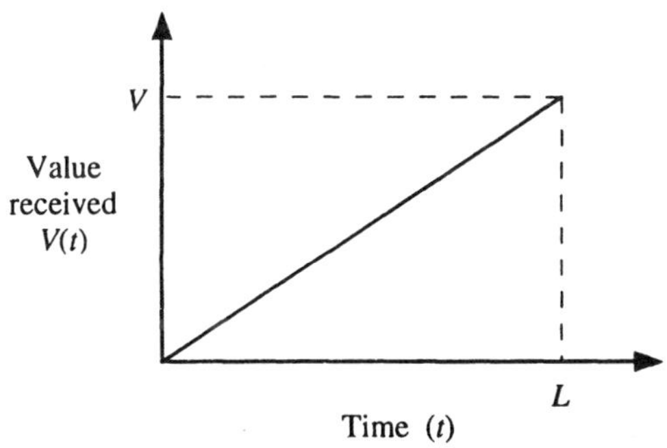

Figure 5.20 Cumulative value received over time during product use, idealized view.

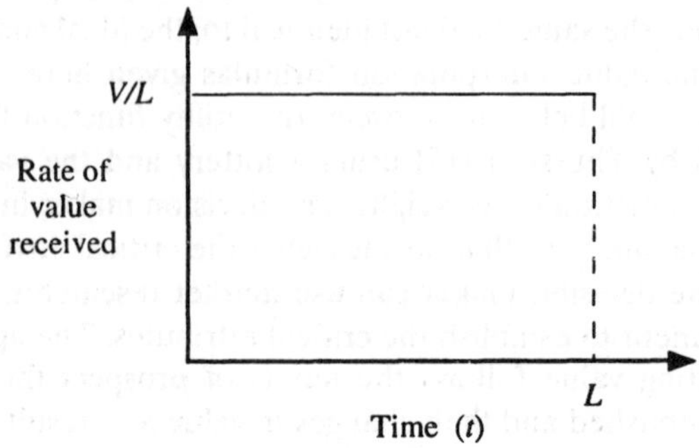

Figure 5.21 Idealized view of a constant rate of value received by customer.

Costs incurred by the customer represent direct losses in value as the result of the expense to repair the product and the loss of use during the period of time that the product is inoperable. Even if the repair costs are covered by warranty the customer suffers some loss when a product malfunctions. Thus, the cumulative value received by the customer at time t is given by [18]:

$$V(t) = \frac{tV'}{L} - C_T(t) + S(t) - \beta\left\{\left[\overline{N_R^*}\,\overline{C_R^*} - \overline{N_R}\,\overline{C_R}\right] + \left[\overline{N_R^*}\,\overline{t_R^*} - \overline{N_R}\,\overline{t_R}\right]\left[\frac{V'}{L} - \frac{\overline{dC_0}}{dt}\right] - \left[\overline{S^*(t)} - \overline{S(t)}\right]\right\} \tag{5.14}$$

The quantity V'/L is the gross rate at which value is generated ignoring the total accumulated operating costs $C_T(t)$ and average resale value $\overline{S(t)}$ considerations. The total accumulated operating costs are equal to the sum of the accumulated normal operating costs, $C_0(t)$ and the accumulated repair costs, $C_R(t)$.

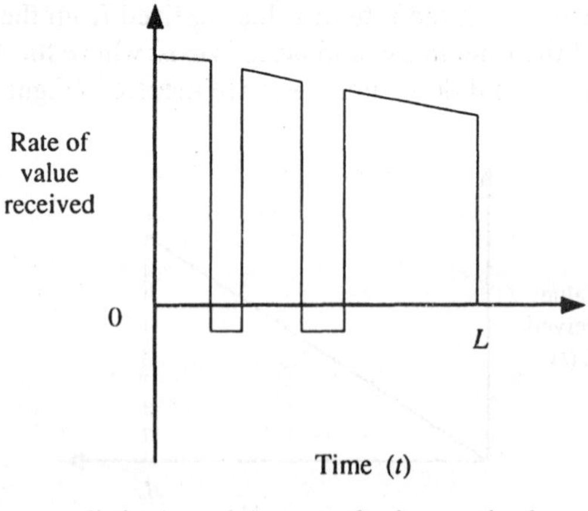

Figure 5.22 A more realistic view of the rate of value received.

The quantity L is the expected lifetime. The terms $\overline{N_R^*}$ and $\overline{N_R}$ are the actual and expected average number of repairs over time t; $\overline{C_R^*}$ and $\overline{C_R}$ are the actual and expected average cost per repair; and $\overline{t_R^*}$ and $\overline{t_R}$ are the actual and expected average downtime per repair. The coefficient β is the multiplication constant required to generate the loss of value attributable to the economic costs associated with unexpected product shortcomings as discussed earlier from prospect theory in Chapter 2. If the net costs in braces are positive (unfavorable), then β is greater than unity, two to three being a reasonable range; if the costs are negative (favorable), β is conservatively taken as equal to one. In using Equation 5.14, the expected repairs, costs, and downtimes can be obtained from mail surveys and the actual amounts from field reports.

If customers on average planned to keep a product for L years and if it performed to expectations in terms of net value generated which results in the terms within braces in Equation 5.14 being zero, then (gross) value received after L years should equal the perceived value at purchase. Perceived value at the time of purchase is connected, therefore and not surprisingly, to the actual cumulative value received. Thus on setting $t=L$, the terms within braces equal to zero, and $V(t=L)=V$, we find on rearranging that the term V' is a function of the expected value at purchase, the expected lifetime costs, and resale value at lifetime:

$$V' = V + C_T(L) - S(L). \tag{5.15}$$

If the net value received is less than perceived value, then the customers on average will lower their perceived value of the product when reconsidering purchasing the product in the future. For simplicity in arriving at Equation 5.15, we did not discount value in future years. This can be remedied by rewriting Equation 5.14 in terms of the value received in a given year instead of the cumulative value, performing a net present worth on the resulting value stream, setting the result equal to perceived value at purchase, and then solving for V'.

In writing Equation 5.14, the product was assumed to be operating at its expected levels of performance throughout its use, except when down for repairs. However, the performance of a product may deteriorate below its expected level before a repair is made. When this condition applies, an additional term multiplied by time t should be added to the expression in braces equal to average gross rate at which value was expected to be generated less the average gross rate at which value was actually generated.

One use of Equation 5.14 is shown in Figure 5.23 for a product whose perceived value at time of purchase was \$218 000 accumulated over its expected seven-year economic life. The resale price was expected to decrease linearly from the purchase price of \$150 000 to approximately zero over the seven-year period and the expected cumulative costs were assumed to be given by $C_T(t) = 100t^3$.

The value received at the instant of purchase is equal to the value of the new but unused product which is simply the price just paid for the product,

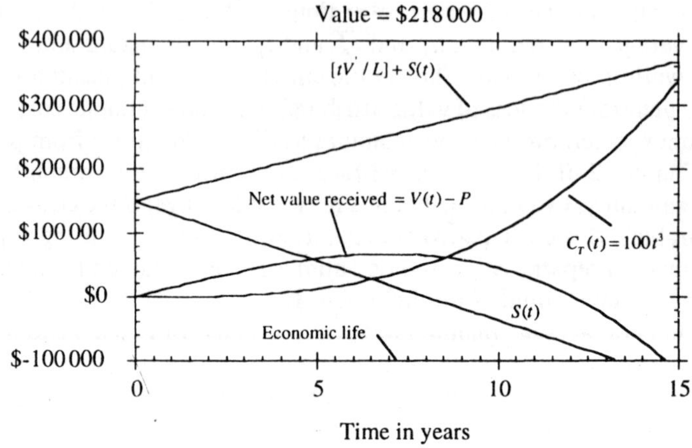

Figure 5.23 Relation of net value received to the economic life of a product.

the assumption being that you are able, if desirous, to immediately sell the product for the price you, just an instant before, paid for it. The net value received from the product at this point (value received less price) is zero. It is through the use of the product that additional value is generated. Economic life is determined by the point at which the time derivative of the net value received, which is given by Equation 5.14 minus purchase price, is equal to zero. At this point the annual costs of keeping the product running begin to exceed the annual value generated by the product. For this example, the expected operational parameters were taken equal to the actual parameters making the sum of the terms in bases equal to zero. This results in the quantity $(V'/L) + \partial S(t)/\partial t$ being equal to the slope of the cumulative cost curve at the economic life point.

If the product is not changed in performance but its operating costs are reduced, the change in value can be computed from the difference between the new and old net value curves as shown in Figure 5.24. In the early life region for the example, the product is delivering net value at the same rate. As use extends beyond five years, the higher rate of accumulation of operating costs for the old product causes its cumulative net value curve to begin to fall below the new product and limits its economic life to twelve years versus twenty-two years for the new product. The old price was used in computing the net value curve of the new product. The added value, δV, of the new product resulting from its improved operating costs is given by the difference between the two maxima which is equal to $130 000 for the example. However, this estimate of the change in value does not include the customer's likely discounting of the future similar to the present worth calculation given by Equation 3.25. When the discounting formula is applied, the increase in value would be reduced.

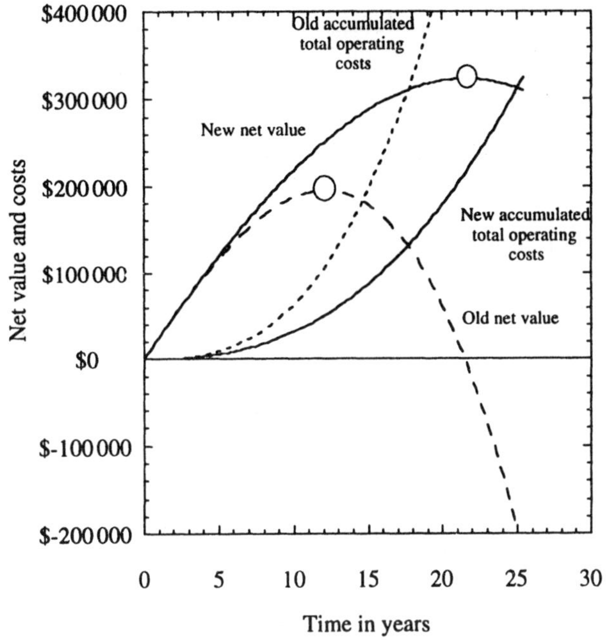

Figure 5.24 Improvements in value received resulting from a reduction in total operating costs.

5.9 Prospect theory implications on value curves

The coefficient β was introduced into the expression for real value received in service (Equation 5.14) to account for the results from prospect theory showing that we are considerably more sensitive to losses than equivalent gains. These considerations should also affect the value curves when changing from a baseline position to an inferior product attribute position as value should fall off initially two to three times as fast as predicted by the value curve. When prospect theory implications are included, the expression for value is defined as $V(g)_{PT}$ and is given by the relation:

$$V(g)_{PT} = V_0 - \beta[V_0 - V(g)] \tag{5.16}$$

where V_0 is baseline value and $V(g)$ is the result given by the regular value curve. β is equal to 1 if $V(g) > V_0$ but approximately equal to 2.5 if $V(g) < V_0$. The behavior of the prospect theory correction is shown in Figure 5.25. A consequence is that an unfavorable change in an attribute for a future product from its current specification for the baseline product can have a negative impact on demand considerably greater than might be expected. For example, the increase in a favorable attribute at some earlier time represented by Δg may have generated an increase in demand of ΔD assuming no change in price. However, the reduction of this attribute by $-\Delta g$ at a later time will likely reduce demand by approximately $-2.5\ \Delta D$.

5.10 Value benchmarking

We have used four different techniques for estimating the value of a product:

1. analysis of demand and price (DP analysis) (Equation 3.11);
2. market survey (DV method of conjoint analysis);
3. value curves (VC) (Equation 5.17);
4. economic value (EV).

There is not a general expression for EV determinations. Each case, such as fuel economy, is special.

The four techniques can be used in concert to arrive at a complete estimation of value for future products including both highly subjective and highly functional attributes. One technique can be used to cross-check the results obtained by another.

The procedure [19] for using the four techniques to forecast the value of a future product is shown in Figure 5.26. First the market shares and transaction prices of the products currently competing in the market segment of interest are used in Equations 3.11 and 3.12 to compute their total values and the differences in value of the products, respectively. The differences found are then analyzed independently using the VC, EC, and CV methods to determine which attributes are responsible for the value differences. (See [7, 9] for applications of the value benchmarking methodology.) A comparison is made between the results of the two procedures for the total value differences as a check of the process and to estimate the errors involved. The process provides a partial test

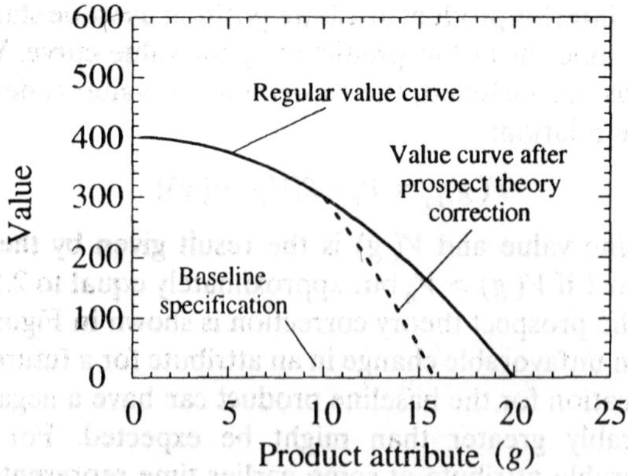

Figure 5.25 Prospect theory correction to a value curve which is needed when an attribute is diminished from customer expectations.

of the value expressions being used for the major attributes and illuminates opportunities for value improvement. We call this process **value benchmarking**. The value of a proposed new product can then be estimated from the attribute changes that are planned for it versus the existing baseline product. We call this process **strategic quality deployment (SQD)** which is described in Chapter 7.

5.11 Linking subsystem and component performance to value

Our discussion to this point has linked value to the performance of the product as measured by its system-level attributes (see, for example, Figures 5.15, 5.16, and 5.17). Thus engineers and planners at the subsystem and component levels need to develop means for linking the performance of their subsystems and components, respectively, to the system-level attributes. If, for example, the weight of the engine block is reduced, how will this affect the value of the vehicle to the customer if no other component attributes are affected? In answering this question, the effect of the weight change on the performance of the engine subsystem must be determined, the effect of the engine subsystem change on other subsystems must be determined, and finally the effect of all the subsystem changes on the system-level attributes of the vehicle must be determined.

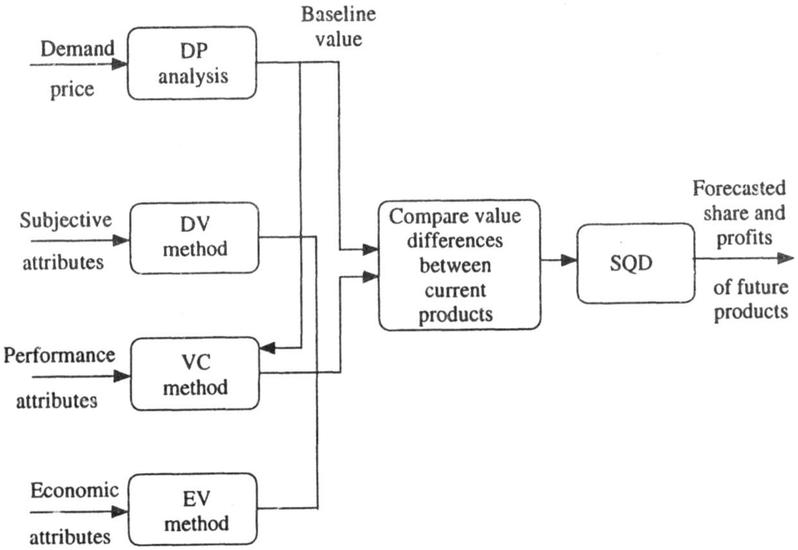

Figure 5.26 The value benchmarking process which is used to verify value versus attribute relationships for strategic quality deployment (SQD). (Source: H.E. Cook (1996) A unified decision support tool for product management, *in Life Cycle Modeling for Innovative Products and Processes*, (eds H. Jansen and F.L. Krause), Chapman & Hall, London, pp. 146–57, © 1996 Chapman & Hall. Adapted with permission.)

For this problem, an important linkage relationship would be the coupling of the weight change to the value for acceleration performance and/or fuel economy. This is not straightforward because neither horsepower nor weight in themselves are attributes which link directly to value. However, the ratio of horsepower to weight couples directly to acceleration performance as shown in Figure 5.27 based upon results compiled from published vehicle performance measurements from *Consumer Reports, Car and Driver*, and *Road and Track*. (In addition to normal experimental error, the scatter observed can arise from differences not accounted for in gear ratios, vehicle aerodynamics, tire adhesion, and weight distribution.) Because 0 to 60 mph acceleration performance has been linked (Figure 5.17) to value, horsepower-to-weight ratio couples to value as shown in Figure 5.28.

If the weight reduction of the engine were major, the front suspension engineer would need to change springs so that the proper ride height would be maintained, and the brake subsystem engineer would probably need to change the characteristics of the line pressure controls to the front and rear brakes to maintain stopping distance performance. The suspension components could also have their cross-sectional properties reduced because of the reduction in vehicle weight. Such action would, of course, require that the added tooling and engineering costs for the redesign not be prohibitive. The accumulated weight change of all of these actions would then be combined to estimate the value change using Figure 5.28. An alternative product scenario could be to reduce the engine displacement to maintain a constant horsepower-to-weight ratio and to improve value through better fuel economy.

Such decisions between alternatives need to be made wisely. Although prototypes of the different alternatives under consideration can be built and evaluated for total costs to manufacture versus their total value to the customer,

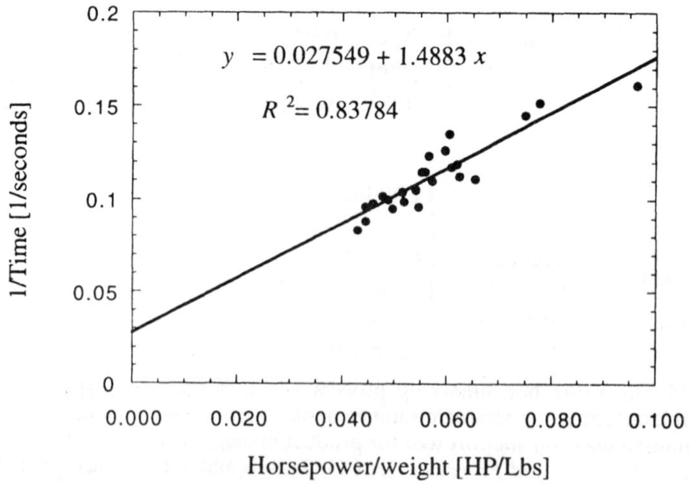

Figure 5.27 Reciprocal of acceleration time from 0 to 60 mph versus vehicle horsepower-to-weight ratio.

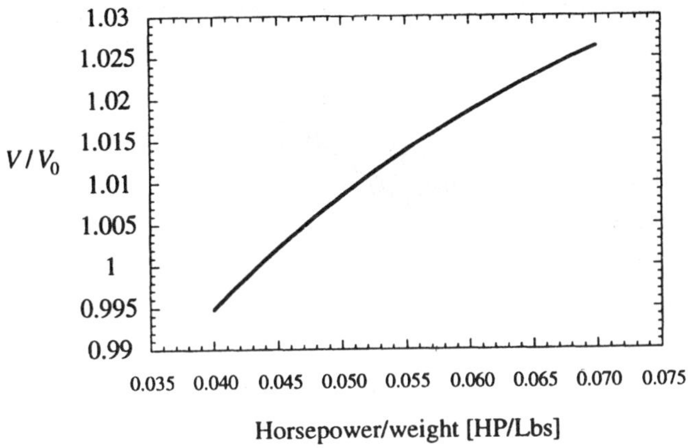

Figure 5.28 Normalized value versus vehicle horsepower-to-weight ratio for family sedans.

this will not always be feasible within the time and costs constraints. As a result, computer simulation of new product alternatives is becoming a very important tool because it is often less expensive and can be done more rapidly than the development and testing of prototypes. The simulations, however, should be structured so that the calculations are taken all the way to the customer/enterprise level in the sense of showing how the product change under consideration affects the system-level attributes, how that in turn translates into value, and how the value and cost changes are expected to impact demand and profits. The need for some prototypes may, however, never completely go away as the actual use and evaluation of real hardware in the hands of potential customers provides fidelity and product insight for certain attributes that computer simulation may never match. Consequently, it is important to develop rapid prototyping methods for components and subsystems that closely approximate the expected performance of the production components and subsystems.

5.12 Value trends

The DP analysis process can be used to track the values of competing products over time and thereby aid in understanding how the market is developing over time. Extrapolation of the value trends show the value levels that should be expected in the future. In conjunction with such plots, it should also be useful for key attributes to show their value history by plotting points on their value curves as a function of time. A hypothetical example of such a trend plot for the value of the horsepower-to-weight ratio of an automobile is shown in Figure 5.29. Such plots can be made even more informative by plotting the same trends for the major competing products.

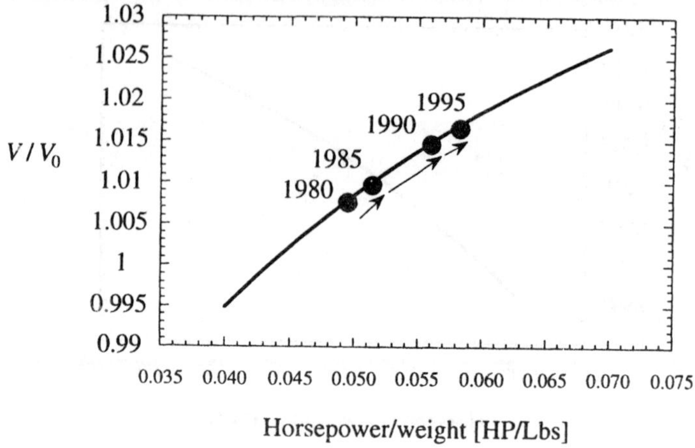

Figure 5.29 Hypothetical value trend plot for an automobile's horsepower-to-weight ratio over the years 1980–1995.

References

1. Green, P.E. and Wind, Y. (1975) New way to measure consumer's judgments. *Harvard Business Review*, July–August, 107–17; Churchill, G.A. (1991) *Marketing Research Methodological Foundations*, 5th edn, Dryden Press, pp. 464–82.
2. Cattin, P. and Wittink, D.R. (1982) Commercial use of conjoint analysis: a survey. *Journal of Marketing*, **46**, Summer, 44–53.
3. Louviere, J.J. and Woodworth, G. (1983) Design and analysis of simulated consumer choice or allocation experiments: an approach based on aggregate data. *Journal of Marketing Research*, **20**, 350–67.
4. Ben-Akiva, M. and Lerman, S.R. (1985) *Discrete Choice Analysis*, MIT Press, Cambridge, MA.
5. Cameron, T.A. and James, M.D. (1987) Efficient estimation for closed-end contingent valuation surveys. *The Review of Economics and Statistics*, **69**, 269–76.
6. McConville, G.P. and Cook, H.E. (1997) Evaluating mail survey techniques for determining the value of vehicle options. *SAE Paper 970764*, Society of Automotive Engineers, Warrendale, PA.
7. Donndelinger, J.A. and Cook, H.E. (1997) Methods for analyzing the value of automobiles. *SAE Paper 970762*, Society of Automotive Engineers, Warrendale, PA.
8. Dillman, D.A. (1978) *Mail and Telephone Surveys The Total Design Method*, Wiley, New York.
9. Silver, R. (1996) *Value Benchmarking to Improve Customer Satisfaction*, M.S. thesis, Department of Mechanical and Industrial Engineering, University of Illinois at Urbana-Champaign, Urbana, IL.
10. Cook, H.E. and Gill, M.R. (1993) On system design. *Research in Engineering Design*, **4**, 215–26.
11. Griffin, M.J. (1978) The evaluation of vehicle vibration and seats, *Applied Ergonomics*, **9**, 15–21.

12. Woodson, W.E., Tillman, B. and Tillman, P. (1992) *Human Factors Design Handbook*, 2nd edn, McGraw-Hill, New York, pp. 676–88.
13. *Automotive News* (1996) May 13, p. 6.
14. McConville, G.P. and Cook, H.E. (1996) Examining the value trade-off between automobile acceleration performance and fuel economy. *SAE Paper 960004*, Society of Automotive Engineers, Warrendale, PA.
15. Simek, M.E. and Cook, H.E. (1996) A methodology for estimating the value of interior room in automobiles. *SAE Paper 960002*, Society of Automotive Engineers, Warrendale, PA.
16. Keeney, R.L. and Raiffa, H. (1976) *Decisions with Multiple Objectives: Preferences and Value Tradeoff*, Wiley, New York.
17. Thurston, D.L. (1992) A formal method for subjective design evaluation with multiple attributes. *Research in Engineering Design*, **3**, 105–22.
18. Cook, H.E. and Kolli, R.P. (1994) Using value benchmarking to plan and price new products and processes. *Manufacturing Review*, **7**, 134–47.
19. Cook, H.E. (1996) A unified decision support tool for product management, in *Life Cycle Modeling for Innovative Products and Processes* (eds H. Jansen and F.L. Krause), Chapman & Hall, London, pp. 146–57.

6 Total quality management

6.1 Introduction

Taken to perfection, total quality management (TQM) has all elements of the enterprise working in harmony and with common purpose. Every action is forecast to improve competitiveness. So-called mistakes are simply part of the ongoing, continuous experiments that guide the learning process. All employees are engaged in and contribute to competitiveness in an active and meaningful way. This may not be the environment in many companies today, but the intense competition of global markets demand that this intensity, harmony, and common purpose be aimed for if the enterprise is to survive and flourish.

TQM includes every bit of management; it is management of the total enterprise. It is applicable to all organizations, public and private, supplying manufactured goods or services to customers. It applies to the individuals and work units within those organizations as each is an internal supplier to others and each is also an internal customer to others. In addition to the practice of TQM at these three levels, Shiba has identified a fourth level of TQM in Japan at the regional/national level where networking is used to transfer and disseminate quality knowledge gained by others and to generate the change in paradigms required to revolutionize quality management [1]. With the emergence of the Malcolm Baldrige National Quality Award, the fourth level is now very much in evidence in the U.S. Moreover, with the signing of Montreal protocols on CFCs and the development of the ISO 9000 standards, TQM is now operating at the international level.

At the enterprise level, TQM is the practice of continuously deploying all means that are legal and ethical in making winning products and profits. The six axioms of total quality management are:

1. Use all means that are legal and ethical.
2. All actions must be in harmony and with a common purpose.
3. Every planned action should be forecast to improve competitiveness.
4. There should be no wasted efforts.
5. Management of the total enterprise must satisfy all stakeholders including customers and society.
6. Emphasis should be on problem prevention.

The purpose of this chapter is to consider in some depth three important questions raised in Chapter 1:

1. What must a company do to assure that its workforce can design and manufacture winning products year in and year out over the long term?
2. How should a manufacturing enterprise operate on a day-to-day basis?
3. What are the rules, goals, procedures, and disciplines that should govern its actions?

Three basic premises underlie our approach to these questions:

1. Structured methodologies are very effective in planning future products and for the execution of those plans.
2. The most fundamental elements of an enterprise's culture should be continuous improvement and prevention of problems.
3. Continuous experimentation followed by actions that support the most favorable outcomes is the basis for continuous improvement.

6.2 The three streams of net value

In considering how to manage total quality, it is important to recognize that quality defined by Equations 4.8 and 4.9 should be viewed as the accumulation on an annual basis of three separate but highly coupled streams of net value (Figure 6.1).

$$Q_i = Q_{1i} + Q_{2i} + Q_{3i}. \tag{6.1}$$

One stream given by:

$$Q_{1i} = A_i \tag{6.2}$$

flows into the enterprise in the form of real dollars as profit. This stream will be called enterprise quality. The second stream is the net value received by the customer which is given by:

$$Q_{2i} = D_i[V_i - P_i]. \tag{6.3}$$

The third stream, Q_{3i}, is equal to the environmental quality impact of the transactions, G_i. It is necessary for the success of the enterprise that both Q_{1i} and the sum of $Q_{2i}+G_i$ be positive. If Q_{1i} is negative over an extended period, the corporation will become bankrupt. If $Q_{2i}+G_i$ is negative because of environmental quality being large negatively, society will curtail the actions of the enterprise through legislation and regulation.

6.3 Special and general forms of the universal metric

Success over the time period δt in a competitive market requires the enterprise to have a positive rate of change of total quality given by:

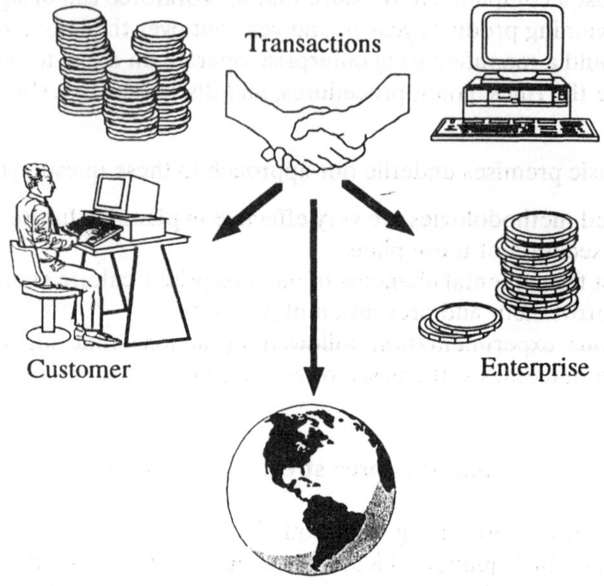

Figure 6.1 Three streams of quality flow from the market transactions. One goes to the customer, one to society, and one to the enterprise.

$$\frac{\delta Q_i}{\delta t} = \frac{\delta A_i}{\delta t} + \frac{\delta Q_{2i}}{\delta t} + \frac{\delta G_i}{\delta t} \tag{6.4}$$

The enterprise will fail in time if it sustains a negative loss in profits from year to year and will also fail in time if the quality it gives to its customers and the rest of society decreases from year to year. Each of the terms on the right-hand side of Equation 6.4 represents a term in the velocity of quality generated by the enterprise. The singular objective of TQM should be to meet demand and to generate and sustain positive unit quality velocities above the level of competitors.

A universal metric, Φ_i^*, can be defined in terms of the time variation of the total quality per unit:

$$\Phi_i^* \equiv \frac{\delta[Q_i / D_i]}{\delta t} = \Phi_i + \frac{\delta\{[G_i - F_i - M_i] / D_i\}}{\delta t} \tag{6.5}$$

where

$$\Phi_i \equiv \frac{\delta[V_i - C_i]}{\delta t}. \tag{6.6}$$

is a special form of the universal metric which applies when fixed costs, investment, and environmental quality per unit are constant over time. The universal metric represents a solution to the metric puzzle described in Chapter 3. The time rate of change of the fundamental metrics given by value, variable cost, fixed costs, investment, and environmental quality is the overall controlling metric. The special form of the universal metric indicates that it is neither value (a surrogate for 'quality' in a more traditional sense), nor cost, nor the pace of innovation (defined as $1/\delta t$) that determines success but the combination given by the time rate of change of value minus cost.

In summary, the following guidelines apply to product realization:

- Total quality must always be positive.
- Over the long term, the velocity of quality represented by the universal metric should be positive and greater than or equal to that of your strongest competitor.
- Demand for the product should be met.
- Profit, as measured over the long term, is the most appropriate objective function for making decisions provided that the externalities associated with environmental quality have been internalized in terms of the product and work place specifications.

6.4 Enterprise order and management

The bottom-line metrics of return on investment, profits, market share, and demand are determined by the fundamental metrics of value to the customer, variable costs, fixed costs, investment level, and the pace of innovation. It is appropriate to ask, 'What generates the fundamental metrics; what are they a function of; what are they connected to; what equations govern them?' The answer, vital to understanding what governs long-term success of the enterprise, is best understood by considering the systems representation of the enterprise (Figure 6.2). Input is shown on the left of the box with finished products as output on the right. Controls such as government regulations, stockholder expectations, price of materials, and customer requirements are shown as input on the top. The bottom of the box lists the tools, described as structure, culture and technology, used by the enterprise to transform the input of raw materials and semi-finished goods into a commercial product. The toolset, which includes the employees and all of the other resources at the disposal of the enterprise, determines the capability of the organization to perform the task. The degree to which any tool is deployed can be defined by a phenomenological variable which we will call an order parameter. The order parameter for a tool is a fraction equal to the number of sites at which the tool is installed divided by the total number of sites at which the tool could, within reason, be installed.

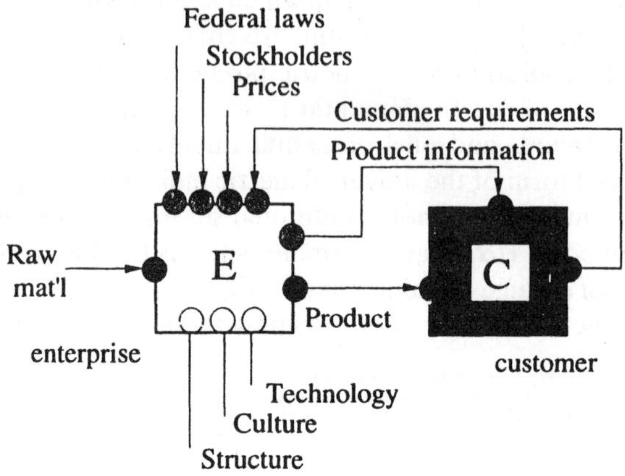

Figure 6.2 Systems view of enterprise showing input, output, controls, and tools. (Source: H.E. Cook (1991) On competitive manufacturing enterprises II: S-model paradigms, *Manufacturing Review*, **4**, 106–15, © 1991 ASME International. Used with permission.)

The concepts of order and its alter ego, chaos, help to understand the workings of an enterprise formed by a collection of interacting elements. When the number of products manufactured by the enterprise is more than a few and when the product has more than a rudimentary level of complexity, a certain degree of discipline or order must exist within the enterprise. Without some order, output from the enterprise would be chaotic and of little value because it would fail to meet customer requirements. The structure, culture, and technology are described in terms of phenomenological order parameters, ρ_1, ρ_2, ρ_3. Enterprise effectiveness as determined by the universal metric is related to the state of order within the enterprise as shown schematically in Figure 6.3. Maximum effectiveness is shown to lie in the figure somewhere between zero (chaos) and unity (complete order). A fully chaotic enterprise has no relationship between input and output and it is consequently ineffective. If a cheese factory were bordering on chaos, an order placed for twenty pounds of cheddar might be shipped as five slices of Swiss. By contrast, a completely rigid enterprise has exacting rules and procedures for everything. Its degree of order is nearly perfect but this enterprise is also ineffective because it can not react quickly to changing consumer tastes and competitive actions. Its pace is glacial and all of its current ideas are rooted in the distant past. If not already dead, last rites are being said.

The diagram in Figure 6.4 describes a situation having local order in which personal computers occupy 37.5% of the available desks. As a result, the degree of order for personal computers is 0.375 and they appear to be arranged in no specific pattern.

The degree of personal computer order has been increased to 50% in Figure 6.5. Is the organization more effective with the higher degree of personal com-

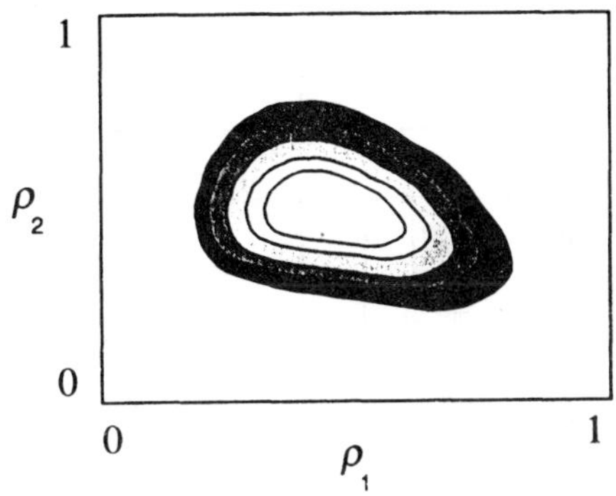

Figure 6.3 Schematic representation of enterprise effectiveness contours as a function of two order parameters.

puter order and do the added benefits outweigh the added costs? The result can only be answered by measuring how the universal metric changed.

Social or cultural order could be described by several order parameters representing the density distributions of important shared values, codes of conduct and dress, ideology, and myths. Technology order could be expressed as the density distributions of skills (design engineers, manufacturing engineers, purchasing agents, lawyers, etc.) as well as tools (personal computers, digital controllers, robots, etc.). A structural order parameter could be the degree to which the enterprise is organized along product versus functional lines. Another could be the distribution of plants and facilities.

Figure 6.4 Example of local order of 37.5 % for personal computers.

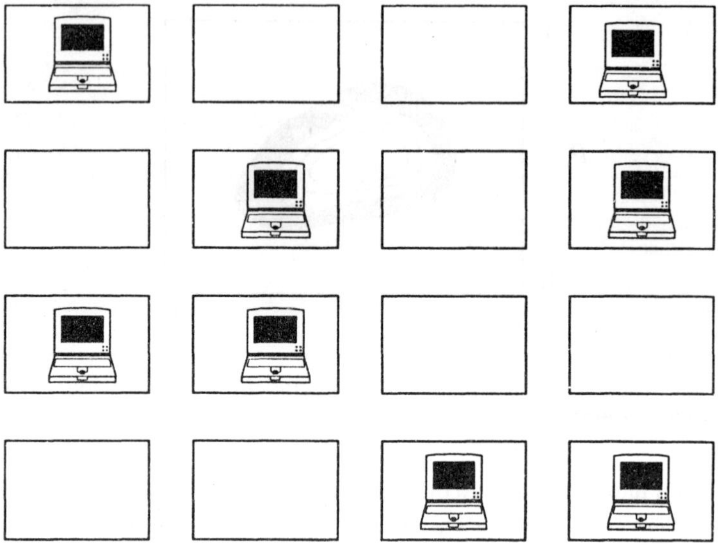

Figure 6.5 Example of local order of 50% for personal computers.

According to the model we are using, the effectiveness of the enterprise is related to the rate at which it is increasing total quality over time. Thus, we can think of output and input not only in terms of the number of products going out and raw materials coming in but also in terms of the universal metrics for input and output as shown in Figure 6.6. An approximate condition for growth of market share is for the special form of the universal metric divided by the average over all competitors to be greater than current share, m_i:

$$\Phi / [N\overline{\Phi}] > m_i \tag{6.7}$$

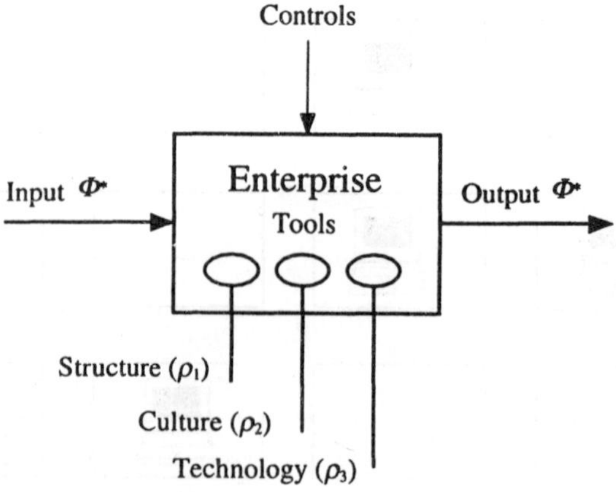

Figure 6.6 Using the universal metric as a performance measure.

The universal metric is managed through adjustments in internal order:

$$\Phi_i^* = f(\rho_1, \rho_2, \rho_3), \tag{6.8}$$

where (ρ_1, ρ_2, ρ_3) are the structural, cultural, and technological order parameters, respectively, for competitor *i*. The job of management, in response to the changing conditions facing the enterprise, is to continuously adjust the internal order vector in a manner that improves the universal metric for its products. This is shown schematically by the arrows in Figure 6.7. The optimum point moves around as conditions change. Management is unable to directly affect the bottom line as the point of control because enterprise order is two steps removed from return and market share. The bottom line can only be managed indirectly by managing internal order to achieve a high rate of continuous improvement in the fundamental metrics (Figure 6.8).

In attempting to meet and better world-class standards, some companies have encouraged greater employee involvement (a change in cultural order), more automation (a change in technological order), and simultaneous engineering (a change in both cultural and structural order). What may be frustrating to management is that there is often a time lag before seeing the improvements in moving internal order from one level to another. Thus management may try for more immediate results by attempting to influence the bottom line directly by changing price and reducing the size of the workforce. This will not be effective when (1) changes in price do not reflect fundamental changes in the cost or value of the product or (2) changes in the level of workforce do not reflect changes in workload or productivity. Management must be able to anticipate the needs of the future and begin making the required changes in its internal order well in advance to avoid the panics associated with crises. The hallmark of good management is proper anticipation and preparation. The art of reacting well to the unexpected is also good management but always having to operate in that fashion is not. A focus on the prevention of problems as opposed to the mindset of fixing problems is the cornerstone of total quality management.

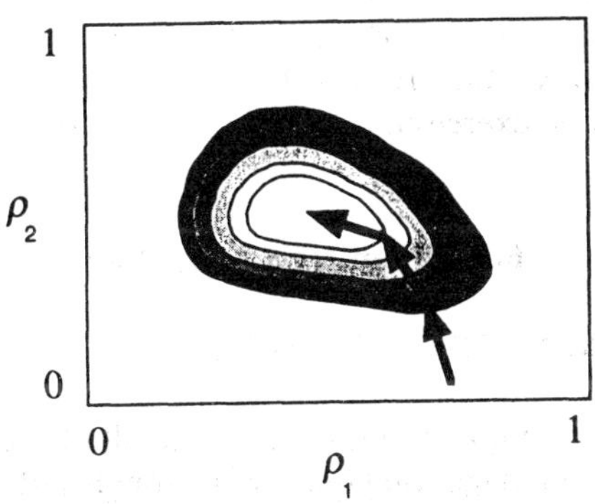

Figure 6.7 Management adjusts internal order to move toward higher effectiveness.

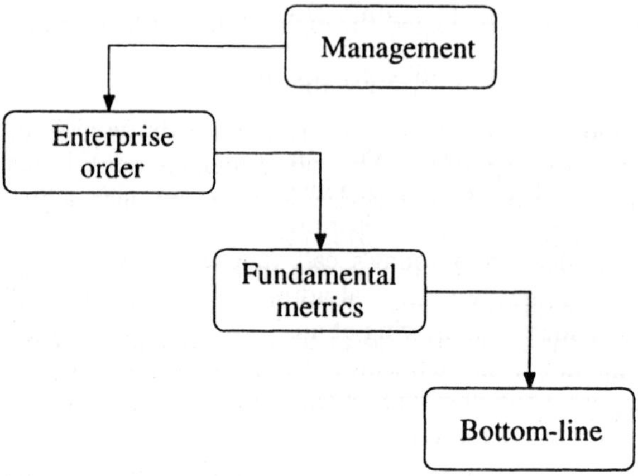

Figure 6.8 Management is two steps removed from managing the bottom line directly. (Source: H.E. Cook (1991) On competitive manufacturing enterprises II: S-model paradigms, *Manufacturing Review*, **4**, 106–15, © 1991 ASME International. Used with permission.)

The enterprise learns how to adjust its internal order to improve effectiveness through continuous experimentation. Shewhart and Deming's PDCA (Plan, Do, Check, Act) cycle is a guide for the process [2]. It is important to experiment with good ideas aimed at important opportunities and to experiment on a scale that is representative but not too large. One idea, for example, would be to place the desks of the design engineer, purchasing agent, and the manufacturing engineer for each component in a particular subsystem side by side in the same room and to compare the performance for product development time and costs for this arrangement versus the prior performance when the three were in separate buildings under separate management. Actions must not be based upon soft, anecdotal information and recollections of important executives but based upon hard, quantifiable information. All potential opportunities can be expressed in terms of changes in the enterprise order vector, $(\delta\rho_1, \delta\rho_2, \delta\rho_3)$, and its relationship to changes in the universal metric, $\delta\Phi_i^*$. Continuous experimentation followed by the rapid dissemination of the ideas proven to be sound generates the basis for continuous improvement but some caution always has to be exercised in scaling up from the experimental level.

6.5 Organization structure

6.5.1 CLASSICAL ADMINISTRATIVE CRITERIA

The management of total quality must permeate all elements of the organization, and the ability to manage total quality is strongly influenced by the nature of the organization structure. In examining the question of how the enterprise should organize itself to be most effective, it is useful to begin by reviewing the guidelines from classical administrative theory, as stated by Urwick (as taken from Brech [3]), Figure 6.9.

Although these guidelines are more than sixty years old, the initiative criterion supports modern ideas of employee involvement such as quality circles and participative management. The continuity criterion's call for 'systematic improvement' is no less than today's call for 'continuous improvement.' It should be understood that the classical administrative criteria are but part of the issues to be considered in arriving at an organizational structure [4]. Other criteria, such as the nature of the business and its goals and work plan, must also be taken into account when deciding on an organization structure. We will consider these factors in what follows.

6.5.2 FUNCTIONALIZATION CRITERION FOR PRODUCT REALIZATION

The functionalization criterion from classical administration theory emphasizes that the successful enterprise has to be 'arranged in logical groups ... as to

Functionalization:

The necessary units of activity involved in the object of the enterprise should be analyzed, sub-divided, and arranged in logical groups in such a way as to secure by specialization the greater results from individual and combined effort.

Correspondence:

Authority and responsibility must be coterminous, coequal, and defined.

Initiative:

The form of the organization must be such as to secure from each individual the maximum initiative of which he is capable.

Coordination:

The specialized conduct of activities necessitates arrangements for the systematic interrelating of those activities so as to secure economy of operation. References from one activity to another should always take the shortest possible line.

Continuity:

The structure for the organization should be such as to provide not only for the activities immediately necessary to secure the object of the enterprise, but for the continuation of such activities for the full period of operation contemplated in the establishment of the enterprise. This involves a continuous supply of the necessary personnel and arrangements for the systematic improvement of every aspect of the operation

Figure 6.9 Classical administrative criteria according to Urwick. (Source: E.F.L. Brech (1958) *Organization, The Framework of Management*, Longmans, London, pp. 371–8.)

secure by specialization the greatest results from individual and combined efforts.' What are the logical groups? What are the specializations?

Concurrent engineering teams have been formed to correct deficiencies in the functional organization that became acute when competition became intense. Nevertheless, the functional organization defined by groups generally noted as marketing, planning, product engineering, purchasing, manufacturing, finance, sales, and service still exists under the overlay of concurrent engineering teams and, in many respects, the teams are attacking the symptoms of the problem with the functional organization – slow, sequential, costly, strong cultural differences, lack of trust between units, and low quality products – but are not correcting the root causes. In other words, concurrent engineering as practiced today by many enterprises may only be a Band-Aid, helpful, but far from complete in attacking the root causes and companies should be seeking ways to improve beyond this limited, albeit improved, state of affairs. Product engineers, purchasing agents and manufacturing engineers have always known that they should, in principle, work together; thus, there must be a fundamental flaw in the functional organization (or in the training of its professionals by our universities) that, over time, has come to inhibit close working relationships. It is very likely that another organization structure – a different set of **logical groups** – and reward system can be formed which does not have the deficiencies often found with functional organizations and will be an improvement over the team structure in use today.

Organizational changes are often implemented all at once company-wide. The resulting disarray that has often resulted from such sweeping change indicates that a more gradual, step-by-step approach may be wiser. This does not suggest that top management should act with timidity but plan bold actions in keeping with the uncertainty of the problem. Experiment on a small scale, broaden the action, and modify the plan as the results of the experiment become clear.

6.6 The natural flow of product realization

The basic premise we will follow in attempting to understand what the logical groups should be is to view the product realization process not as organizational elements interacting but at a higher level of abstraction in terms of systems, subsystems, and components being planned and manufactured in response to customer needs. With this perspective, the natural flow of product realization and the fundamental activities that support product realization become clear and can be scrutinized.

The flow of requirements begins with the customer stating his or her needs and ends with a product being delivered to the customer by the manufacturing enterprise. The final product is a system which is assembled from subsystems which in turn are assembled from components manufactured by the enterprise and its suppliers. Three methods, as discussed by Ulrich and Eppinger [5], are generally used to identify customer needs: interviews, focus groups [6], and observing the product in use. Generally, needs are stated quite simply. In the case of an automobile, for example, a potential customer may state that it should handle well, be quiet and reliable, carry five passengers comfortably, and be priced under $20 000. However, the amount of detailed information that must be generated by the enterprise to define the thousands of parts that need to be manufactured and assembled to meet this rather simple statement of needs is enormous.

This explosion in complexity is managed through a series of steps in which an increasing number of persons and skills are brought to bear as the effort unfolds. An outline of the requirements flow from the customer through the steps of the product realization process for the enterprise and its suppliers is shown in Figure 6.10. The boxes in the chart refer to the logical (natural or fundamental) activities needed to design, develop, manufacture, and assemble a complex product. As drawn, the chart shows the requirements flow. The logical parts flow is exactly in the opposite direction. Together we denote these two processes as the logical requirements and parts (LRP) flow.

Each task in Figure 6.10 has an array of inputs and outputs as shown in Figure 6.11. Parts and requirements are the inputs and outputs of each LRP task. The transformation of input to output uses the tools of structure, culture, and technology. The requirements passed from an internal customer to an internal supplier should include the following:

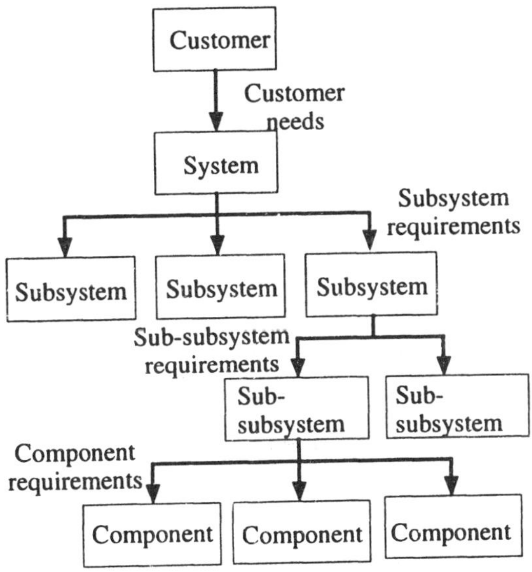

Figure 6.10 Logical requirements flow from customer through the manufacturing enterprise.

- piece cost
- investment
- weight
- performance
- material(s)
- hard point geometry
- assemble to
- package with
- reliability
- receive date
  - prototype
  - production
- service time and cost
- process plan
  - input
  - process route
  - ship to
- disposability

Hard point geometry refers to the dimensions of the part or subsystem that are mandated by the activity sending the requirement and unfolds from the system to component level as requirements flow through the enterprise. The complete set of dimensions needed to fully specify the system or component is only arrived at when the requirements flow reaches the component/stock activities. The dimensional requirements for the packaging of subsystems is set by the systems activity and the requirements for the packaging of sub-subsystems and components within a subsystem are set by the appropriate subsystem activity. In the output from the system level, piece cost, for example, refers to the cost of the entire system, whereas piece cost for the requirements to a subsystem refers to the cost of that subsystem. The same unfolding applies to each of the other categories.

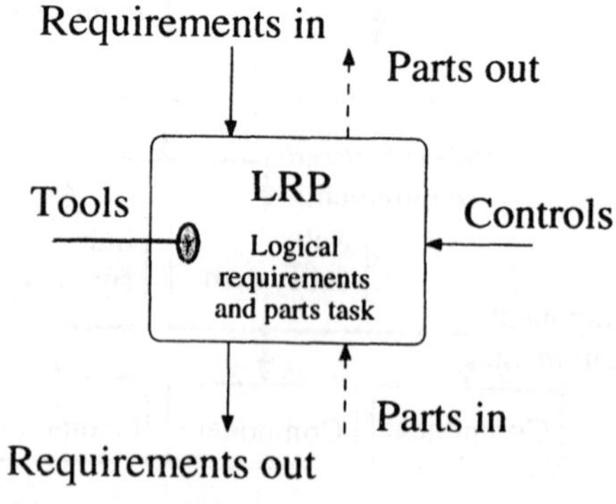

Figure 6.11 Input, output, and tools for a task in the logical requirements and parts flow.

At the subsystem task level, for example, 'requirements in' refers to the performance of the subsystem. If the system were a car and the subsystem of interest an engine, one subsystem requirement would be engine horsepower. System performance requirements influenced by horsepower would be acceleration and fuel economy but these involve not just horsepower but a host of variables, e.g. weight and drag, involving many other subsystems. The challenge to systems engineers is to establish performance requirements for the subsystems which ensure that the system performance requirements will be met including the requirements for costs, investment, and weight.

6.6.1 SYSTEM TASK

The task for those having system responsibility – marketing, planning, design, engineering, purchasing, final assembly, sales, and service – in the overall LRP flow is as follows:

1. to discover, understand, and meet customer needs;
2. to develop the total system specifications for the product based upon:
   - customer needs
   - variable costs
   - total fixed costs
   - total investment costs
   - expected actions of competitors
3. to translate the system specifications into key performance specifications for each subsystem, including the cost, weight, and reliability;
4. to procure subsystems that meet those requirements;
5. to establish the start date for delivery of the subsystems, the rate of subsystem delivery, and total number of subsystems to be delivered;
6. to assemble the final product which meets all customer requirements;
7. to deliver the final product to the customer;
8. to establish the service requirements and service procedures for the product;
9. to ensure that the system and the workplace for the system meets all enterprise and governmental requirements.

The system task can be factored into a generalized lead planning activity and a generalized manufacturing activity as in Figure 6.12. Customer needs are discovered by the activities responsible for system planning and translated first into system performance specifications and then into the subsystem requirements believed necessary to generate the system performance. Those responsible for system assembly must review the subsystem requirements to ensure that the subsystems can be assembled before the requirements are transmitted.

6.6.2 SUBSYSTEMS TASKS

The tasks for those having subsystem responsibility – planning, engineering, purchasing, and subsystem assembly – in the overall LRP flow are as follows:

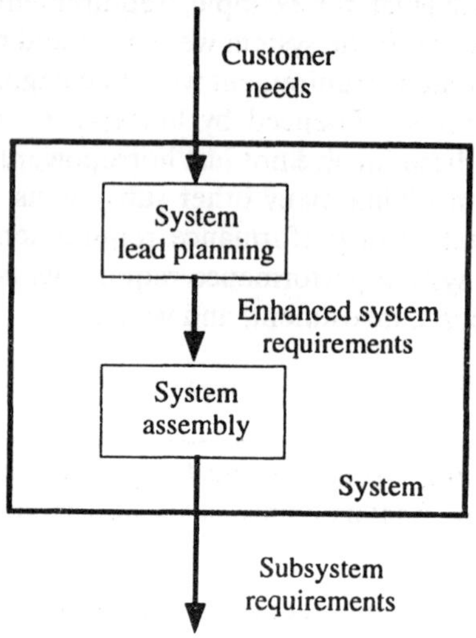

Figure 6.12 System task factored into system lead planning and system assembly.

1. to understand and meet the needs of the system;
2. to develop the total subsystem specifications for the product based upon system requirements, variable costs, investment costs, and expected actions of competitors;
3. to translate the subsystem specifications into key performance specifications for each component in the subsystem, including the cost and weight;
4. to procure components that meet those requirements;
5. to establish the start date for delivery of the components, the rate of delivery and the total volume to be delivered;
6. to assemble the subsystem which meets all requirements;
7. to deliver the subsystem to the point of final assembly;
8. to establish the service requirements and service procedures for the subsystem and transmit those requirements to the system task meeting all system requirements;
9. to ensure that the subsystem and the workplace for the subsystem meets all enterprise and governmental requirements.

Each subsystem task can also be divided into a lead planning and a subassembly activity as shown in Figure 6.13. The responsibilities are similar to those within the system activity except that they relate to the design, manufacture, and assembly of a subsystem.

6.6.3 COMPONENTS TASKS

The tasks for those having component responsibility – engineering, purchasing, and manufacturing – in the overall LRP flow are as follows:

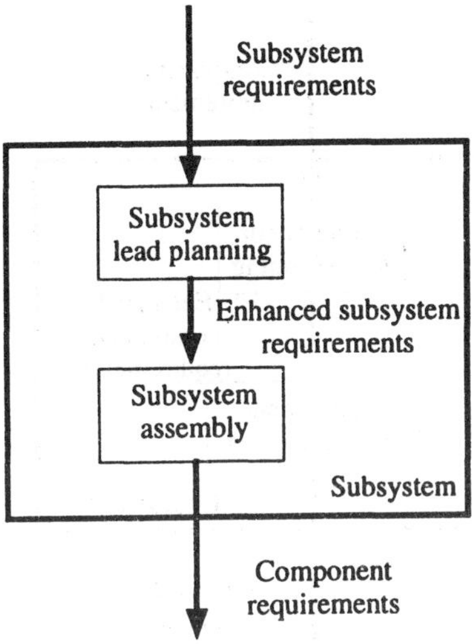

Figure 6.13 Subsystem task factored into subsystem lead planning and subassembly.

1. to understand and meet the needs of the subsystem and system;
2. to develop the total component specifications for the product based upon subsystem requirements, variable costs, investment costs, and expected actions of competitors;
3. to translate the component specifications into key performance specifications for the input stock and raw materials used to manufacture the component;
4. to procure stock and raw materials that meet those requirements;
5. to establish the start date for delivery of the stock and raw materials, the rate of delivery and the total volume to be delivered;
6. to manufacture the component from input stock and raw materials so that it meets all requirements;
7. to deliver the component to the point of subsystem assembly;
8. to establish the service requirements and service procedures for the component and transmit those requirements to those responsible for the subsystems task;
9. to ensure that the component and the workplace for the component meets all enterprise and governmental requirements.

Finally each component task has a lead component planning function and a component manufacturing function which generates requirements for stock and raw materials (Figure 6.14).

Each manufacturing or assembly activity also has a logistics planning activity for day-to-day operations and for the machines required for production and the persons needed to maintain and operate those machines. If any manufacturing or assembly activity is exploded for more detail, it should only display the

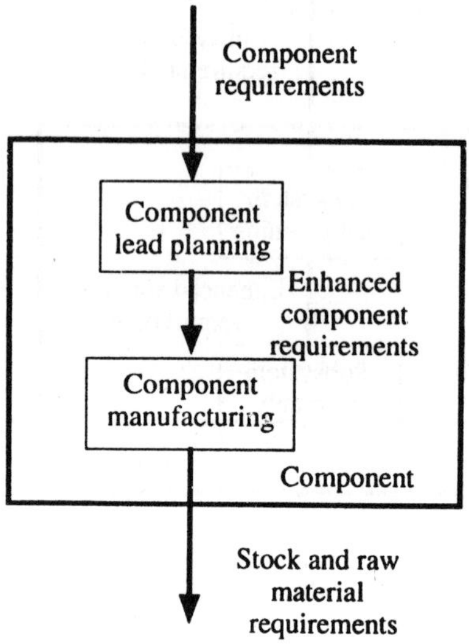

Figure 6.14 Component task factored into component lead planning and component manufacturing.

logistics planning activity and the requirements and parts flow between the machines on the factory floor. All other persons (design engineers, manufacturing engineers, purchasing agents, marketing, etc.) and tools such as computers and product development test equipment not used directly in day-to-day production are grouped under the lead planning activity umbrella but may, in fact, be in different divisions or departments.

6.7 Total quality performance against objectives

6.7.1 TRADITIONAL PRACTICES

The enterprise, to be successful, must set challenging and well-defined objectives and measure its performance against them. The setting of objectives based upon traditional accounting practices is now believed to be counterproductive for enterprises operating in competitive markets [7, 8, 9].

One reason for this is that traditional accounting practices have focused on direct labor costs and direct labor productivity with indirect costs allocated to operating units as burden or overhead in proportion to their direct costs. High machine utilization which has also been a goal of traditional accounting and control practices has serious flaws as a performance measure because, if not matched to sales rate, costs associated with excess work in process and inventory are generated [10]. Direct labor is often a small fraction of total costs today as opposed to the turn of the century when traditional accounting practices were developed. The environment has changed but the outdated paradigm lingers.

Another problem with traditional performance measurements is that control of the individual local elements by them does not assure that the performance goals of the overall enterprise are met. Simply put, the macro performance of the enterprise is not equal to the summation of its traditional micro performance measures. The chief executive raised this question to his top operational managers during the early 80s: 'How can it be that the performance requirements of each division have been met but the enterprise itself has failed to meet its objective?' Although the answer to this important question is complex, a major factor is that traditional performance measures stress local optimization at the expense of optimizing the effectiveness of the full enterprise.

6.7.2 WHAT SHOULD PERFORMANCE MEASUREMENTS DO?

Although the above establishes much of what performance measurements should not do, what then should they do? Highlighted below are five guidelines for performance measurements which have direct strategic implications for the enterprise:

1. Track sources of competitiveness.
2. Encourage improvements that benefit the enterprise, its customers, and society.
3. Benchmark against your major competitor.
4. The sum of the local measures should equal the performance of the enterprise.
5. Be in harmony with the organizational structure.

The first guideline is due to H. Thomas Johnson [11]. The second arises from the fact that competitiveness mandates continuous improvements which benefit all that are engaged in the market transaction. Continuous improvement focuses not so much on where you are today but the rate at which you are improving versus the pace of your competitors. Performance measurements should thus be set against the benchmark of your major competitor (guideline 3). Practices that result in local optimization should be changed so that the performance of the total enterprise can be obtained by summing up the performance of each individual unit (guideline 4). Each action needs to be measured by how it improves the value or reduces the cost of the full system.

It is inescapable that a close relationship exists between what can be used as an effective performance measure and the nature of the organization structure. The organization structure must be harmonious with the performance measurement you want to use and vice versa (guideline 5). As noted by Kaplan [12], 'the new role for aggregate cost measures would seem to require financial summaries that cut across machines, cost centers and other organizational units.' The fifth guideline can, in principle, be met if the organization structure is such that it supports the correspondence of authority and responsibility for system, subsystem, and component design.

6.7.3 SYSTEM/SUBSYSTEM ORGANIZATION

The S-model formalism developed in Chapter 3 provides a basis for developing performance and thus accounting measures which encompass the five guidelines stated above provided that the operating units follow the system rules for managing the LRP flow and organization structure:

1. Each organizational unit within the enterprise must have well-defined goals based upon improvements in the fundamental metrics for the total system.
2. Performance should be measured against these goals.
3. Authority and responsibility for setting input requirements should reside within the unit receiving input.
4. Authority, responsibility, and capability for transforming input to required output should reside within the unit making the transformation.
5. Subtasks having a large number of transactions between each other should be the paradigm for grouping task units into departments.

As a first step, let's apply these system rules to the system lead planning activity for a functionally organized automotive manufacturer, Figure 6.15. Also shown for completeness are several functions (abbreviated) outside of the system activity. The requirements flow is shown by solid lines and the flow of parts is shown as dashed lines. Only the major channels of flow are shown. Each requirement line is also a conduit for feedback regarding the requirements.

The system responsibility is seen to be shared in a diffuse manner by the functional organization as each activity shown within the system box is generally headed by a vice president. Responsibility is diffuse in the functional organi-

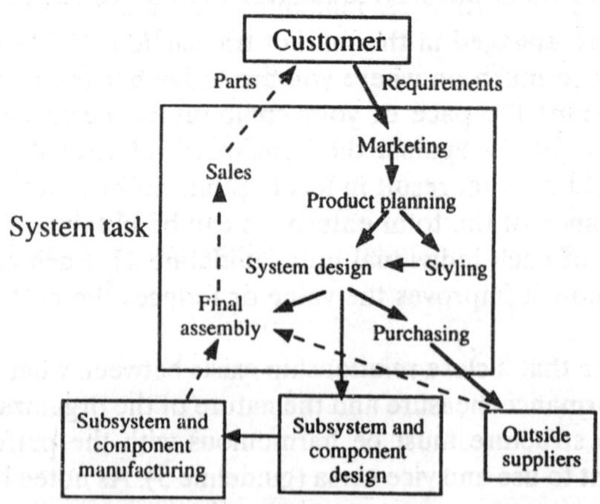

Figure 6.15 System task mapped over functional organization showing diffuseness of system responsibility between organizational units.

zation because design, purchasing, manufacturing, and assembly are separate units. This diffuseness of responsibility and authority within the functional organization violates the correspondence criterion of classical administrative theory. This factor and the glacial pace of decision making associated with it has led to the formation of concurrent engineering teams having members from each of the respective functions shown. Ideally, the teams are empowered with the responsibility and authority to proceed with the development of new products with minimal interference from the parent organizations.

Another way in which to organize the systems activity would be to have the organizational boxes defined by the LRP flow itself. The resulting organization structure – a system/subsystem structure – would be identical to the boxes shown for the LRP flow given earlier for the total enterprise. In the system/subsystem organization, each unit receiving parts is empowered with the full authority and responsibility for setting the requirements for its input and is also empowered with the full authority and responsibility for adding the appropriate value to the input received in order to meet the requirements placed on it by the unit downstream which receives its output [13, 14, 15].

The system unit has full system responsibility which includes marketing, systems planning, systems engineering, packaging, final assembly, sales, and service (Figure 6.16). Each subsystem can be divided into more and more subordinate systems until the individual components are reached. Each subsystem unit would have all the necessary personnel and skills – purchasing, engineering, manufacturing, financial control, etc. – to make the transformation from input to output and to set the requirements on incoming material. The purpose of the requirements flow is to maximize the quality of the full system. This prevents the subsystems and components units from optimizing on their own which would lead to a sub-optimization of the full system.

An important feature of the system/subsystem structure is that each unit on average should have the same culture because each unit has manufacturing, engineering, and purchasing personnel as well as planning. The strong cultural differences in the functional organization likely arise from the sharp organizational differences which often have distinctly different goals and sets of performance criteria. For example, manufacturing tends to have a short-term view and design engineering has a longer-term view because this is often how their performances are measured. In the system/subsystem organization, each unit has a broad view of the importance of time with goals for near and longer term.

For an automobile company, the main subsystem units would be likely to be engine, transmission, body, interior, and chassis–suspension–brakes. The customer is aware of these critical subsystems and it is important that the responsibility for them is well-defined in the organization. Because responsibility and authority for them will be very diffuse in a functionally organized automobile company, finger pointing will be likely to be one of the more intense activities when something goes wrong.

An example of how the system activity might be organized in system/subsystem organization is shown in Figure 6.16. The functional units of sales, marketing, product planning, system design engineering, and styling have been combined into a single unit entitled 'system design for customer.' The system design for final assembly function and final assembly operations make up the remaining functions. This organization structure closely parallels the LRP flow within the systems activity. In the functional organization, marketing manages the interface between the customer and the enterprise keeping the enterprise abreast of changing customer needs. Sales manages the customer interface with regard to selling the resulting product. Purchasing provides the interface between the system activity and outside suppliers. A difficulty is that each of these functions will often stress their own interest and not present a balanced system perspective for product value, costs, and lead time. They also inhibit direct lines of communication between the system engineering function and the customer, for example.

The arrangement of functions shown in Figure 6.16 is more open and flexible in interfacing with the customer and suppliers. The responsibility and authority for each activity is also clearly defined. This is displayed within the LRP flow chart by the fact that the requirements and parts flow link the same organizational pairs and are in equal and opposite directions. The final assembly operation includes a purchasing function in the LRP structure. Each different system (e.g. vehicle platform for an automotive enterprise) produced by the manufacturer would have a similar type of structure. The system and component activities can be organized using similar considerations.

This process of mapping the natural LRP flow can be extended to any activity and level within the enterprise to study a variety of product realization issues

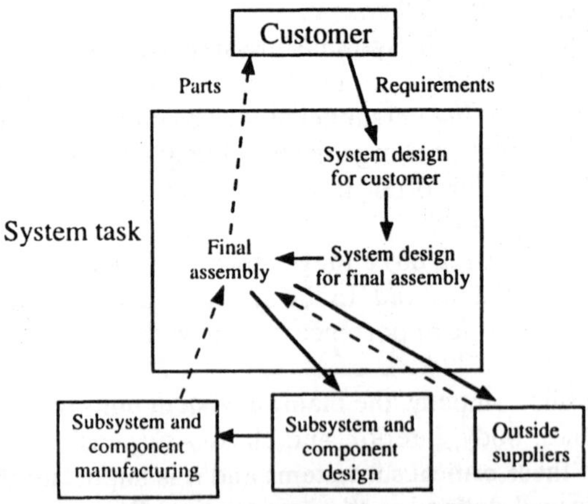

Figure 6.16 System task mapped over system/subsystem organization where one person, the system leader, is responsible and has full authority for setting system specifications and key subsystem requirements.

including database needs, computer networks and integration, operations research, just-in-time delivery, manufacturing cell design, concurrent engineering planning, organization design, and process planning. A key advantage to using the LRP flow charting method to study organizational and other changes is that it links the current and proposed actions at any level within the enterprise to its impact on customer needs. It is an enabler for continuous improvement throughout the enterprise which stimulates and guides ongoing positive change in manufacturing effectiveness and product competitiveness. The process of using flow charts to document what is done today and to plan better processes for tomorrow is a form of as-is and to-be analysis, discussed in Chapter 1.

The natural flow of work does not mean that there is a natural reporting hierarchy with the system unit at the top. The reporting relationship could be one in which the leaders of the system and subsystems groups report to a common leader. As there could be several distinct products and thus several systems groups, the reporting hierarchy could be as shown in Figure 6.17.

6.8 Activity-based strategic accounting

6.8.1 CONCEPT

With an organization structure in place which follows the system rules listed earlier, the system/subsystem organization being one example, it is then possible to realistically consider implementing activity-based strategic accounting (ABSA) practices to measure the improvements in profitability that each activity makes to the enterprise.

Consider an enterprise organized into four levels as shown in Figure 6.18a. Starting from level 1, these could be system, subsystem, sub-subsystem and component. Each organizational box in Figure 6.18a lists the costs that are added by that organization. The total cost of the product equals the sum of the costs over all of the boxes which is 100 for the example shown. A product innovation originating within the component organizational unit represented by the center box in level 4 is considered. As this innovation progresses up the chain,

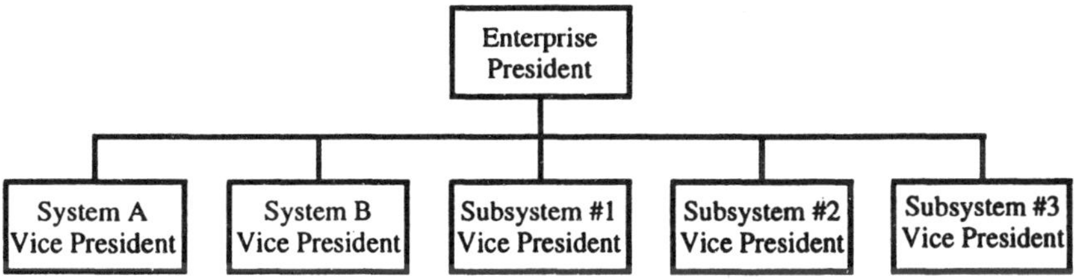

Figure 6.17 A possible reporting hierarchy for system/subsystem organization at highest level.

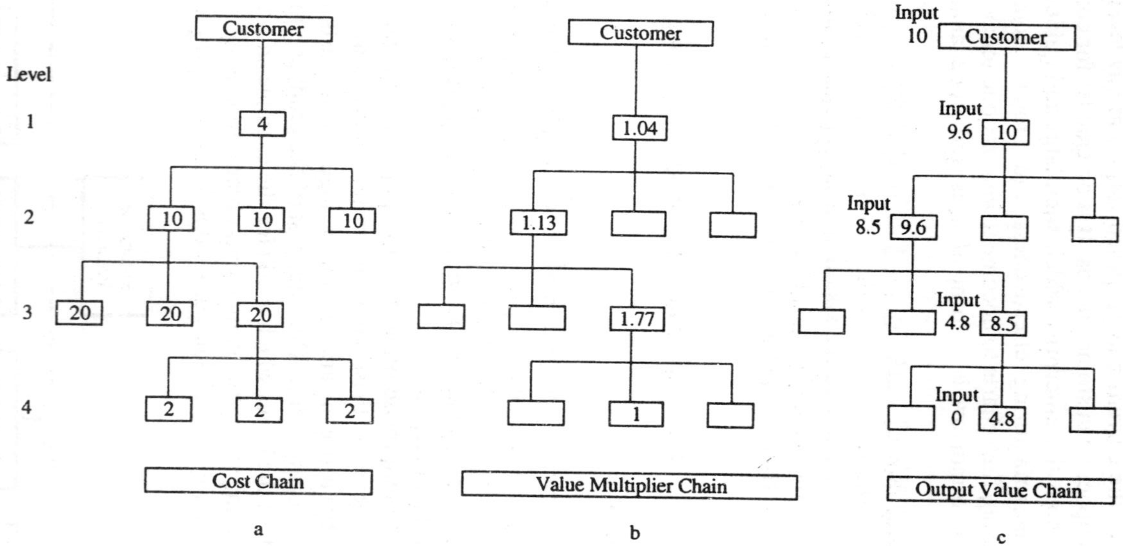
Level
1
2
3
4
Customer
4
10
10
10
20
20
20
2
2
2
Cost Chain
a
Customer
1.04
1.13
1.77
1
Value Multiplier Chain
b
Input 10
Customer
Input 9.6
10
Input 8.5
9.6
Input 4.8
8.5
Input 0
4.8
Output Value Chain
c

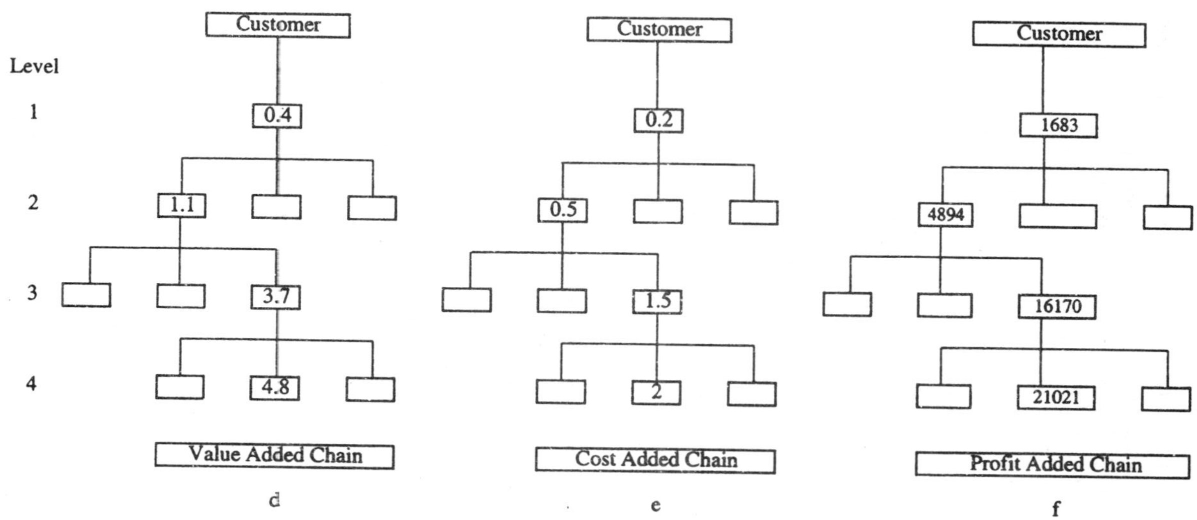

Figure 6.18 Managing performance objectives using activity-based strategic accounting (ABSA).

each affected unit will add value to it. For example, the unit in level 3 may paint the part and attach it to the sub-subassembly. The level 2 unit may assemble the sub-subsystem to a subsystem and the system organization in level 1 may distribute and advertise the final product. All of this activity adds value to the component as it moves through the chain.

The numbers in the boxes of Figure 6.18b represent the value output by each organizational unit on receiving one unit of value as input, the organization responsible for the original innovation in such a chain always having an output multiplier equal to unity. For example, the level 3 unit sends out \$1.77 of increased value on receiving \$1.00 of value from the component unit shown in level 4. The level 2 unit sends out \$1.13 of value on receiving \$1.00 of added value as input. Finally the system level unit sends out \$1.04 units of increased value for \$1.00 of value received as input.

The question begged in all of this is 'How are the value multiples determined?' The answer is that they should be determined either through negotiations between the organization units or as directed by top management. A good starting point for the negotiations is the cost structure shown in Figure 6.18a because the value multiplier for a unit should be approximately equal to one plus the fraction formed by the operating cost of the unit divided by the sum of its operating cost and those of all of the other units sending input to it. The value multipliers shown in Figure 6.18b were determined in this manner. For example, the 1.77 in level 3 is equal to $1+(20/(20+3\times2))$; the 1.13 in level 2 is equal to $1+(10/(10+3\times20+3\times2))$; and the 1.04 in level 1 is equal to $1+(4/(4+3\times10+3\times20+3\times2))$. If the organizational units shown as not having input actually had input, then their operating costs would also have to be included in computing the value multipliers.

Consider the example shown in Figure 6.18c of a total value improvement of the product equal to 10 at the customer which originated as an innovation in level 4. The value chain can then be worked backwards from the customer, to see how much added value was made by each affected unit. The system organization in level 1, which sent out to the customer a total value improvement of \$10 from the enterprise, received as input a value improvement of \$9.60 (= 10/1.04). Working down the chain in this manner, we find that the originator in level 4 produced \$4.80 of the total \$10 ultimately delivered to the customer. The amount each unit added is shown in Figure 6.18d. Each unit affected may also have incurred added costs in implementing the innovation. Hypothetical added costs are listed in Figure 6.18e.

If we assume that competitors do not make any changes and that the enterprise has chosen to price the new product using the scenario:

$$P = P_0 + \alpha_V \delta V + \alpha_C \delta C,$$

then the resulting price change can be computed from the changes in total value and total cost per unit. Moreover, the change in profit by activity (Figure 6.18f) can also be forecast using Equation 3.22. If we take $K = 100$ for this problem and the baseline quantities listed in Table 6.1, the overall resulting improvements due to the product innovation are shown in the last column of Table 6.1 where we have used $\alpha_V = \alpha_C = 1/2$ in the pricing scenario expression and the assumption that competitors will not make any price or value changes. The exact choice for the coefficients in the pricing scenario should be set by those responsible for the system level of the product. If transfer prices were being used to compute the profitability of business units within the firm, then the transfer price offered by a unit should equal the baseline transfer price before the change plus the added output value and cost from the unit multiplied, respectively, by α_V and α_C.

6.8.2 TOWARD ABSA PRACTICE

Although the ABSA concept described above is straightforward, the development of an accounting practice for it would require some important changes to traditional practice. First, the organization structure should be along the system/subsystem design. The systems unit would need to specify what the key system level attributes, $\tilde{g}_i$, were and how value is expected to depend on changes in these attributes. Methods for doing this were described in Chapter 5. Next the subsystem units in conjunction with the system unit would need to establish how key subsystem attributes, , influenced multiple system-level attributes. Finally, the sub-subsystem and component units would need to work with their respective subsystem units to establish how the sub-subsystem and component attributes in turn influenced the subsystem attributes. The major accounting practice change would be to monitor and track key attributes at the system, subsystem, sub-subsystem, and component levels. This implies a cultural change and could thus be difficult for many enterprises to implement. Nevertheless, if the true sources of competitiveness were to be tracked as already voiced, then the tracking of key product attributes becomes mandatory.

6.9 Managing environmental quality

With the rapid growth in population and number of manufactured goods on a finite earth, the production of wastes and effluents associated with the manu-

Table 6.1 Baseline metrics and added contributions to the baseline metrics due to the cost and value chain processes shown in Figure 6.18

Metric	*Baseline*	*Added*
Demand	5000	292
Cost	\$100	\$4.16
Value	\$300	\$10.00
Price	\$200	\$7.08
Investment	\$100 000	\$0.00
Profit	\$400 000	\$43 768

facture, use, and disposal of goods and services is of great concern. These byproducts of market transactions impact environmental quality and are grouped under the term 'externalities' in microeconomic theory [16]. The losses to environmental quality resulting from pollution of the air, water, and soil by effluents from manufacturing plants are one example of externalities. Noise pollution from factories, aircraft, trucks and cars is another. Externalities have no direct markets; clean air, clean rivers, or low noise decibels cannot be bought or sold in some market. The change in environmental quality associated with a given transaction, however, does not always have to be negative. One example of a transaction with a positive externality is the inoculation of a person against a communicable disease. The purchase of liability insurance is another.

The loosely regulated market has poorly internalized environmental quality losses [17] leading to the London Green Smog, Love canal, Dioxin, DDT, etc. There are two reasons for this. First, environmental losses are often not seen or fully understood at the time the transaction takes place. Second, losses – usually insignificant for a single, market transaction – arise from the aggregate, cumulative effects of the manufacture, use, and disposal of many products or services leading to degradation over the long term of common property resources such as air and water.

Thus, in response to societal needs, regulatory agencies such as EPA, OSHA, DOT, FDA, etc. are formed by the government to assess the real or projected environmental quality losses [18]. Because regulatory decisions represent a means for incorporating externalities into the market for goods and services, they have far-reaching impact and must be made carefully and wisely. Just as manufacturing and service industries have implemented TQM practices to improve customer satisfaction, regulatory agencies need to employ rigorous TQM practices to better serve the nation and to lessen their chances of making flawed decisions. However, the practice of TQM, whether it is by a manufacturing firm or a regulatory agency, can only be effective when the mission or responsibility of the organization is understood by all stakeholders, when the mission can be described in terms of current goals, when the goals can be divided and deployed as specific tasks assigned to every element of the organization, and when the authority and capability necessary to carry out those tasks is vested in each element.

6.9.1 MISSION OF THE REGULATORY AGENCY

Ideally the mission of a regulatory agency should be (1) to determine what the important environmental factors are under its jurisdiction, (2) to quantify the relationship between the level of each environmental factor and the losses generated in environmental quality, and (3) to act in a manner that effectively and properly introduces this information into the marketplace in terms of an economic incentive for reducing environmental losses.

Within this ideal mission, the responsible agency has several ways in which to introduce environmental quality information into the marketplace:

1. **Public awareness:** the agency can make potential customers aware of the losses generated by various environmental factors. With concerns raised in the minds of potential customers, the value of the product is reduced. This in turn reduces demand for the product and encourages manufacturers to find ways of reducing the levels of environmental factor present. Warning labels on cigarettes act in this manner, encouraging buyers to reduce consumption or quit smoking entirely and manufacturers to reduce the amount of tar and nicotine, all of which has occurred.
2. **Set mandatory maximum levels:** the approach used perhaps most often by regulatory agencies is to set legal limits on environmental factors. Exhaust emissions on automobiles are a well-known example of this type of environmental factor control.
3. **Effluent fees:** the agency can require the manufacturer or user to pay a fee in proportion to the level of the environmental factor. The example considered in Chapter 4 was of this type.
4. **Develop a marketplace for environmental quality credits:** a combination of approaches 2 and 3 can be used in which organizations that have environmental factors below a given level can sell its excess environmental quality to those who exceed the level.

It is possible that any one of these three approaches could arrive at the optimum level for the environmental factors that would balance the interests of all of society including buyers and sellers. However, this is unlikely. When a common good such as air or water is involved, public awareness (approach 1) will not ensure that it is properly cared for [19]. Awareness is very important but not sufficient.

There is also a fundamental difficulty with approach 2, setting maximum levels, because it requires that the regulating agency not only understand the losses caused by the environmental factor but also know the manufacturing costs needed to reduce the level of the environmental factor. The agency faces considerable difficulty in determining the manufacturing costs as it is not a manufacturer. The agency not only needs to know the impact of control technology on variable costs but the size of the investments needed as well. Moreover, these costs are often unknown until a manufacturer begins to experiment in reducing the levels of the environmental factor. Furthermore, the costs can change, usually reduced, as experience is gained by the manufacturer on the learning curve and technology advances. Because of the considerable uncertainty in costs, it is unlikely that the regulating agency will arrive at the optimum level of effluents using the second approach.

Finally the agency also needs an understanding of the value to the customer if it attempts to set a legal limit that balances all interests. The requirement in the early 1970s of having the seatbelt locked before starting an automobile is a good example of where an agency failed to understand the sensitivities of the customer. The resulting uproar suggested that the loss in value to the customer of having to deal with seatbelt interlocks greatly outweighed the externality caused by the absence of such devices.

The legal limit has considerable favor because it appeals to strong interest groups. A mandated level appeals to environmentalists because the manufacturer cannot buy out of it by paying an effluent fee. Moreover manufacturers may prefer a legal limit because its replacement by an effluent fee would increase the overall level of competition in the marketplace. However, the fundamental flaw in the legal limit approach is that it suggests that there is a large negative effect on environmental quality as the limit is just exceeded and no effect if the effluent is below the limit, neither of which is generally true.

Approach 4, the establishment of a marketplace for environmental quality credits, requires that the agency first set a target level for the environmental factor which must be met either in actual practice or through the use of purchased credits to offset the amount of environmental factor in excess of the target. Setting of the target level for this approach is no different in theory from the setting of the target level for approach 2 entailing the same problems of uncertain manufacturing costs and value assessments. But because it gives extra flexibility in the overall market it is preferable to approach 2.

The use of an effluent fee should be more effective in protecting the interests of all of society than a legal limit. Ideally the fees would be used to directly compensate those suffering losses generated by the effluent. In instances where the losses are diffuse and borne by all of society, the fees would supplement the general fund. The task of the regulating agency is also easier in that it only has to deal with a part of the problem that it should come to know best, the accurate and timely assessment of environmental factors and their associated environmental quality losses. Use of an effluent fee to regulate the market externalities causes the loss of environmental quality to be internalized by the manufacturers and, in principle, results in the optimum effluent level always being satisfied vis-à-vis each individual manufacturer, reflecting its specific manufacturing costs and level of technical expertise. It also encourages manufacturers to continue the search for cost-effective ways to reduce the level of the environmental factor, whereas the legal limit offers a weaker incentive. In its study of decision making at the EPA, a National Research Council panel strongly recommended that effluent fees be considered as a method of regulation by the agency [20]. By establishing the costs of externalities associated with an environmental factor, the regulating agency becomes an important contributor to the market forces and to the market's overall effectiveness in providing for the balanced interests of all of society. However, when the agency sets a mandated legal limit, it reduces the efficiency and effectiveness of the marketplace. When the evidence is clear that an environmental factor generates environmental losses that are greater than the benefits, the responsible agency should act to ban outright those processes and products that generate the factor. DDT, for example, is recognized as making such a dire assault on environmental quality.

Regulatory agencies should and do find themselves often in conflict with both business and elected government officials because their actions to improve environmental quality will generally affect both manufacturing costs and jobs in a way which will be unfavorable to those that are directly affected. However, the agency should not be allowed to compromise its role if it has accurately assessed, which the free market fails to do, the losses to environmental quality resulting from market externalities. It is in the interest of society as a whole, business, labor, consumers, and our elected representatives, that those losses be entered into the overall microeconomic equations that should govern the marketplace. These costs are as real and as important as other more traditional factors such as labor rates, inventory, and the price of energy. Simply stated, all of the costs and values must be understood and weighed to arrive at intelligent policy decisions.

Thus, society needs the agency to correctly assess the environmental quality losses of existing and impending markets and to find effective ways of working with elected officials to ensure that those losses are internalized in the marketplace. Going beyond this simple mission, for example engaging in full market analyses of the costs of control, not only interferes with those market forces that are proper and efficient, it makes the agency engage in matters where it does not have the right expertise, assessing the cost of control technology to the manufacturer and the value of the product to the customer.

6.10 Unemployment and fair trade

At the national level, a full description of total quality should include a summation of the microeconomic contributions from all of the products from the individual enterprises less a factor proportional to the number of persons that are unemployed. A conflict arises between this macroeconomic expression for total quality at the national level and the microeconomic equations for profitability and quality at the enterprise level because higher quality accrues when the equivalent quantity and value of goods are produced by fewer persons which makes for fewer jobs in a given company or industry.

Although there may be several ways of resolving or at least accommodating this conflict, the most effective way is to have the federal government exercise its responsibility at the national level to ensure that interest rates and the infrastructure defined by the national networks for communication, transportation, education, and training be sufficient to sustain a robust, growing economy in which the creation of new enterprises and the growth of existing enterprises results in a low level of unemployment.

We also need to apply TQM concepts to trade and environmental concerns at the international level. In the competition between manufacturers in the global marketplace, the issue often arises as to whether or not trade in certain products is fair. If a company in country A exports its products to B but uses practices that yield a cost advantage which are unlawful in country B, is this cost advantage fair or unfair? If slave labor were being used in country A, it would probably not be allowed to sell products in a country that abhorred such a practice. But what if country A does not employ slave labor but simply has less stringent rules for smokestack emissions and toxic wastes? What if the standard minimum wages are different between the two countries? What if a company in country A simply has special access to certain capital markets at a significantly lower rate than companies in country B? These last three questions may be more difficult to answer than the first. In addressing them, it is necessary to understand that the sale of a product involves more markets than just the direct market for the finished product (Figure 6.19). The finished products market is interconnected to and supported by the markets for raw materials, technology, labor, and capital. These allied markets are necessary for a functional market in finished products. If any of the allied markets are closed or restricted to some companies, then trade in the finished product can become unfair as judged by Equation 6.1 for the net value of the product to society.

Societal issues are a proper subject for governmental concern, review and action. For example, if certain capital markets are open to some competitors but not others or subsidies are available to some but not all, then the effective trade in goods is not fair. All companies that sell in an open market for finished goods should also participate in an open market for capital if competition is to be fair according to Equation 6.1. Likewise, the environmental and safety requirements should be common between companies if trade in their finished products is to be truly fair. Fair trade demands that total quality management be practiced at an international level between countries.

6.11 Metric taxonomy

A useful way of summarizing the TQM approach presented here is to outline the taxonomy of metrics for an enterprise (Figure 6.20). The product metrics

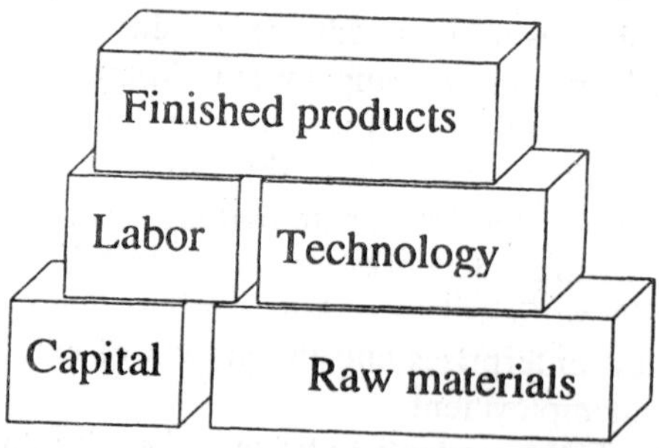

Figure 6.19 The market for final products is inseparable from the allied markets for labor, technology, capital, and raw materials. If trade in finished goods is open for a group of companies, then the allied markets for each should be open to all competitors in the group independent of what the home base is for the competitors.

are represented by the attributes of the components, subsystems, and system. They are generated by the work of the people within the enterprise using the tools provided. The product's system-level attributes define its fundamental metrics of value, variable cost, fixed cost, investment costs, environmental quality, and the pace of innovation. The product's fundamental metrics determine its bottom-line metrics given by:

- return on investment
- market share
- demand
- price
- total quality

Total quality as measured by the product's net value to society is the most encompassing of the bottom-line metrics. It includes contributions from cost, investment, profits, customer satisfaction, and environmental impact. All actions that favorably influence quality also favorably influence one or more of the other bottom-line metrics. For example, the product specifications that generate the maximum in quality can also yield the maximum in profits. Total quality and the other bottom-line metrics are managed by adjusting, as the business environment changes, the degree of order within the enterprise which includes the skills and size of the workforce, the levels of technology used in all aspects of the business, the organization structure, and the culture of the workplace so that the enterprise improves the total quality of its products at a rate equal to or better than its competitors.

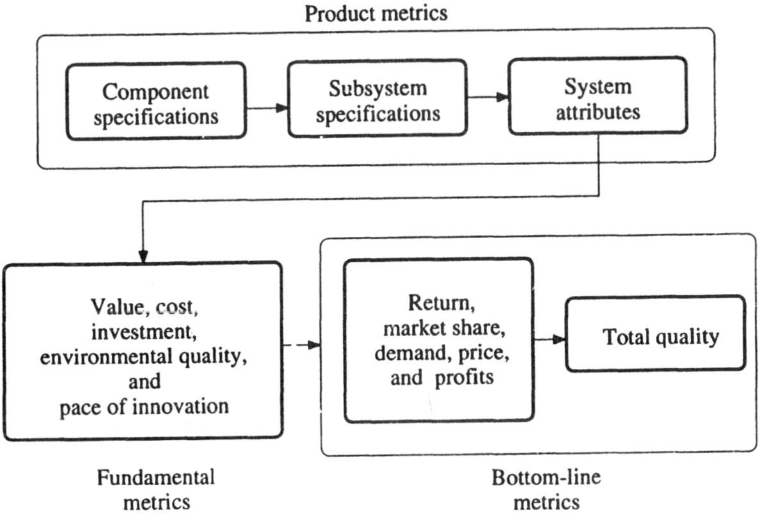

Figure 6.20 Taxonomy of metrics from product, to fundamental, to bottom line.

References

1. Shiba, S. (1990) Total quality control in Japan, total quality management, in *A Report of Proceedings from the Xerox Quality Forum II*, July 31–August 2, 1990, (ed. L.J. Harris), Xerox Corp., Rochester, NY.
2. Scherkenbach, W.W. (1988) *The Deming Route to Quality and Productivity*, Mercury Press, Rockville, MD, p. 35.
3. Brech, E.F.L. (1958) *Organization, The Framework of Management*, Longmans, Green and Co., London, pp. 371–8.
4. Simon, H.A. (1976) *Administrative Behavior*, Free Press, New York.
5. Ulrich, K.T. and Eppinger, S.D. (1995) *Product Design and Development*, McGraw-Hill, New York, p. 38.
6. Greenbaum, T.L. (1993) *The Handbook for Focus Group Research*, Lexington Books, New York.
7. Kaplan, R.S. (1990) Limitations of cost accounting in advanced manufacturing environments, in *Measures for Manufacturing Excellence*, (ed. R.S. Kaplan), Harvard Business School Press, Boston, MA, pp. 15–38.
8. Goldratt, E.M. and Cox, J. (1986) *The Goal*, North River Press, Croton-on-Hudson, NY.
9. Johnson, H.T. (1990) Performance measurement for competitive excellence, in *Measures for Manufacturing Excellence*, (ed. R.S. Kaplan), Harvard Business School Press, Boston, MA, pp. 63–89.
10. Goldratt, E.M. and Cox, J. (1986) *The Goal*, North River Press, Croton-on-Hudson, NY.
11. Johnson, H.T. (1990) in *Measures for Manufacturing Excellence*, (ed. R.S. Kaplan), Harvard Business School Press, Boston, MA.
12. Kaplan, R.S. (1990) in *Measures for Manufacturing Excellence*, (ed. R.S. Kaplan), Harvard Business School Press, Boston, MA, p. 37.
13. Cook, H.E. (1992) Organizing manufacturing enterprises for customer satisfaction. *Manufacturing Systems: Foundations of World-Class Practice*, (eds J.A. Heim and W.D. Compton), NAE Press, Washington, DC, pp. 116–27.
14. Cook, H.E. (1991) On competitive manufacturing enterprises II: S-model paradigms. *Manufacturing Review*, **4**, 106–15.
15. Cook, H.E. (1992) New avenues to total quality management. *Manufacturing Review*, **5**, 284–92.
16. Salvatore, D. (1991) *Microeconomics*, HarperCollins, NY, pp. 591–612.
17. Jasanoff, S. (1990) *The Fifth Branch*, Harvard University Press, Cambridge, MA.
18. Seneca, J.J. and Taussig, M.K. (1979) *Environmental Economics*, Prentice-Hall, Englewood Cliffs, NJ, pp. 103–4.
19. Seneca, J.J. and Taussig, M.K. (1979) *Environmental Economics*, Prentice-Hall, Englewood Cliffs, NJ, pp. 93–5.
20. (1977) *Decision Making in the Environmental Protection Agency Vol. II*, National Academy of Sciences, Washington, DC, p. 20.

7 Product planning and quality deployment

7.1 Product decision making

Product decisions are made within a demanding business environment shaped by the intense pressures of short development time, limited budget, incomplete information, and strong competitors. Yet certain persons, as if they have some magical, intuitive talent, have demonstrated the ability to quickly evaluate a situation and decide upon a successful course of action. If the source of their talents were truly magical and if there were a ready supply of persons so blessed, decision support tools would not be needed.

Reality, however, is that their intuition is not magical but comes from the insight provided by a profound knowledge of the product's marketplace and technology which can only be gained from experience and a constant, ongoing reassessment of the relevance of that experience to the changing environment. Moreover, apart from the few celebrated, high-level decision makers, there are many other persons – design and manufacturing engineers, product planners, marketing specialists, purchasing agents, and financial analysts – who regularly make important decisions having make-or-break implications in the ongoing development of complex products such as a car, road scraper, or computer. Some have developed profound knowledge, but many with less experience have not yet done so. It is the decision support needs of these persons that has led to the popularity and use of tools such as conjoint analysis, value engineering, statistical process control, Taguchi methods, and quality function deployment (QFD) [1].

In his discussion of analysis versus intuition, Mintzberg [2] cited the findings of Peters, Hammond, and Summers [3] who considered the distribution of errors in an experiment designed to contrast the two approaches to decision making. They found that analysis gave the exact answer much more frequently but it also led to a wider distribution of wrong answers than intuition. Thus use of both perspectives should be beneficial with intuition used to arrive at quick, ball-park estimates and to rule out analytical results that are judged far off the mark.

The product plan is the guide for the enterprise to achieve its desired level of profitability. The belief here is that all persons involved in the product realization process at all levels should think and plan strategically using relatively simple but reliable decision support tools whenever possible. If persons have been specifically designated as product planners, they should, nevertheless, reside within tasks having broad design and manufacturing responsibility because the leader of each task has the ultimate responsibility for planning the task to meet the requirements of its customer. This view is in contrast to having a product planning group that resides outside day-to-day operations and profit–loss responsibility.

7.2 Quality function deployment (QFD)

The QFD structured methodology was developed to address four key product management questions:

1. What is needed by customer?
2. Why is it needed?
3. How much of an improvement needs to be made?
4. How can it be achieved?

QFD links product decision making from design through manufacturing by an unfolding of what's-needed versus how matrices – the so-called 'houses of quality' – and has proven to be an insightful tool for planning of new products [4]. The purpose of this chapter and the one which follows is two-fold: the first purpose is to make a direct, quantitative connection between product decision making and the strategic matters of market share and profit which is not done in QFD. The second is to structure deployment, not in the 'house of quality' manner, but more rigorously as a waterfall of linked experiments from the system to the component levels of product design and manufacturing. On making these refinements, quantitative forecasts can be made for demand and profits and the crucial quality tools of design of experiments; Taguchi methods; and statistical process control become seamlessly integrated into the overall product planning formalism which has been called SQD or **strategic quality deployment** [5].

The curves shown in Figure 7.1 illustrate how SQD addresses the four questions listed above. The system attributes, represented by the parameter g along each of the horizontal axes, are the 'what is needed.' The position of the strongest competitor, benchmarked by the point g_{Cmpt} (Figure 7.1b), reflects 'why it is needed' because this competitor is either ahead of you or will be ahead if you do not act. The 'how much' is given by the difference between the current and future target specifications defined by the maxima, g_T and g_T', respectively, for the current and future (projected) total quality. The 'how' is achieved by lowering the cost curve for the future product which allows the target specification to be moved to g_T' cost effectively and by reducing variance, the width of the distribution function. A limited set of subsystem alternatives is generally evaluated and the best from the group is chosen, a process known as satisficing [6].

(a)
(b)
(c)

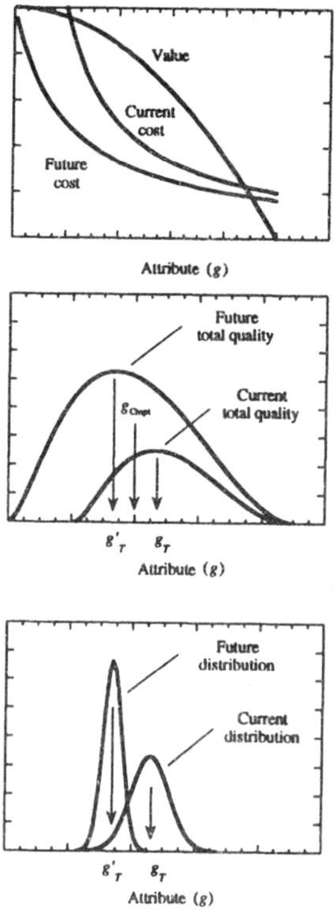

Figure 7.1 (a) Value and cost curves, (b) total quality curves proportional to $[V - C]^2$, and (c) frequency functions of the system level attributes. (Source: H.E. Cook (1992) New avenues to total quality management. *Manufacturing Review*, 5, 284–92, © 1992 ASME International. Adapted with permission.)

7.3 Strategic planning within the systems perspective

Before work on a new product begins, agreements should be reached within the enterprise as to:

1. the definition and measurement of product value, variable cost, fixed cost, and investment;
2. what the bottom line is and how it is to be computed for forecasting purposes;
3. the general nature of the tasks of product realization including their input/output relationships and timing;
4. who has responsibility and authority for each task;
5. the capability of each team for performing its task.

In the absence of these agreements, a consensus does not exist on the definition of key product terms, lines of authority, and the ability of the organization to do the job. Each shortcoming jeopardizes the development of a successful product. For example, many design engineers today define value as performance divided by cost [7], whereas marketing specialists define it in terms of part worths [8] or utiles. The connection between the two definitions is seldom made because it is not straightforward to do which leads to an impedance in the processing of information between marketing and engineering. Also concern for the bottom line should be shared by all as the output of each task influences it. But for all involved to understand and project how their decisions and actions will impact profits, more rigor must be applied to profit forecasting than is done today in most corporations. Forecasts should be developed from algorithms based upon projections of the variable cost, value, and investment level for the proposed future product and for the impact that future competing products are expected to have on demand. The algorithms used for forecasting the profitability of the future product should be available to all persons involved in the product realization process as all need to make decisions based upon it.

The guide for strategic planning from system to subsystem to component is the customer/company needs loop shown replicated at each of the levels in Figure 7.2. At the system level, subsystem alternatives are evaluated for their impact on the system-level attributes. At the subsystem level, component alternatives are evaluated for their impact, again, on the system-level attributes. Finally, at the component level, dimensional, material, and processing alternatives are evaluated for their impact, once again, on the system-level attributes. For the enterprise to be effective, there needs to be, in addition to the formal flow of information shown in Figure 7.2, an ongoing, informal flow of information between all levels on a continuing basis with each informing the others of potential new opportunities and risks.

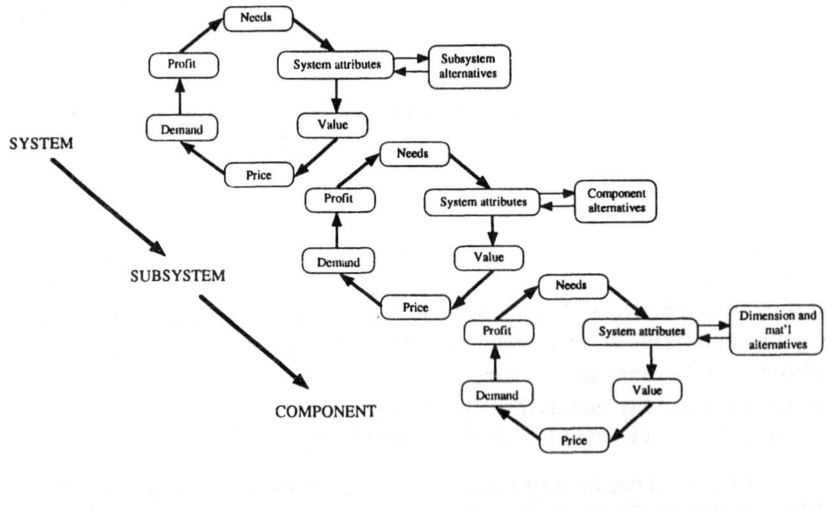

Figure 7.2 The structured methodology for product planning follows repeats of the customer/company needs loop at the system, subsystem, and component levels.

The product requirements flow, as discussed in Chapter 6, results in a highly structured and synchronized flow of hardware from component manufacturing, to subassembly, to final assembly (Figure 7.3). Components are shipped to the requirements of the subassembly planners and subassemblies are shipped to the requirements of the system planners. Finally, the full system is shipped in accord with the requirements of the final customer. Each task in Figure 7.3 is performed by a team made up of members from all the functional disciplines whose skills are needed to complete the task. Decisions are made within a common strategic context for all tasks at all levels.

Planning at all levels can be divided into the six key steps shown in Table 7.1. The objective of situation analysis, the first step in planning, is to understand the product's current profitability and the opportunities and risks to profits in the future. This requires that the fundamental metrics of value, cost, and the pace of innovation be determined for the competing products including their current position and where they will likely be in the future.

On receiving the key requirements from a customer, suppliers must factor them into their planning processes. At each level, as already discussed, the planning process is guided by the relevant customer/company needs loop in Figure 7.2. Provided that societal interests regarding environmental quality have been internalized, the use of profit as the objective function at every level – or related quantities such as present worth (Equation 3.25); internal rate of return; or time to break even (Equation 3.26) – makes the deployment of requirements and actions taken by suppliers strategic in nature.

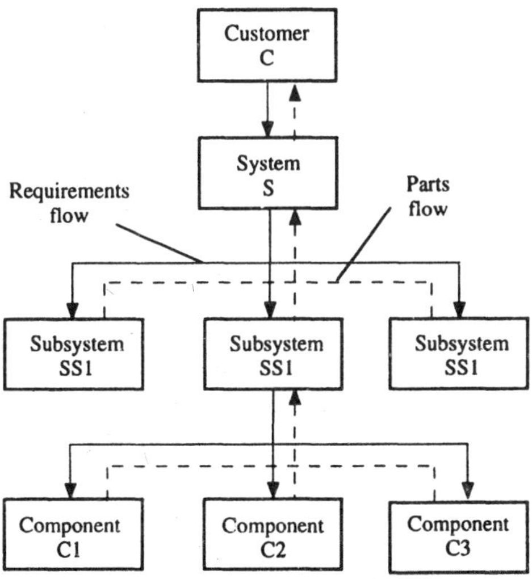

Figure 7.3 The requirements flow and the resulting parts flow through the enterprise. (Source: R.P Kolli and H.E. Cook (1994) Strategic quality deployment. *Manufacturing Review*, 7, 148–63, © 1994 ASME International. Used with permission.)

Table 7.1 The six key planning steps

Step	*Activity*
1.	Situation analysis (as-is).
2.	Brainstorming for possible alternatives.
3.	Reduction of possible alternatives to a manageable number.
4.	Testing alternatives.
5.	Selecting alternatives.
6.	Transmitting requirements to suppliers of task (to-be).

7.4 SQD planning at the system level

7.4.1 SITUATION ANALYSIS

A key element of situation analysis (Table 7.1) is gaining an understanding of customer requirements or needs which are obtained, not surprisingly, by talking to customers and observing how they use and misuse the current product. It is important to establish how their expectations are being met with the product they are currently using and to determine what they would like that is not currently available and what things need to be improved. It is also important to experience the product in the customer's environment. In this way, new and better ways of developing the product should become apparent which may not have been stated by the customer. If the product under consideration is markedly different from any product the customer is currently using, it will be difficult to determine how the potential customer will like or dislike the product without being able to use a prototype version over a trial period. If there is not sufficient budget or time to have such prototypes available to potential customers, then the value of the proposed product must be estimated by a jury evaluation of persons within the company who have had access to the limited number of prototypes available.

The requirements, as stated by the customer, need to be linked to measurable system attributes of the product. For example, if customers are being asked about their satisfaction with a personal computer, a statement that the disk drive is 'a bit noisy' needs to be translated into a noise level measured at a point that corresponds to the distance, on average, between the customer and the disc drive. The desire that a customer wants 'more pep' in an automobile can be translated into several measurable attributes including acceleration time from 0 to 40, from 0 to 60, and from 40 to 70. Each of these acceleration time attributes correspond to different environments for vehicle usage that can be important to the customer. More specifically, the customer should be asked if the pep is needed to enter and merge with high-speed traffic on an interstate, or needed to get moving from stops in city traffic, or needed in passing slow vehicles on a two-lane highway. Perhaps acceleration in all three environments is in need of an upgrade. Once customer needs are translated into measurable system-level attributes, then it is possible for the system team to define system specifications for these attributes and engineering objectives for subsystems which, if met, should result in the customer being well-satisfied with the performance of the new product.

Product specifications are seldom, if ever, exactly at their ideal points because costs are usually prohibitive. Thus, every specification decision involves a cost versus benefit (value) trade-off. Each system-level attribute can have its relationship to value determined by the value benchmarking methodology described in Chapter 5. At the system level, ideal and critical specification points can vary with market segment. For example, the weighting exponent for vehicle acceleration for a family sedan is roughly 0.2 [9] but it might be as large as 0.7 or more for persons who drive sports cars or personal luxury cars.

An important part of situation analysis within the system task is the setting of financial targets for the product. Using the S-model, these are set in order as follows:

1. Select desired level of annual profits, A.
2. Select desired level of annual demand, D, and market share, m.
3. Select target price, P.
4. Select fixed cost, F_C, investment level, M, and upper bound on variable cost:

$$C \le P - \frac{F_C + [M / Y_{RS}] + A}{D} \tag{7.1}$$

5. Select lower bound on value (for $E_2 = 1$):

$$V \ge P + \frac{N\bar{P}}{N+1}[m+1] \tag{7.2}$$

where $\bar{P}$ is the average price of all products competing within the segment.

Sanity checks need to be made on the bounds on variable cost and value. Is the variable cost reasonable and is the point given by the value and price targets within the range established for the market segment, as shown by a plot such as the one for the automobile market in Figure 3.16? If the cost and value limits are not reasonable, then the financial targets need to be adjusted accordingly. Instead of using Equation 7.1 and its simplifying approximation of an average yearly investment, M/Y_{RS}, for arriving at the target for variable cost, a present worth calculation can be made (Table 10.10).

7.4.2 BRAINSTORMING

Once customer needs have been assessed and overall cost, value, demand, investment, and profit targets established to jointly meet customer and company needs, the system team should meet with experts from the subsystem teams to see if subsystem alternatives are available 'off the shelf' that will generate the required level of value in the product. If not off the shelf, what can be developed in terms of new subsystem characteristics to obtain the performance improvements? Such brainstorming sessions between the system and subsystem teams should be an ongoing process so that the system team is fully aware of the state of subsystem technology at all times and that the subsystem teams are likewise aware of changing customer needs.

In evaluating alternatives derived from brainstorming, planners and engineers do not have the luxury of an unlimited time for study. As tests are made to better define the confidence level of the promising alternatives targeted for the future product, the uncertainty in the performance of a product relative to a baseline should decrease with time roughly as $1/\sqrt{t/t_0}$, where t is the time equal to nt_0 for testing n replications of an alternative under consideration. However, the chance that someone beats you to the marketplace increases during the development period which causes the uncertainty in the success of the proposed product to increase proportional to t. When the two competing effects are combined, the overall uncertainty as a function of time has the form shown in Figure 7.4. The minimum indicates that there is an optimal time period for planning the product, as discussed in Chapter 1.

Although the curve in Figure 7.4 given by the elementary model is an idealization, it is shaped by the two overarching risks – the risk of acting without sufficient testing and the risk of not acting soon enough – that lead to an optimum time for product development. The actual development time needed for products contains not only the time for testing but also the time needed for planning, designing, and tooling the product. There have been significant reductions in product development time as result of doing more operations in parallel and by the reductions in bureaucratic waste through teamwork [10].

The process of brainstorming alternatives can be aided by using a list of generic product system attributes such as those shown in Figure 5.10 to stimulate the search for novel, groundbreaking ideas. A key concept to keep in mind

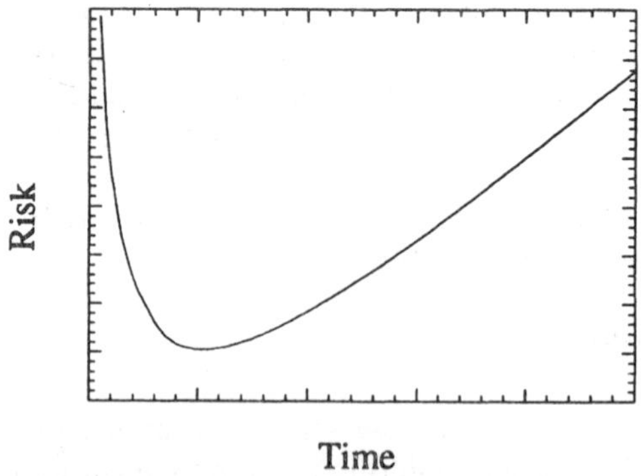

Figure 7.4 Risk versus time has a minimum reflecting a balance of the time needed to reduce new product risk through testing (planned experimentation) and the increase in risk with time for not acting due to the increasing likelihood of a competitor coming to the marketplace first.

when planning at the system level is the way or ways that the product can be differentiated favorably from competition over the short term as well as the long term. The opportunities for long term differentiation will, of course, always be fewer because viable competitors in time will catch-up to most advancements. The sliding door added by Chrysler to the driver's side of its minivans is an attribute, for example, which yielded an important but short term advantage. In addition to the product attributes listed in Figure 5.10, the sales and service experience can be an important element in differentiating a product as shown by the large value which Saturn customers give the so-called 'hassle-free' buying experience. This may prove to be an advantage for Saturn over the long term as other competitors must generate a difficult cultural change at their dealerships to provide the same type of environment.

Favorable product differentiation must be wisely managed, however, to be effective over the long term. Since its introduction, the Macintosh computer has had the friendliest operating system but Apple chose not to license the system and not to compete broadly in the marketplace, strongly targeting educators but not stressing most business and home users. Its competitors, however, targeted all of the markets using a less advanced PC technology based upon Microsoft's DOS operating system and Intel's chip technology. But the widespread use of the Microsoft and Intel technologies has caused software developers to write mainly for PC users, making it a *de facto* standard, to the detriment of Macintosh users. Moreover, Microsoft has greatly improved the friendliness of its operating system and Intel has continuously improved its chip technology to support the ravenous appetite for memory and speed demanded by the growing software. Earlier, in a similar vein, the much greater market penetration of the VHS format led to its becoming the *de facto* standard for VCRs versus Sony's protected and now defunct Betamax format which had superior visual quality. The lesson from these two examples is that, if your technology could become a *de facto* standard for an industry in need of a standard, you should not limit its use by holding it as a proprietary, in-house technology, even though it may be superior to the competition, but should work at making it become the standard by aggressively licensing it to competitors. You retain control of the technology that set the standard and thereby retain a competitive lead in implementing upgrades.

7.4.3 REDUCE ALTERNATIVES TO A MANAGEABLE NUMBER

Usually there will be more potential alternatives identified than time available to consider. These have to be reduced to a manageable number by using expert judgment before proceeding.

7.4.4 TEST ALTERNATIVES

A rigorous way of exploring the merits of the alternatives chosen for consideration is to design an efficient experiment using an orthogonal array (Appendix

B). The experiment can involve prototype hardware, computer simulations, or a combination of the two. It is important before beginning the experiment to assess whether some alternatives should have strong interactions and to allow for those interactions in the design of the experiment.

The performance of the product should be evaluated, in principle, against every system attribute for each experimental trial. A schematic representation of a spreadsheet format suitable for guiding the experiments and for analyzing the results at the system level is outlined in Table 7.2. The first column lists the trials by number. Next is the column of ones used in defining the performance level of the baseline product. The columns labeled 11, 21, 31, 41, ... represent those subsystem modifications or alternatives that are being considered as replacements for subsystems in the baseline vehicle. If a particular subsystem is being considered for trial q, the number 1 will appear in its column for this trial and if this subsystem is not being considered, the number 0 will appear.

In terms of an 'artificial life' viewpoint [11], the string of 0s and 1s under the subsystem alternatives heading for each trial form the genetic code for the particular prototype being evaluated. The overall experiment represents a selection process in which each prototype is evaluated for its fitness to survive in the marketplace using profitability as the objective function. The experiment allows the merits of each subsystem to be evaluated and those found to be most fit for generating profits are joined thereby forming the genetic code of the final product.

Based upon the quantities entered into the spreadsheet shown in Table 7.2, system attributes 1, 2, 3, etc. for a vehicle could be vehicle interior noise, acceleration time from 0 to 60 mph, stopping distance, seat vibration in dB, top speed, fuel economy, turning radius, etc. With the value curve equations for the attributes included in the spreadsheet, the value of the prototype for each trial is automatically computed in column V when the measurements for the system attributes are entered. The variable costs are entered in column C, fixed costs and investments are summed together in column W, and forecast annual profits in the last column are computed from Equation 3.24.

The use of annual profits, shown in the last column, as the objective function requires that a strategy be invoked for pricing. A conservative strategy is to compute the projected price of the future product by adding to the baseline price one-half of the value increase for the new product but adding nothing to price for cost increases. When the cost increases are due to inflation, they can be priced for and covered if all competitors choose to pass these along. The general expression for demand given by Equation 3.8 includes an evaluation of the projected impact of the actions of competitors which requires projecting their new value targets and pricing.

7.4.5 SELECT ALTERNATIVES

The importance of each particular subsystem alternative ij to the objective function is defined by a coefficient λ_{ij} (determined using matrix operations in the spreadsheet as described in Appendix B and the conjoint analysis problem con-

Table 7.2 System-level experimental design evaluating subsystem alternatives

	Subsystem alternatives							System attributes								Value	Cost	Investment	Profits
Trials	*0*	*11*	*21*	*31*	*41*	*51*	* * *	*1*	*2*	*3*	*4*	*5*	*6*	*7*	* * *	*V*	*C*	*W*	*A*
1	1	0	0	0	0	0	* * *	85	11	140	4	110	22	36	* * *	#	#	#	#
2	1	0	0	0	1	1	* * *	80	9	135	3	115	20	42	* * *	#	#	#	#
3	1	0	1	1	0	0	* * *	82	9	140	4	120	28	42	* * *	#	#	#	#
4	1	0	1	1	1	1	* * *	75	9	138	3	135	24	42	* * *	#	#	#	#
5	1	1	0	1	0	1	* * *	77	8	125	3	135	22	42	* * *	#	#	#	#
6	1	1	0	1	1	0	* * *	82	7	125	4	140	22	40	* * *	#	#	#	#
7	1	1	1	0	0	1	* * *	80	7	190	4	145	26	40	* * *	#	#	#	#
8	1	1	1	0	1	0	* * *	75	7	120	3	140	24	42	* * *	#	#	#	#
*	*	*	*	*	*	*	* * *	*	*	*	*	*	*	*	* * *	#	#	#	#
*	*	*	*	*	*	*	* * *	*	*	*	*	*	*	*	* * *	#	#	#	#
*	*	*	*	*	*	*	* * *	*	*	*	*	*	*	*	* * *	#	#	#	#

sidered in Chapter 5) which forecasts the increase in profits on using subsystem ij versus the baseline subsystem. The subscript i denotes the type of subsystem and j denotes its level, $j=0$ being the baseline level. Subject to the constraints in the form of the price, cost, investment, and value targets, level j which yields the largest λ_{ij} should be chosen for each subsystem i for the new vehicle. The coefficient λ_0 is equal to the profitability for the baseline vehicle. If the constraints represented by the targets are not met, then the system team must either develop other alternatives, revise the targets, or cancel the development of the new product as Chrysler, for example, did in the mid-90s with their LX luxury sedan.

It is vital that the subsystem alternatives selected for implementation be verified by constructing real prototypes as opposed to virtual to determine if the predicted performance is realized. If the predicted performance is not found, it is likely, apart from experimental blunders, that interactions are present which were not accounted for in the design of the experiment.

7.4.6 TRANSMIT REQUIREMENTS

If the system team chooses to proceed, their next step is to review their findings for the system-level targets with the subsystem teams and to specify the key subsystem targets needed for the product. Once the subsystem teams concur that they can meet their respective requirements, the system team can formally transmit the key performance requirements to the subsystem teams (Chapter 6).

7.5 SQD at subsystem level

When those responsible for subsystem i receive the requirement that they are to incorporate level j of the subsystem for the new product, there will be several alternative ways of designing, manufacturing, and assembling subsystem i within the broad requirements given for level j. The alternatives at the subsystem level are also indexed with pairs of dummy variables which now denote the type and level of the component alternatives. If, for example, those responsible for the system task have chosen subsystem 11, a V6 engine, then one component alternative considered might be for the valve action mechanism. Component 10 would be the baseline (e.g. push rod) and component 11 would represent an alternative (e.g. an overhead cam). Component alternative 20 could be the baseline steel piston pin and component 21 could represent a ceramic piston pin. The steps for SQD at the subsystem level closely follow the steps at the system level:

1. Get full system requirements and key subsystem-level requirements from system team and seek out opportunities for improving the product through

subsystem modifications. Transmit to the system team and other subsystem teams the field variables that affect the subsystem.

2. Develop the complete set of subsystem level attributes that bear directly on the customer requirements as expressed by the system level attributes.
3. Using the value versus full system attribute expressions established by the system team, estimate how the total value of the product measured in dollars changes as each subsystem attribute is varied by itself and in concert with the others. Arrive at a tentative set of subsystem performance specifications that include the key requirements transmitted by the system team.
4. Meet with knowledgeable persons to decide what component alternatives can be modified or developed to help meet the subsystem requirements and to improve the full system through improvements in the subsystem.
5. Using DOE methodology and procedures, arrange the evaluation of component alternatives as real or simulated hardware to support an efficient and effective experiment.
6. Using DOE methodology and procedures, determine the amount that each component alternative contributes to the objective function and select the best combination of alternatives.
7. Construct a prototype subsystem in real hardware having these selected alternatives to verify the prediction.
8. If verified, transmit requirements to the component teams.

The simulated spreadsheet SQD chart for a typical subsystem is shown in Table 7.3. Note that the labels for the columns for the system attributes, as well as value, cost, W (fixed costs plus investment), and the profit objective function are identical to those at the system level. This continuity from Table 7.2 reflects the fact that measures of value, costs, and investment associated with design changes, at any level, are only meaningful when determined from their impact on the total system. Customers care only about the full system behavior. One consequence of this important rule is that the total value of a given subsystem cannot generally be determined unless it were an option or unless there is a separate market for it. Fortunately, it is usually not important to know the value of a particular subsystem or a component. What can be determined is the change in value for a change in a subsystem or component and that is very important.

The same type of DOE process used at the system level is followed in determining the best component alternatives for each subsystem. The resulting λ coefficients are for component alternatives at the subsystem level and those most preferred should again be for the level j which has the largest λ_{ij} for each component i. Once those responsible for the subsystems have selected the best component alternatives, they are transmitted to those responsible for component design and manufacture. This is not a complete design definition for each component; it contains only the key characteristics that are important from the point of view of the subsystem.

Table 7.3 Subsystem-level experimental design evaluating component alternatives

	Component alternatives							System attributes								Value	Cost	Investment	Profits
Trials	*0*	*11*	*21*	*31*	*41*	*51*	* * *	*1*	*2*	*3*	*4*	*5*	*6*	*7*	* * *	*V*	*C*	*W*	*A*
1	1	0	0	0	0	0	* * *	75	7	135	3	110	26	36	* * *	#	#	#	#
2	1	0	0	0	1	1	* * *	75	7	135	3	110	27	36	* * *	#	#	#	#
3	1	0	1	1	0	0	* * *	75	7	135	3	110	27	36	* * *	#	#	#	#
4	1	0	1	1	1	1	* * *	77	7	135	3	115	28	36	* * *	#	#	#	#
5	1	1	0	1	0	1	* * *	75	7	135	3	110	27	36	* * *	#	#	#	#
6	1	1	0	1	1	0	* * *	72	7	130	3	110	26	36	* * *	#	#	#	#
7	1	1	1	0	0	1	* * *	72	7	135	3	100	25	36	* * *	#	#	#	#
8	1	1	1	0	1	0	* * *	72	7	135	3	110	28	36	* * *	#	#	#	#
*	*	*	*	*	*	*	* * *	*	*	*	*	*	*	*	* * *	#	#	#	#
*	*	*	*	*	*	*	* * *	*	*	*	*	*	*	*	* * *	#	#	#	#
*	*	*	*	*	*	*	* * *	*	*	*	*	*	*	*	* * *	#	#	#	#

7.6 SQD at component level

The basic elements of the process at the component level mirror those at the subsystem and system levels:

1. Get full system requirements, full subsystem requirements and key component-level requirements from the subsystem team and seek out opportunities for improving the product through component modifications.
2. Develop the complete set of component level attributes that bear directly on the customer requirements as expressed by the system level attributes.
3. Using the value versus full system attribute expressions established by the system team, estimate how the total value of the product measured in dollars changes as each component attribute is varied by itself and in concert with the others. Arrive at a tentative set of component performance specifications that include the key requirements transmitted by the subsystem team.
4. Meet with knowledgeable persons including component manufacturing personnel (who are assumed to be part of the component team) to decide what component features can be modified or developed to help meet the component and subsystem requirements and to improve the full system through improvements in the component.
5. Using DOE methodology and procedures, arrange the evaluation of component alternatives as real or simulated hardware to support an efficient and effective experiment.
6. Using DOE methodology and procedures, determine the amount that each component feature contributes to the objective function and select the best combination of alternatives.
7. Construct a component prototype in real hardware having these selected features to verify the prediction.
8. If verified, transmit the requirements to component manufacturer and supplier of raw material stock.

The process of experimental design and analysis is repeated once more at the component level (Table 7.4) where dimensional, raw material, and manufacturing alternatives are evaluated. Again, the right half of the spreadsheet program is identical to what it was at the system and subsystem levels.

For the overall product realization process to work efficiently and effectively, open channels of communication are required between those responsible for the system, subsystems, and components from the early stages of the product realization process to its completion. In other words, far from the impersonal description given in Chapter 6 of the LRP flow, there must be considerable prior discussion and understanding generated before the requirements are officially passed and accepted. Moreover, those responsible for the system should have had experience at both the subsystem and component level and those responsible at the subsystem level should have had experience at the component level.

Table 7.4 Component-level experimental design evaluating dimensional, material, and processing alternatives

	Dimensional, material, and processing alternatives							*System attributes*								*Value*	*Cost*	*Investment*	*Profits*
Trials	*0*	*11*	*21*	*31*	*41*	*51*	* * *	*1*	*2*	*3*	*4*	*5*	*6*	*7*	* * *	*V*	*C*	*W*	*A*
1	1	0	0	0	0	0	* * *	75	7	130	3	110	28	36	* * *	#	#	#	#
2	1	0	0	0	1	1	* * *	75	7	130	3	110	30	36	* * *	#	#	#	#
3	1	0	1	1	0	0	* * *	75	7	130	3	110	28	36	* * *	#	#	#	#
4	1	0	1	1	1	1	* * *	75	7	130	3	110	29	36	* * *	#	#	#	#
5	1	1	0	1	0	1	* * *	75	7	130	3	110	30	36	* * *	#	#	#	#
6	1	1	0	1	1	0	* * *	75	7	130	3	110	30	36	* * *	#	#	#	#
7	1	1	1	0	0	1	* * *	75	7	130	3	110	28	36	* * *	#	#	#	#
8	1	1	1	0	1	0	* * *	75	7	130	3	110	29	36	* * *	#	#	#	#
*	*	*	*	*	*	*	* * *	*	*	*	*	*	*	*	* * *	#	#	#	#
*	*	*	*	*	*	*	* * *	*	*	*	*	*	*	*	* * *	#	#	#	#
*	*	*	*	*	*	*	* * *	*	*	*	*	*	*	*	* * *	#	#	#	#
																#	#	#	#

As the development of the plan waterfalls from the system through the subsystem and component levels, overall system performance – projected value, cost, and investment – should improve. Continuous communication is needed to follow the process and in assessing the strategic aspects in terms of projected market share, investment levels, and costs as well as an understanding of the overall risks. The system group should upgrade the specifications and the profit forecasts as design and manufacturing improvements are developed at the subsystem and component levels. A preliminary waterfall of requirements should also begin well in advance of the approval date for taking the new product to production. Moreover, several iterations of the requirements pass-through should be anticipated before a program is solidified and ready for final approval. At the systems level, certain simplifying assumptions will generally be made at the outset regarding the performance of future subsystems and components and the actions of competitors. These will need to be refined as the program develops and as the performance of new subsystems and components and future competitive products are better understood.

7.7 Interactions

The simulated spreadsheets for the SQD process were idealized in that they did not show, at least explicitly, possible interactions between the alternatives being considered. In the QFD process, interactions form the 'roof' in the 'house of quality.' With SQD, interactions are represented in the classical DOE manner as additional columns in the array of alternatives. For example, if alternatives for subsystems i and k interact, then a column headed $ijkl$ should appear for each level j and l considered in the system-level experimental design. (See Appendix B for a more complete description of how interactions are treated.)

When considering interactions in subsystem-level experiments, there are two distinct types: interactions between components within one subsystem and interactions between different subsystems. Interactions between components within a given subsystem are handled in the classical DOE manner as additional $ijkl$, etc. columns. Those between different subsystems, but being evaluated at the subsystem levels of SQD, can be evaluated using fields and boundary conditions in the following context:

Each subsystem is designed under the assumption that it will operate under certain constraints. For example, the brake system is designed under the assumption that the maximum vehicle weight and the weight distribution between front and rear wheels will be a specified amount and that the cooling air to each brake will be a certain amount as a function of vehicle speed. Microprocessors are designed to operate within a certain electromagnetic and temperature environment. These 'fields' should be established at the outset of the design process for each subsystem by those responsible for the systems task and are transmitted as part of the requirements flow. The establishment of fields should not be made in an arbitrary manner by those responsible for the systems task but in concert with those responsible for the subsystems tasks and with the overall plan for the product in mind. Each subsystem is also physically packaged into the internal and external product boundaries by those responsible for the systems task. The package defines the extent of the spatial boundaries that each subsystem can occupy and the attaching points between subsystems. This information is transmitted as part of the requirements flow.

When those responsible for a subsystem propose a change that will affect any of the fields and/or boundary conditions of any other subsystem, they should get approval by those affected and by those responsible for the systems task. If approved, those responsible for the systems task must transmit the new fields and revised boundary conditions to all subsystems.

The interaction between subsystems evaluated at the subsystem level of SQD is shown schematically in Table 7.5. The fields and boundary conditions for the other subsystems are noted as F1, F2, ... If a change is needed by subsystem *ij* in field for subsystem *kl*, for example, its impact on value, cost, and investment on subsystem *kl* must be included in computing the λ coefficients for profits for the proposed change in subsystem *ij*.

7.8 Application example of system-level planning

To illustrate how certain tools and concepts of the previous chapters can be integrated into a system-level planning methodology, let's consider the hypothetical example of an automobile manufacturer, noted as C, wanting to improve the position of its midsized vehicle against its competitors, manufacturers A, B, D, and E. The vehicles by each manufacturer also carry the names A, B, C, D, and E.

7.8.1 SITUATION ANALYSIS APPLICATION

For situation analysis, C used the value benchmarking tools of demand/price analysis, value curves, and the teardown cost analysis of competing vehicles. The demand/price analysis findings are listed in Table 7.6 and shown in graphical form in Figure 7.5, the values for the products in the fourth column of Table 7.6 being determined from the sales and transaction prices in columns two and three, respectively, using Equation 3.11. (Actual plots for family vehicles across four product segments are shown in Figure 3.16.) The hypothetical plot in Figure 7.5 shows that there is an approximately linear relationship between value and price with price changing at a rate of 0.862 times the change in value. Although vehicle C has a cost advantage of $1000 with respect to four of the competing vehicles as determined from teardown studies, it has a value disadvantage of over $3000 with respect to vehicle A.

By studying product trends over the last two model introductions, C found that their average value increased by about 5% between model introductions with the

Table 7.5 Subsystem-level experimental design with component alternatives and interactions between subsystems noted as changes in fields

	Component alternatives							Fields				System attributes								Value	Cost	Investment	Profits
Trials	*0*	*11*	*21*	*31*	*41*	*51*	* * *	*F1*	*F2*	*F3*	*F4*	*1*	*2*	*3*	*4*	*5*	*6*	*7*	* * *	*V*	*C*	*W*	*A*
1	1	0	0	0	0	0	* * *	#	#	#	#	75	7	135	3	110	26	36	* * *	#	#	#	#
2	1	0	0	0	1	1	* * *	#	#	#	#	75	7	135	3	110	27	36	* * *	#	#	#	#
3	1	0	1	1	0	0	* * *	#	#	#	#	75	7	130	3	110	27	36	* * *	#	#	#	#
4	1	0	1	1	1	1	* * *	#	#	#	#	77	7	135	3	115	28	36	* * *	#	#	#	#
5	1	1	0	1	0	1	* * *	#	#	#	#	75	7	135	3	110	27	36	* * *	#	#	#	#
6	1	1	0	1	1	0	* * *	#	#	#	#	72	7	130	3	110	26	36	* * *	#	#	#	#
7	1	1	1	0	0	1	* * *	#	#	#	#	72	7	135	3	100	25	36	* * *	#	#	#	#
8	1	1	1	0	1	0	* * *	#	#	#	#	72	7	135	3	110	28	35	* * *	#	#	#	#
*	*	*	*	*	*	*	* * *	#	#	#	#	*	*	*	*	*	*	*	* * *	#	#	#	#
*	*	*	*	*	*	*	* * *	#	#	#	#	*	*	*	*	*	*	*	* * *	#	#	#	#
*	*	*	*	*	*	*	* * *	#	#	#	#	*	*	*	*	*	*	*	* * *	#	#	#	#

Table 7.6 Demand/price analysis for the values of five competing products

Model	*Annual sales*	*Transaction price*	*Value*	*Value difference from average*	*Variable cost*
A	300 000	\$21 000	\$41 222	\$2 022	\$16 000
B	280 000	\$20 000	\$39 963	\$763	\$16 000
C	260 000	\$18 000	\$37 704	(\$1 496)	\$15 000
D	220 000	\$19 000	\$38 185	(\$1 015)	\$15 000
E	200 000	\$20 000	\$38 926	(\$274)	\$16 000
Averages =	252 000	\$19 600	\$39 200	\$0	

slope of value versus price between vehicles being as shown in Figure 7.5. New models were introduced roughly every five years by each manufacturer.

Historically, C had invested \$1 billion to develop a new model paid in installments of M/Y_{RS} = \$200 million each year over five years and fixed costs were F = \$200 million per year. Current annual profits were A = \$380 million or \$1462 per vehicle. (These were computed using Equation 3.14.) The goal was to move closer towards the higher profitability of vehicle A by achieving a more favorable combination of value and price. This, of course, would have to be done cost effectively. The vehicle was to be assembled in a plant having a maximum capacity of 300 000 units annually and needed to sell the current volume of D = 260 000 to maintain two-shift capacity.

The company was willing to invest \$1.5 billion prorated over five years (M/Y_{RS} = \$300 million annually) to develop the new product with fixed costs

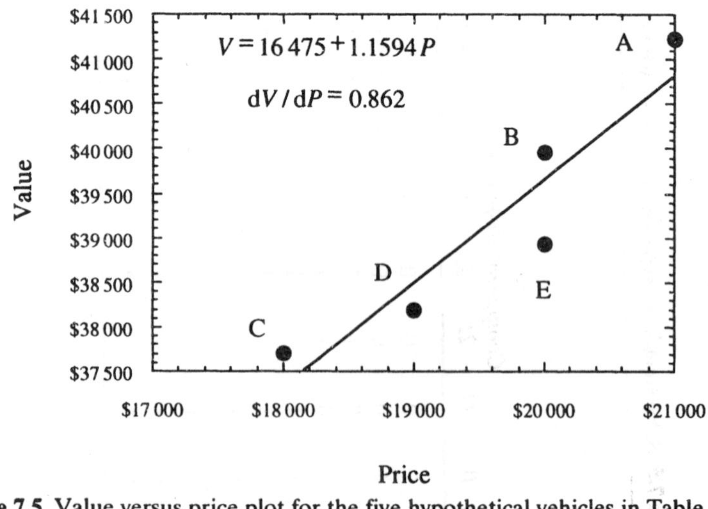

Figure 7.5 Value versus price plot for the five hypothetical vehicles in Table 7.6.

remaining at current levels and it wanted to make a profit of A = \$1 billion annually. Using a target price of P = \$21 000, the system planners computed the target variable cost of the new vehicle at \$15 230 using Equation 7.1. The target value for the new vehicle was computed at V = \$42 000 using Equation 7.2; the average price, $\bar{P}$, of all competing vehicles was taken as \$21 000; and C's minimum market share was taken as m = 0.21, an increase of 0.004 over its current market share of 0.206.

7.8.2 SEARCH FOR AND SELECTION OF OPPORTUNITIES

Although brainstorming for opportunities to reach these goals should begin as already noted, with a review of the list of generic product attributes (Figure 5.10) to stimulate a search for groundbreaking product initiatives, C skipped this step as being too risky for its limited objectives (as opposed to Planes Inc. in Chapter 1) and went directly to a consideration of which of the following specific system-level attributes could be moved closer to their ideal positions in a cost-effective manner.

- acceleration
- egress and ingress
- exterior design
- head room
- image
- interior design
- interior noise
- lateral acceleration
- leg room
- name value
- operating costs
- options
- overall length
- overall width
- range
- reliability/durability
- seat vibration
- shoulder room
- stopping distance
- thermal comfort
- time to start
- top speed
- turning radius
- view of the road

In this regard, C found it particularly helpful to plot value curves and to show on each curve where its product resided vis-à-vis its strongest competitor, vehicle A. The value curve for acceleration time from 0 to 60 mph [9] is shown in this way in Figure 7.6 with the positions of vehicles A and C superimposed on the curve as determined from their (hypothetical) acceleration performances of 10.5 and 12 seconds, respectively. Similar value curve plots are shown in Figures 7.7 and 7.8 for interior noise at 70 mph [12] and front leg room [13] with the (hypothetical) positions of vehicles A and C noted. Opportunities for value improvement are seen in each.

Spreadsheets were also very useful to C in assessing attribute opportunities because they could readily be used to compute the amount of improvement needed in a particular attribute to gain a desired value increase. Table 7.7 shows the fifteen attributes identified by C as having a moderate to high likelihood for cost-effective improvements. Those for which three-point value curves had been generated have their estimated critical and ideal specifications shown along with their exponent and nominal baseline.

A comparison of attribute levels between vehicles C and A is shown in Table 7.8. Vehicle C is seen to have a $217 advantage in fuel economy value but vehicle A is seen to have a large name value advantage. Fuel economy value was computed from the amount of fuel saved over an operating life of 100 000 miles at a fuel price of $1.20 per gallon. The equations used for computing the value of fuel economy and vehicle range are listed in Table 7.9. Company C found its name value of minus $1000 in line 4 of Table 7.8 relative to company A by asking respondents to a DV survey (Chapter 5) their willingness to pay for a vehicle built by manufacturer C versus an identically specified baseline vehicle built by manufacturer A. (Although this again is a hypothetical situation, name values of real vehicle manufacturers have been determined in this manner and differences found were as large as $2000 [14].) The value coefficients for lines 5 through 15 were computed by substituting their respective parameters in Table 7.7 into the exponentially weighted parabolic approximation expression given by Equation 5.7.

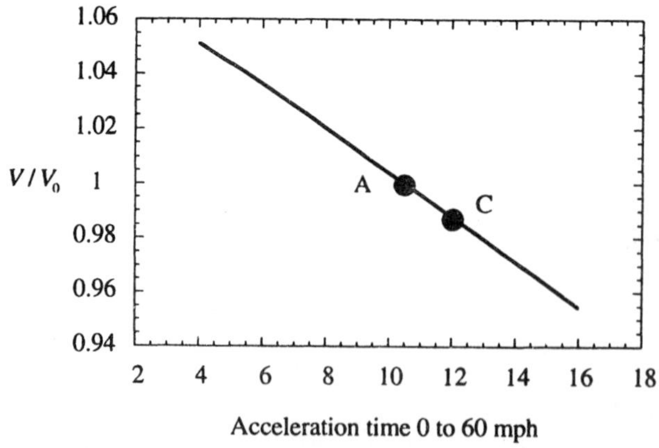

Figure 7.6 Value of acceleration time from 0 to 60 mph for hypothetical vehicles. (Value curve segment taken from results given in [9].)

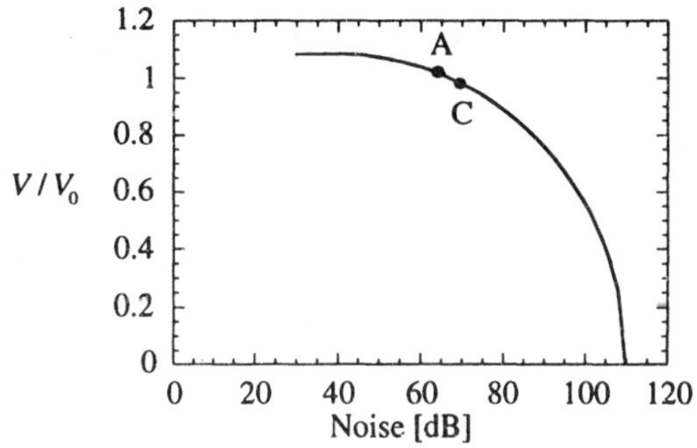

Figure 7.7 Value of interior noise at 70 mph for $\gamma = 0.5$ for hypothetical vehicles. (Value curve segment taken from results given in [12].)

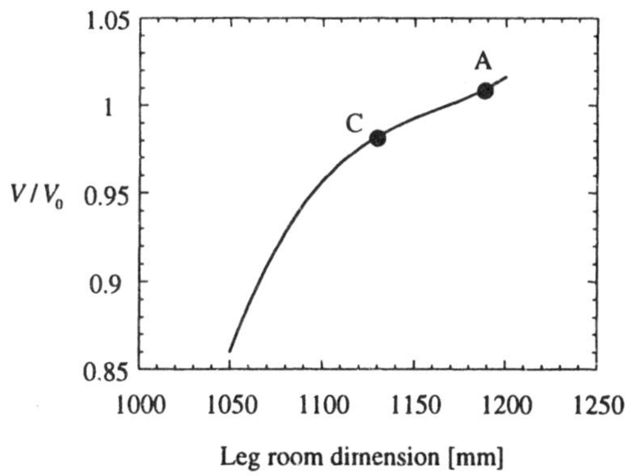

Figure 7.8 Value of front leg room (combined men and women) for hypothetical vehicles. (Value curve segment taken from results given in [13].)

The last two columns in Table 7.8 show the exploratory actions as determined by C's system-level planners as being part of the strategic plan for improving the competitiveness of the new model. The improvement in fuel economy to 27 mpg was targeted to at least maintain if not increase vehicle C's leadership in this attribute. Leadership would be sacrificed if C chose to increase engine horsepower to improve acceleration performance. C believed that its inferior name value was caused by A's reputation for very high reliability. Using $300 for the value of one less repair [14], C set an objective of reducing lifetime repairs (repairs over 100 000 miles of operation) for the new model by four for a $1200 improvement in value. Major improvements in front leg room and interior noise at 70 mph were also targeted.

The analysis in Table 7.8 shows that vehicle C would gain over $4000 in value from these changes and exceed A's current value by over $500. However, over the development period of five years, competition would not be standing still

Table 7.7 Attributes targeted for possible improvement for application example. (Source: reprinted with permission from SAE Paper 970762 © 1997 Society of Automotive Engineers, Inc.)

Line	*Attributes*	*Type*	*Attribute units*	*Nominal baseline*	*Critical*	*Ideal*	*Exponent*
	Fuel economy and range						
1	Overall mpg	Economic	mpg	24.00	N/A	N/A	N/A
2	Volume fuel tank	Economic	gal	17.00	N/A	N/A	N/A
	Options						
3	Major options	Economic	N/A	Vehicle A	N/A	N/A	N/A
	Name Value						
4	Stated willingness to pay	Survey	N/A	Vehicle A	N/A	N/A	N/A
	Value curves						
5	Front leg room	LIB	millimeters	N/A	N/A	N/A	N/A
6	Turning circle	SIB	meters	12.0	20.54	1.83	0.125
7	Overall length	SIB	millimeters	4800	7620	762	0.125
8	Overall width	SIB	millimeters	1780	2743	762	0.15
9	Door-top to ground	LIB	millimeters	1280	1128	1499	0.03
10	Luggage capacity	LIB	cubic meters	0.45	0.00	0.85	0.08
11	Idle noise	SIB	dBA	43.0	110	40	0.20
12	Max. 1st gear noise	SIB	dBA	71.0	110	40	0.05
13	70 mph noise	SIB	dBA	67.0	110	40	0.50
14	Time for 0–60 mph*	SIB	seconds	10.5	40	2.00	0.16
15	Skidpad lateral accel.	LIB	Gs	0.72	0.25	1.00	0.20

*Times are transformed to Log(1/time) which becomes a LIB psychometric force variable.

		Vehicles					
		C		A			
Line	Attributes	Dimension	Value from baseline	Dimension	Value from baseline	Exploratory new model actions	Value change
	Fuel economy and range						
1	Overall mpg	24.00	0.00	23.00	($217)	Improve FE to 25 mpg	**$556**
2	Volume fuel tank	17.00	0.00	17.00	$0	No change	
	Options						
3	Major options		0.00		$300	No change	
	Name value						
4	Stated willingness to pay		($1000)		$0	Reduce average lifetime repairs by 4	$1200
	Value curves		*Coefficient*		*Coefficient*		
5	Front leg room	1130	0.982	1190	1.010	Add 60 mm	$1097
6	Turning circle	12.00	1.000	12.00	1.000	No change	
7	Overall length	4870	0.998	4800	1.000	No change	
8	Overall width	1800	0.998	1790	0.999	No change	
9	Door-top to ground	1250	0.994	1250	0.994	No change	
10	Luggage capacity	0.50	1.005	0.40	0.994	No change	
11	Idle noise	42.00	1.000	42.00	1.000	No change	
12	Max 1st gear noise	70.00	1.001	70.00	1.001	No change	
13	70 mph noise	69.00	0.986	64.00	1.018	Reduce to 64 dB	$1242
14	Time for 0–60 mph*	12.00	0.988	10.50	1.000	No change	
15	Skidpad lateral accel.	0.70	0.995	0.70	0.995	No change	
16	Net value coefficient		0.948		1.011		
17	Net change vs baseline		($2029)		$446		
18	Value versus baseline		($3029)		$529		$4095
19	(Value C – Value A)	Current =	**($3558)**			After changes =	**$537**

Table 7.9 Equations used for computing the value of fuel economy and range (source: M.E. Simek (1994) *Human Factors Value Modeling Applied to Vehical Analysis and Development*, MS Thesis, Department of Mechanical and Industrial Engineering, University of Illinois at Urbana-Champaign, pp. 45–9)

Attribute	*Value equation*
Fuel Economy	$V_i = \dfrac{1.20 \times 10^5}{FE_i}$
Range	$V_i = \dfrac{10^5 V_{Time}}{3600 FE_i}\left[6.8 + \dfrac{214}{0.8T_i}\right]$

Symbol	*Definition*
FE_i	Fuel economy in mpg of vehicle *i*.
V_{Time}	Value of one hour of a person's time.
T_i	Volume of fuel tank in gallons.

and, if historical trends of value growth continued, the competing vehicles could be expected to increase in value by 5%. A forecast was then made of C's competitive position under the scenario of it increasing its value by $4095 and each of its competitors increasing their value by 5%, the results being shown in Table 7.10 and the value–price plot of Figure 7.9. The added improvements by C over the 5% norm should move vehicle C from the lowest position on the current value versus price plot to a point very close to second position.

C's system-level planners reviewed the strategy with the planners for each of the subsystems. They concluded that a 12% reduction in vehicle weight generated by more intensive use of lightweight materials and use of an engine with 12% less horsepower would meet or exceed the desired improvement in fuel economy at no loss in acceleration performance. They also concluded that the reliability targets would be met through the redesign of several subsystems,

Table 7.10 Forecast of sales, price, and values for new product (application example)

Model	*Annual sales*	*Transaction price*	*Value*	*Value difference from average*	*Variable cost*
A	300 633	$22 777	$43 283	$1 681	$16 000
B	279 963	$21 722	$41 961	$359	$16 000
C	282 285	$21 530	$41 799	$197	$16 000
D	219 017	$20 646	$40 094	($1 508)	$15 000
E	199 411	$21 678	$40 872	($730)	$16 000
Averages	256 262	$21 670	$41 602	$0	

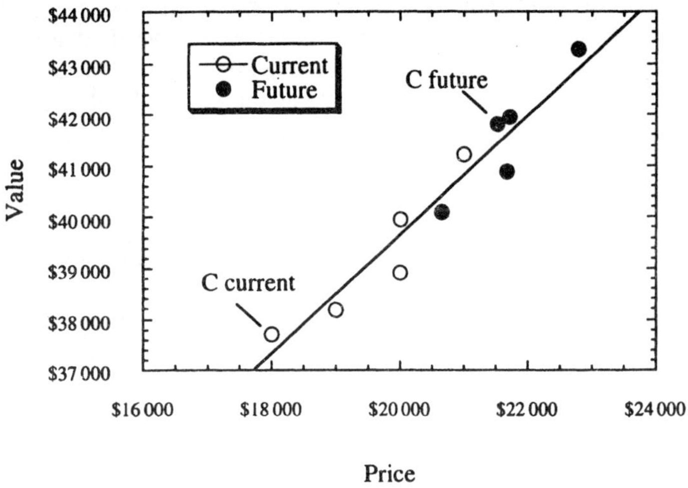

Figure 7.9 Hypothetical value versus price relationships forecast for future vehicles compared to their current relationships.

improvements in key manufacturing processes, and intensive application of design of experiment and statistical process control methods. The smaller engine and improved package design would allow the added front leg room to be generated without sacrificing rear leg room and trunk space. The smaller engine plus 50 lbs of added sound insulation, improved sealing of the interior against tire and engine noise, and improved aerodynamics to reduce wind noise was projected to reduce the noise level at 70 mph to 64 dB. The added weight needed for the insulation was to be offset by secondary weight reductions in brakes and suspension parts.

The added variable cost for the proposed changes were estimated at $1000 and the investment level was estimated at $1.5 billion ($0.5 billion more than required for the previous model) with no increase in fixed costs. The forecast profitability of the future vehicle based upon the demand, price, and variable cost forecasts in Table 7.10 was slightly over $1 billion annually or over $3700 per unit, a marked improvement over the profitability of the current vehicle. All of the objectives for the new vehicle program were forecast to be met with these attribute changes. Before transmitting a detailed list of requirements to the subsystem tasks, several prototypes were to be built with these changes so that they could be evaluated against vehicles C and A as currently manufactured to verify the anticipated improvements and to confirm that other attributes such as crash worthiness, corrosion, etc. would not be adversely affected.

The process used for system-level planning in this example did not follow the exact system-level DOE formalism envisioned in Table 7.2 in which competing subsystems were evaluated. Instead, those responsible for the subsystem tasks had evaluated the competing subsystem alternatives on their own and had put them 'on the shelf' for future consideration at the system level. Thus all that was necessary for the subsystem team to do was to evaluate and verify that the proposed vehicle with the new subsystems recommended by the various teams would, in fact, meet the objectives. This made the system-level experiment much simpler as only one kind of prototype needed to be evaluated against the current baseline vehicle.

7.9 Making decisions for new products under uncertainty

The influence of scatter or variance in the product attributes on value and profits was not examined in the above discussion nor was the significance of the outcomes evaluated versus the baseline. The question is 'How confident can you be that an action, which on average was favorable based upon tests of a small number of prototypes, will remain favorable when evaluated across the large population needed to serve the marketplace?' What is the risk to profits on investing a large sum of money for a new product based upon the information obtained from tests on a limited number of prototypes? In other words, those attributes indicated as favorable need to be tested for significance. Although we will look at these issues in great detail in Chapter 9 for experiments at the component or parameter level of design, we need to consider here how such considerations can be made at the system level of design.

In addition to the uncertainty in the system attributes, g_i, such as acceleration times, noise levels, etc., there is also uncertainty in the relationships between value and the attributes. Moreover, in any real product development situation, there is also uncertainty due to the unforeseen changes over time in labor costs, material costs, interest rates, investment costs, tax rates, and Federal regulations. Unforeseen developments in the prices and values of competitive products and changes in consumer tastes can have a major impact on profitability. Finally, there will always be uncertainty as to what the values and prices of competitive products will be in the future.

Many of these uncertainties can be formally included in the profit equation by replacing each parameter x_j that appears in it by the expression:

$$x_j = \bar{x}_i + t_s\sqrt{\frac{s_i^2}{n_i}}$$

where $\bar{x}_j$ and s_i^2 are the sample average and variance, respectively, of parameter x_j; t_S is a random variable defined by Equation A.22 and determined from Monte Carlo simulations; and n_i is the number of terms used to compute the average. The scatter found by repeatedly simulating the outcomes generates the range of profitability forecast based upon the combined variances associated with the parameters. This behavior closely approximates the real scatter expected for a t-distribution (Appendix A). With the Delphi method, estimates used for the economic parameters are averages obtained from experts in fields such as finance, purchasing, and economics. Marketing specialists and engineers provide estimates of the coefficients that affect value, the slope of the demand curve, and the number of competitors. The distributions of the estimates of each term are then used to compute the sample averages and sample variances of the term. Once the statistical expressions for the terms have been entered into the spreadsheet, repeated computations are used to arrive at the uncertainty in profits expected for each combination of subsystems.

References

1. Akao, Y. (1990) *Quality Function Deployment*, Productivity Press, Cambridge, MA; Hauser, J.R. and Clausing, D. (1988) The house of quality. *Harvard Business Review*, May–June, 63–73.
2. Mintzberg, H. (1994) *The Rise and Fall of Strategic Planning*, Free Press, New York, 1994, p. 327.
3. Peters, J.T., Hammond, K.R., and Summers, D.A., (1974) A note on intuitive versus analytic thinking. *Organizational Behavior and Human Performance*, **12**, 125–31.
4. *The Fifth Symposium on Quality Function Deployment*, The American Supplier Institute, Dearborn, MI, June 20–2, 1993.
5. Kolli, R.P. and Cook, H.E. (1994) Strategic quality deployment. *Manufacturing Review*, **7**, 148–63.
6. March, G. and Simon, H. (1967) *Organizations*, Wiley, pp. 194–9.
7. Fowler, T.C. (1990) *Value Analysis in Design*, Van Nostrand Reinhold, New York, p. 19.
8. Churchill, G.A., Jr (1991) *Marketing Research Methodological Foundations*, 5th edn, Dryden Press, Chicago, IL, pp. 464–70.
9. McConville, G.P. and Cook, H.E. (1996) Examining the trade-off between automobile acceleration performance and fuel economy. *SAE Technical Paper 960004*, Society of Automotive Engineers, Warrendale, PA.
10. Clark, K.B. and Fujimoto, T. (1991) *Product Development Performance*, Harvard Business School Press, Boston, MA, pp. 76–82.
11. Levy, S. (1992) *Artificial Life the Quest for a New Creation*, Pantheon Books, New York.
12. McConville, G.P. (1996) *Developing Value Relationships for Automotive Attributes*, MS Thesis, Department of Mechanical and Industrial Engineering, University of Illinois at Urbana-Champaign, pp. 26–33.
13. Simek, M.E. and Cook, H.E. (1996) A methodology for estimating the value of interior room in automobiles. *SAE Technical Paper 960002*, Society of Automotive Engineers, Warrendale, PA.
14. Donndelinger, J.A. and Cook, H.E. (1997) Methods for analyzing the value of automobiles. *SAE Paper 970762*, Society of Automotive Engineers, Warrendale, PA.

8 Parameter and tolerance design

8.1 Introduction

Parameter design, the third step of Taguchi's robust design process, improves quality by reducing variation in the attributes of a product and moving the means of the attributes toward their ideal specifications. As such, it is a very important refinement of the system design and represents a form of 'off-line' quality control in the sense that its purpose is to design quality into the product up-front so that its performance is relatively insensitive to changes in the manufacturing and in-use environments. The methodology is powerful and has also been used to guide redesigns of components and processes to correct known product deficiencies in the field. Tolerance design, the step following parameter design, sets the maximum allowable excursion for a given attribute about its target value.

The concept of a quality loss function (Chapter 4) was introduced by Taguchi to guide the parameter design process, the objective being to minimize the loss function. This is equivalent to maximizing total quality in the SQD process just described. There are no fundamental differences between Taguchi's four-step process for robust design and the SQD process described in Chapter 7. However, several highly useful concepts and viewpoints – the quality loss function, signal-to-noise ratios, and the deliberate, controlled introduction of experimental noise through outer arrays to duplicate the range of conditions found in production – were developed by Taguchi for parameter design. We will explore these concepts here and blend them into the SQD formalism. The condition, often invoked, that alternatives to improving quality involving cost increases cannot be considered as part of parameter design will be relaxed here as there can be many instances where the best action to take may involve a cost increase that is more than offset by a large improvement in quality and customer satisfaction.

8.1.1 SHAFT DESIGN EXAMPLE

The relationships between the different elements of the design process can be understood by considering the design of a driven shaft turning in a bearing as shown in Figure 8.1. The first step is to arrive at a target or nominal size for the diameter of the shaft. On ignoring dynamical considerations, this involves balancing the improved product value associated with increasing the shaft diameter against the additional costs. Value improves because the chance of fatigue failure is reduced with increasing size. Costs increase because of the need for larger bearings and for the material added to the shaft.

If the cumulative distribution that the shaft has failed at time t is given by $\wp_C(t)$, then the value of the system at time t' is given by:

$$V(t') = \frac{V_0}{T}\left[t' - \int_0^{t'} \wp_C(t)dt\right] \tag{8.1}$$

where T is the expected lifetime of the system and V_0 is the total value delivered by the system in the absence of shaft failure over its lifetime. For the expected lifetime, value is thus given by:

$$V(T) = V_0\left[1 - \frac{1}{T}\int_0^{T} \wp_C(t)dt\right]. \tag{8.2}$$

The assumption here, for simplicity, is that a shaft failure results in the loss of the total system with no form of repair possible. (More realistic considerations would require that the consequences of a shaft failure be evaluated with regard to safety, collateral damage, and repairability, all of which are beyond the scope of our immediate interest.) The cumulative distribution of shaft failure decreases as its diameter increases but the total system cost increases with the square of the diameter and will overtake the value curve at some point (Figure 8.2).

The target diameter, $d_{S,T}$, for the shaft is the diameter which gives the maximum difference between value and cost. In considering the problem to this point, the bearing diameter, d_B, has been assumed to be at its correct nominal relationship to the shaft diameter. However, once the nominal shaft diameter has been selected, the impact of the manufacturing variances in shaft and bearing diameters about their nominal dimensions must be considered. The resulting value relationship is given by:

$$V = V_T - k_d(d_B - d_S - \Delta_I)^2 \tag{8.3}$$

Figure 8.1 Shaft and bearing illustrating the setting of targets for the shaft's diameter and the standard deviation of the shaft's diameter.

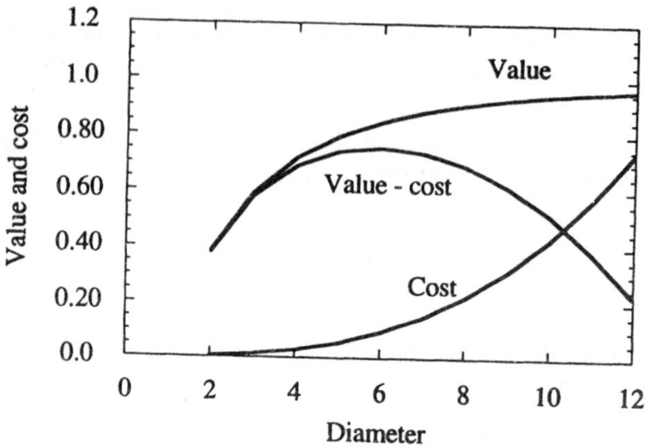

Figure 8.2 Schematic behavior of shaft value and cost as a function of the shaft diameter. The target diameter is set at the maximum in value minus cost.

where k_d is a constant and Δ_I is the ideal difference in shaft and bearing diameters for proper function. The size of k_d in absolute units will be much larger than one-half of the second derivative of the value function shown in Figure 8.2 because deviations of thousandths of an inch can markedly affect the operation of the shaft in the bearings but will not have an important impact on the value computed from Equation 8.2. Averaging value over the full distribution of shaft and bearing diameters gives:

$$\bar{V} = V_T - k_d\left[\sigma_S^2 + \sigma_B^2 + (\mu_B - \mu_S - \Delta_I)^2\right] \tag{8.4}$$

where σ_S^2 and σ_B^2 are the variances and μ_S and μ_B are the averages of the shaft and bearing diameters, respectively. If we assume that the variance of the bearing is fixed and that the difference between the average shaft and bearing diameters is equal to the ideal difference, Δ_I, then $\bar{V}$ decreases in quadratic fashion with σ_S, the standard deviation of the shaft diameter (Figure 8.3).

Also shown in Figure 8.3 is the cost for reducing the standard deviation of the shaft diameter. This cost will be singular as the standard deviation approaches zero. Just as there was a target diameter for the shaft there is also a target standard deviation for the shaft diameter $\sigma_{S,T}$, which yields maximum quality. Traditional design practice uses the concept of tolerance ϵ instead of a target standard deviation. For three-sigma control, the two are connected by the relationship:

$$3\sigma_{S,T} \leq \epsilon \tag{8.5}$$

If the shaft sizes were normally distributed, then it is expected that 99.7% of the diameters will lie within the tolerance range of $\pm\epsilon$.

In this sample design problem, target specifications were found for the nominal shaft diameter and also for the standard deviation of the diameter.

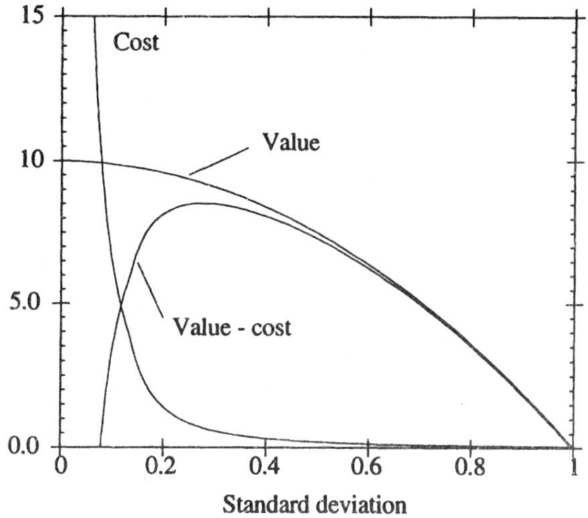

Figure 8.3 Schematic representation of shaft value as a function of the shaft standard deviation. The target standard deviation is set at the maximum in value minus cost.

For both cases, value and cost were evaluated for the total system. In terms of the SQD process, all of these considerations would fit under the component design classification of SQD. However, we note in passing that in terms of Taguchi's robust design classification, the first step, finding the nominal shaft diameter, was a system design effort because it involved the determination of a nominal target specification. The second step, finding the best specification for the variance of the shaft diameter, was a combination of parameter and tolerance design.

8.2 Latitude and robustness

8.2.1 SHIRT COLLAR EXAMPLE

Missing from the shaft diameter and variance problem just reviewed were considerations of latitude and robustness which have an important place in Taguchi's methodology. Consider the situation in Figure 8.4 of three nominal-is-best (NIB) value curves as a function of shirt collar size for a man with a 15.5 inch diameter neck. Material A has little stretch, material B has moderate stretch and material C has considerable stretch. Otherwise the materials look and feel the same. Thus, if the collar size were manufactured slightly small, material A quickly loses value because it is too tight whereas material C has a considerable degree of what is known as latitude. Shirts with collars that are too large lose value as a result of being unsightly. This means that shirt C can be tighter fitting in the manner of a collar on a turtle neck sweater and yet not be uncomfortable. The cost of materials and the nominal cost to cut and sew are assumed to be constant for the three fabrics. The machines used to cut and sew, however, generate a distribution $f(g)$ of collar sizes g about the mean of 15.5.

The fraction $\delta\phi$ of shirt sizes produced between g and $g+\delta g$ is given by

$$\delta\phi = f(g)\delta g,$$

for the frequency distribution in Figure 8.5. The integral of $f(g)$ over all sizes is equal to unity. If value in Figure 8.4 is approximated by its parabolic form in the vicinity of the maximum at $g = \bar{g} = 15.5$:

$$V(g) = V(\bar{g}) - k_T(g - \bar{g})^2 \quad (8.6)$$

The coefficient k_T, equal to one-half of the second derivative of value with respect to shirt size, is greatest for shirt A and least for shirt C. It follows from the definition for standard deviation that the change in value expected for the average shirt from the distribution shown above versus a shirt having nominal value $V(\bar{g})$ is given by:

$$\overline{\Delta V} = -k_T\sigma^2, \quad (8.7)$$

where σ^2 is the population variance of shirt size. Thus, for this example, as for the shaft design problem, the value of the product is compromised by variance even though the mean of the distribution is on target.

If the product is robustly designed in the manner specified by Taguchi, then it will have latitude or reduced sensitivity to the natural sort of variance that occurs in the field. For the above example, it means searching for materials like C. An additional consideration of the robustness of the shirt design is its resistance to the normal forces of deterioration that it is subjected to in the field, the wear and tear it receives during use and during washing and ironing.

8.2.2 SCALING FACTOR: LAWN MOWER SHEAR PIN

Lawn mowers connect the blade to the drive shaft with a shear pin (Figure 8.6) so that a torque overload from the blade hitting a root or other obstacle will simply shear the pin (which is easily replaced) and not damage or break the

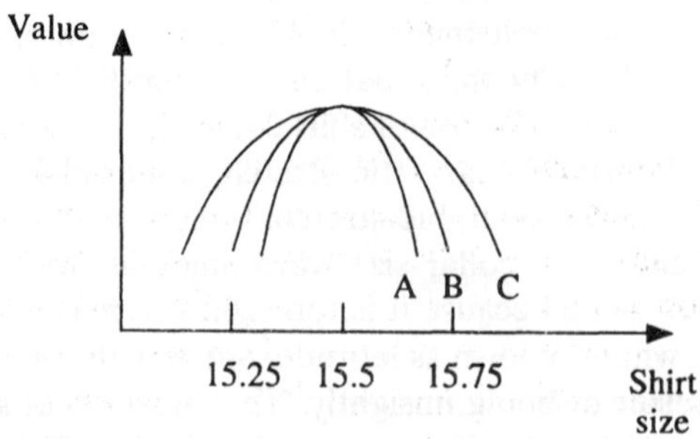

Figure 8.4 Increasing latitude for shirts from A to B to C.

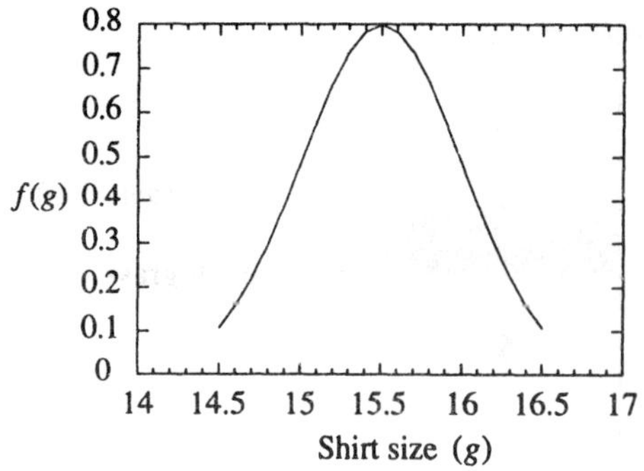

Figure 8.5 Schematic frequency distribution for shirt collar sizes.

crankshaft in the engine. The pin acts as a mechanical fuse and must shear if the torque reaches the upper level τ_2 to prevent crankshaft damage but must not shear at the torque τ_1 required to cut grass (Figure 8.7).

If the area of the shear pin is A and the diameter of the small shaft is d, then the shear strength S of the pin's material must lie between the limits:

$$\frac{2\tau_1}{Ad} < S < \frac{2\tau_2}{Ad}. \tag{8.8}$$

If the pin's shear strength varies with annealing temperature as shown in Figure 8.8, variations in the temperature ΔT of the heat treat furnace will lead to variations in shear strength. The lower annealing temperature for this hypothetical example will yield smaller variations in strength as a result of the temperature excursions. Thus, the temperature control factor should be set at the lower level thereby avoiding the expense of a more accurate control mechanism for the furnace.

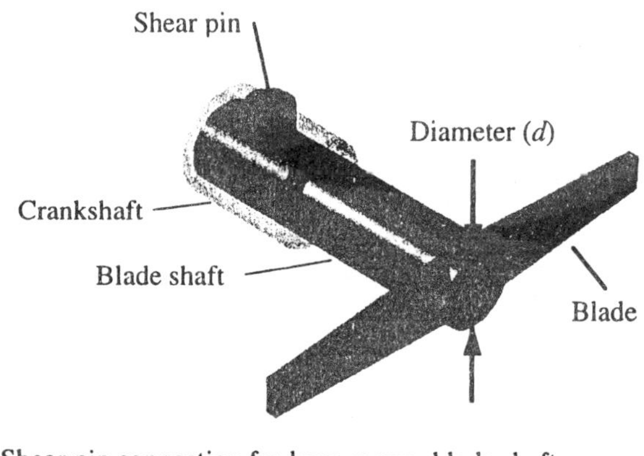

Figure 8.6 Shear pin connection for lawn mower blade shaft as an example of a scaling factor.

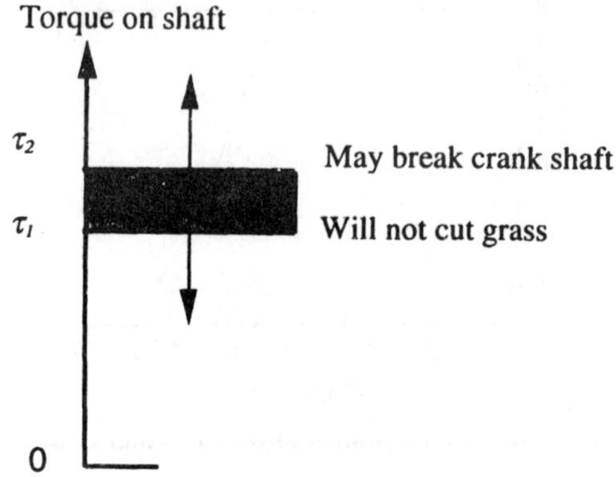

Figure 8.7 Operational range for torque on blade shaft.

The design was made robust by choosing a heat treatment temperature that gave latitude to variations in the control factor. This of course could result in the pin being too strong if its diameter had been designed for the lower shear strength material given by the higher annealing temperature. But the pin's diameter represents a scaling factor which can readily be modified so that the specification can be returned to the NIB target. Unfortunately, not all NIB problems have a suitable scaling factor. For these problems, repeated iterations of combinations of factors are evaluated to find the best combination for bringing the specification close to target and reducing variance simultaneously.

8.3 Objective functions for parameter design

8.3.1 PROFITABILITY

Parameter design considers various alternatives at the component level and follows a planned experimental design, generally an orthogonal array, to deter-

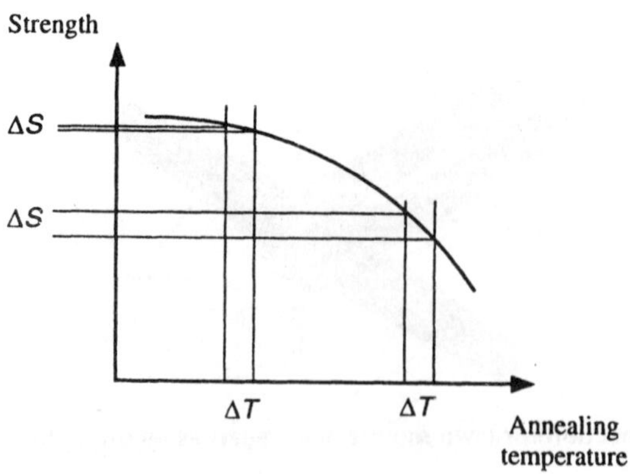

Figure 8.8 Schematic representation of material shear strength variation with heat treatment temperature.

mine which set points (control factors) are to be evaluated for each trial q (Appendix B). The fundamental objective function or output for the alternatives under consideration should be sustained profitability, as measured by their forecast present worth. (We again assume that the enterprise has internalized the costs for meeting all governmental regulations.)

For simplicity, we will use profit (Equation 3.24) as the objective function with the investment prorated over the Y_{RS} years that the alternatives are expected to be in production:

$$A(q)=D(q)[P(q)-C(q)]-F(q)-[M(q)/Y_{RS}]. \tag{8.9}$$

The forecast demand for trial q is given by:

$$D(q)=D_0+K\{\delta V(q)[1-\alpha_V]-\alpha_C\,\delta C(q)\} \tag{8.10}$$

where D_0 is the demand at trial $q=1$, the baseline trial, $\delta V(q)$ and $\delta C(q)$ are the differences between the alternatives being considered at q and those at $q=1$ for value and variable cost, respectively. Following Equation 3.20, price is assumed to be given by:

$$P(q) = P_0 + \alpha_V\,\delta V(q) + \alpha_C\,\delta C(q) \tag{8.11}$$

where P_0 the baseline price and we have assumed, for subscript simplicity, that competitors are not changing cost or value. Consequently, Equation 8.9 can be written as:

$$\begin{aligned} A(q) = &\left\{D_0 + K\left[\left[1-\alpha_V\right]\delta V(q) - \alpha_C\delta C(q)\right]\right\} \\ &\left[P_0 - C_0 + \alpha_V\delta V(q) - \alpha_C\delta C(q)\right] - F(q) - \left[M(q)/Y_{RS}\right]. \end{aligned} \tag{8.12}$$

The average forecast value for the product's critical attributes at trial q is given by:

$$\overline{V(q)}=V_0\overline{v_1(q)\,v_2(q)\,v_3(q)\,....} \tag{8.13}$$

which follows from Equation 5.8 on averaging.

8.3.2 LOSS FUNCTION

Instead of using profit, Taguchi and Wu [1] introduced the concept of a loss function for parameter design which is the sum of two terms, one being variable cost as described above and the other being the cost of inferior quality, CIQ (Chapter 5). If we add to the loss function the fixed costs and amount of investment needed per unit produced, we can write the loss function for trial q as:

$$\Lambda(q) - \Omega(q) + C(q) + W(q)/D_0, \tag{8.14}$$

where the CIQ for trial q is given by:

$$\Omega(q) = V_I - V(q), \tag{8.15}$$

and:

$$W(q) = F(q) + (M(q) / Y_{RS}).$$

The loss function is equal to zero if all costs are equal to zero and if the product is at its ideal specification $g = g_I$ where value equals V_I.

Only when demand is given by:

$$D = K[V - P] \\ = K[V - C]/2 \tag{8.16}$$

which is the expression for a monopoly pricing to maximize profits, does Taguchi's loss function become equivalent to using profit (Equation 8.9) as the objective function for parameter design. For this condition, the expression for profit becomes:

$$A(q) = [K / 4][V(q) - C(q)]^2 - F(q) - [M(q) / Y_{RS}]. \tag{8.17}$$

Using a Taylor's expansion, we find that the variation in profit from the baseline trial (trial 1) per unit produced at the baseline condition is given by:

$$\delta A(q) / D_0 \equiv - \delta \Lambda(q) \tag{8.18}$$

where

$$\delta \Lambda(q) = [\delta C(q) - \delta V(q)] + [\delta F(q) + (\delta M(q) / Y_{RS})] / D_0 \tag{8.19}$$

is the difference between the loss, $\Lambda(q)$ for trial q and the baseline loss, $\Lambda(1)$, for trial 1.

When scrap losses are present, the loss function becomes:

$$\Lambda(q) = \Omega(q) + C_S(q) + \{C_{SP}(q) + W(q)/D_0\} \tag{8.20}$$

where $C_S(q)$ is the cost of scrap per unit and $C_{SP}(q)$ is noted as 'set point cost' which equals variable costs less warranty and scrap costs. Warranty costs, $C_W(q)$, result when the manufacturer chooses to repair off-target products. The warranty repair adds $\Delta V_R(q)$ value to $V(q)$ and the loss function becomes:

$$\Lambda(q) = \Omega(q) - \Delta V_R(q) + R(q) + C_S(q) + C_W(q) + \\ \{C_{SP}(q) + W(q) / D_0\} \tag{8.21}$$

A residual loss of value, $R(q)$, is incurred because the product was off target during a certain time of use by the customer and because a loss of use was experienced during repair. Braces are used in Equations 8.20 and 8.21 to denote that the last two terms do not depend upon the statistical properties of the outcomes of trial q. The averages of the other terms in the expression, excluding V_I, do, however, depend upon the statistical distribution.

As amply shown by practitioners of Taguchi's methodology, the loss function is very useful for interpreting single-attribute parameter design experiments and Elsayed and Chen have developed a multiattribute formulation of the loss function and explored its properties to a limited extent [2]. Nevertheless, the loss function approach has shortcomings in properly assessing the strategic impact of quality improvements which is particularly limiting for system-level considerations. The fundamental difficulty is that it is only through the profit equation itself can the appropriate coupling between value and cost be obtained. The reason is that value directly influences demand and changes in costs directly influence the profit per unit and the two only connect in the expression for profit.

8.4 Single-attribute parameter design

We will consider only single-attribute problems in this chapter and in Chapter 9, for simplicity. This is not a serious shortcoming because most parameter design studies to date have been single-attribute problems. However, the system through component level considerations in Chapter 7 are for a large multiattribute problem with nested experiments. Case Study 4 is an example of a multiattribute parameter design problem in which tests of significance are included for completeness.

8.4.1 SIGNAL-TO-NOISE RATIO

Taguchi does not use his loss function directly as the objective function for parameter design but transforms it to a signal-to-noise (S/N) ratio given by:

$$Y(q)=-10\log_{10}(\bar{\Lambda}(q)) \tag{8.22}$$

to improve the additivity of the coefficients determined from the linear model for the outcomes, $Y(q)$. Because changes in variable, fixed, and investment costs are usually not considered in parameter design experiments, the expression most often used for the signal-to-noise ratio is given by:

$$Y(q)=-10\log_{10}(\bar{\Omega}(q))\,. \tag{8.23}$$

For a single attribute, the cost of inferior quality for replication y of trial q is assumed to be a continuous function of a basic variable $x(q, y)$:

$$\Omega(q,y) = k_T[x(q,y)-x_I]^2$$

where x_I is the basic variable when the attribute g is at its ideal specification. The basic variables are defined in terms of the attribute $g(q, y)$ in Table 8.1 for the three common situations: SIB, NIB, and LIB.

The expected value of the cost of inferior quality in Equation 8.23 is evaluated as a sample average by most practitioners of Taguchi's robust design:

$$\bar{\Omega}(q) \approx [1/n]\sum_{y=1,n}\Omega(q,y)$$

where the summation is over y = 1, ..., n replications of a trial. (For simplicity, we assume throughout that for a given experimental design the number of replications for each trial is constant and equal to n.) The fundamental quantity of interest, however, is the CIQ for the entire population. The fact that we need to estimate the population statistics in terms of the sample statistics leads to uncertainty in the population CIQ which we manage using the tests for statistical significance developed in Chapter 9.

8.4.2 NIB1

For the NIB condition there are, as stated earlier, two possibilities which we denote as NIB1 and NIB2. NIB1 has a scaling factor, such as the diameter of the lawn mower shear pin discussed earlier, but NIB2 does not. In an experiment designed to reduce the variance in the thickness of a film, for example, the scaling factor, α, is the deposition time needed to generate a certain thickness for a thin film. The experiments, however, would be run using a nominal, constant time for deposition given by α_0. The mean film thickness $\mu(q)$ for the population estimated by the sample average, $\bar{g}(q)$, can be used to develop a new time or scaling level, $\alpha(q)$, for each trial (q) which should generate the ideal mean thickness, g_I, using the expression:

$$\alpha(q) = \frac{\alpha_0 g_I}{\mu(q)} \approx \frac{\alpha_0 g_I}{\bar{g}(q)} \tag{8.24}$$

In this fashion, it is possible, in principle, with any NIB1 problem to change the scaling level from α_0 to $\alpha(q)$ thereby shifting the distribution from $g(q, y)$ as determined from the experiment to a new distribution $\hat{g}(q, y)$ whose mean is at the ideal specification, g_I. This ability to place the mean at the ideal position results in a monotonic dependence of the loss function on $\mu(q)$ for the NIB1 problem, whereas the NIB2 loss function without a scaling factor is not monotonic. Because of NIB1's monotonic dependence, it is easier to analyze NIB1 experimental results versus NIB2 to determine which set points (factors) are favorable and which are not.

Because the scaling level $\alpha(q)$ is used to shift the mean to the ideal position, the resulting average loss of value from ideal, $\overline{\Delta V}(q) = \Omega(q)$, for the shifted distribution, $\hat{g}(q, y)$, is equal to the sum of $k_T [\hat{g}(q, y) - g_I]^2$ over the entire population, n_p:

Table 8.1 The basic variables $x(q, y)$ and ideal value of basic variable x_I used in Taguchi's SIB, NIB, and LIB loss functions

Condition	$x(q, y)$	x_I
SIB	$g(q, y)$	0
NIB	$g(q, y)$	g_I
LIB	$1/g(q, y)$	0

$$\Omega(q)=\frac{k_T}{n_P}\sum_{y=1,n_P}[\hat{g}(q,y)-g_I]^2. \tag{8.25}$$

When multiplied by $\alpha_0/\alpha(q)$, the shifted distribution in Equation 8.25 becomes the original unshifted distribution $g(q,y)$ and likewise g_I becomes the population mean, $\mu(q)$:

$$\begin{aligned}\Omega(q)&=\left[\frac{\alpha(q)}{\alpha_0}\right]^2\left[\frac{k_T}{n_P}\right]\sum_{y=1,n_P}\left[\frac{\alpha_0\hat{g}(q,y)}{\alpha(q)}-\frac{\alpha_0 g_I}{\alpha(q)}\right]^2\\&=\left[\frac{\alpha(q)}{\alpha_0}\right]^2\left[\frac{k_T}{n_P}\right]\sum_{y=1,n_P}[g(q,y)-\mu(q)]^2\end{aligned} \tag{8.26}$$

On replacing the summation in Equation 8.26 by the population variance of the unshifted distribution given by:

$$\sigma^2(q)=\left[\frac{1}{n_P}\right]\sum_{y=n_P}[g(q,y)-\mu(q)]^2 \tag{8.27}$$

we see that the CIQ can be written as:

$$\Omega(q)=k_T[\alpha(q)/\alpha_0]^2\sigma^2(q). \tag{8.28}$$

Using Equation 8.24, we arrive at Taguchi's celebrated form for the NIB1 CIQ written here in terms of the population statistics:

$$\Omega(q)=\frac{k_T g_I^2\sigma^2(q)}{\mu^2(q)} \tag{8.29}$$

When costs are a function of the attribute, the scaling factor should be used to shift the attribute to the point where the loss function is at a minimum, the loss function replacing the CIQ in Equation 8.29. However, the manufacturing costs for a process are generally not influenced by variation if all of the product is shipped in which case the coefficient k_T in Equation 8.29 should still be set equal to one-half of the second derivative of the CIQ even when considering the loss function itself.

8.4.3 SIB, NIB2, AND LIB

The CIQ (value loss) relations for four types of loss functions are given in Table 8.2 in terms of the population mean, μ, and population variance, σ^2, of the basic variable x for each condition. The quantity V_0 is the value for the baseline condition at $\bar{x}(q=1)\equiv x_0$. The quantity g_C is the critical specification where value goes to zero and γ is the weighting exponent (Equation 5.13). Taguchi has provided an approximate estimate of the coefficient k_T in terms of warranty losses divided by the square of the tolerance [3]. Using the value model given in

Chapter 5, k_T is determined, as expressed in Table 8.2, by the value of the product at the baseline specification and the partial derivatives, evaluated at x_0, of the exponentially weighted functions $f_V(x)$ from Table 5.2 with respect to the basic variable, x. Taguchi's estimate of k_T is approximate because there is not always a direct relationship between warranty repair costs and the value loss being experienced by the customer. For example, a glove box rattle costing $10 to repair may generate as much loss of value to the customer as brake squeal costing $100 to repair. Moreover, the loss of value for either problem may be considerably more than $100.

A major assumption in Taguchi's theory of the loss of the quality due to variation is that the partial derivatives of the loss function or the CIQ exist at the baseline point g_0 (see the derivation of the loss function in Chapter 4). However, as a result of the findings from prospect theory, if all persons have the same baseline expectation for the attribute, there is likely to be a kink in the value curve precisely at this point as indicated schematically in Figure 5.25. Further illumination of this important point is the subject of current research. At this juncture, the view is that there may be a distribution of kinks which are related to the distribution of customer expectations for the attribute. This should result in a renormalized value curve having an effective curvature in the vicinity of the baseline that is greater than that given by the unkinked value curve and dependent upon the sensitivity by which persons can perceive a change in the attribute from baseline. The losses associated with product vari-

Table 8.2 Single-attribute value formulas for the cost of inferior quality CIQ, $\Omega(q) \equiv V_I - V(q)$

Condition	$\Omega(q)$
SIB	$k_T\left[\sigma^2(q)+\mu^2(q)\right]$
NIB1	$\dfrac{k_T g_I^2 \sigma^2(q)}{\mu^2(q)}$
NIB2	$k_T\left[\sigma^2(q)+\left[g_I-\mu(q)\right]^2\right]$
LIB	$k_T\left[\sigma^2(q)+\mu^2(q)\right]$

$$k_T = \frac{-V_0}{2}\left\{\gamma[\gamma-1]\left[\frac{\partial f_V(x)}{\partial x}\right]^2\Bigg|_{x_0} + \gamma \frac{\partial^2 f_V(x)}{\partial x^2}\Bigg|_{x_0}\right\}$$

See Table 8.1 for x and Table 5.2 for $f_V(x)$.

ation are thus likely to be larger than calculated based upon a smooth value curve or a smooth loss function curve. For example, if, as in Figure 5.25, all customers had the same baseline expectation, g_0, and it was not at or near the ideal point, the leading term in the CIQ loss due to product variation as a result of the prospect theory kink is proportional to the absolute value of the first derivative of the smooth (not kinked) value curve at g_0 times the standard deviation of the distribution. The loss contributed by the second order term, which in the standard formulation is the leading non-zero term and proportional to the second derivative of the smooth curve times the variance, is increased by roughly 50%. The developments which follow use the same assumption made by Taguchi which is that the CIQ curve is smooth. Based upon the above, the results provide only a lower bound for the CIQ losses due to variation.

Although the expressions in Table 8.2 are written in terms of the population mean and variance of the basic variables for each condition, these quantities will not be known exactly because the sample sizes used in experiments are relatively small compared to the population of products that are to be manufactured. When the population is normally distributed, the expectation of the population variance, $\sigma^2(q)$, for trial (q) based upon a sample measurement of variance is, from Equation A.21, given by:

$$\begin{aligned} E(\sigma_S^2(q)) &= \nu s^2(q) E\left(\frac{1}{\chi^2}\right) \\ &= \frac{\nu s^2(q)}{\nu - 2} \end{aligned} \tag{8.30}$$

where ν is the number of degrees of freedom (df) which is equal to $n - 1$ if variances have not been pooled and

$$s^2(q) = \left[\frac{1}{n-1}\right] \sum_{y=1,n} [x(q,y) - \bar{x}(q)]^2$$

is the measured unbiased sample variance for trial (q). (The pooling of variances is discussed in Chapter 9.) Equation 8.30 should be interpreted as follows: many different population variances can yield the same value for a single, unbiased sample measurement of the variance for ν df. If we consider the distribution of possible population variances for the normally distributed attribute and average them, the average will be that given by Equation 8.30. (This result can be readily demonstrated using computer simulation.) If, on the other hand, we repeatedly make ν df sample measurements taken from a specific normal population and compute the unbiased sample variance for each and average them, we find from Equation A.17 that, in the limit of an infinite number of sample measurements, the average will be given by:

$$E(s^2(q)) = \frac{\sigma^2 E(\chi^2)}{\nu} = \sigma^2.$$

This is a specific example of a general theorem which states that the expectation of the unbiased sample variance is equal to the population variance, the proof appearing in most textbooks on statistics. However, when the df is small, use of the unbiased sample variance as a point estimate, as seen from Equation 8.30, will, on average, yield a significant underestimate of the average population variance computed from the distribution of possible population variances.

The point estimate for the population average $\mu(q)$ is the sample average $\bar{x}(q)$ obtained from the expectation of Equation A.20. For a normal distribution, the point estimate for $\mu^2(q)$, which appears in the expressions for the CIQ in Table 8.2, is given by:

$$E(\mu_S^2(q)) = \bar{x}^2 + \frac{\nu s^2(q)}{[\nu - 2]n} \tag{8.31}$$

Now having discussed the issues in arriving at point estimates, suffice it to say that the importance of point estimates is much less than the attention usually given to them. Given the sample measurements, it is the distribution or range of possible population outcomes that is most important. Methods for arriving at the distribution of possible population outcomes are discussed in Chapter 9.

8.4.4 SINGLE-ATTRIBUTE CIQ AND LOSS FUNCTION

Using the linear set point model (Appendix B) for describing the outcomes of experimental trials, the CIQ given by Ω_{ij} for set point *ij* acting alone is obtained by first computing the CIQ phi or lambda coefficients which are primed to denote the log transformation for improving additivity has been used:

$$[\varphi'(\Omega)] = [\mathbf{X}^{\mathrm{T}}\mathbf{X}]_{\varphi}^{-1}[\mathbf{X}^{\mathrm{T}}]_{\varphi}[-10\log_{10}\Omega] \tag{8.32a}$$

or

$$[\lambda'(\Omega)] = [\mathbf{X}^{\mathrm{T}}\mathbf{X}]_{\lambda}^{-1}[\mathbf{X}^{\mathrm{T}}]_{\lambda}[-10\log_{10}\Omega] \tag{8.32b}$$

depending on the form chosen for the **[X] matrix** for describing the experimental design. The matrix form $[-10\log_{10}\Omega]$ represents a column matrix or vector in which the elements from the top are $-10\log_{10}(\Omega(1))$, $-10\log_{10}(\Omega(2))$, ... The solutions given by Equations 8.32a and 8.32b represent regular least square solutions. After the coefficients have been computed and when the number of trials is greater than the number of unknown coefficients, the outcomes predicted for each trial by the coefficients determined from Equations 8.32a and 8.32b are equal to the measured outcomes plus a least square error term. For example, the predicted model outcome for the coefficients determined from Equation 8.32b is given by:

$$[\mathbf{X}][\lambda'(\Omega)] = [-10\log_{10}\Omega] + [err].$$

The CIQ point estimate for each trial (which exists only for $\nu > 2$) is computed by substituting, into the appropriate expression in Table 8.2, Equation

8.30 for the population variance and Equation 8.31 for the population mean. When the df is 2 or 1, the range for the CIQ for a specified level of confidence must be used instead of a point estimate. This should not be seen as a problem as the range is a more complete representation of a statistical quantity than a point estimate. Monte Carlo methods are used, for example, in Chapter 9 to compute distributions (ranges with confidence levels) for the uncertainty in the CIQ for arbitrary df.

Each lambda coefficient $\lambda'_{ij}(\Omega)$ computed from Equation 8.32b (or phi coefficient determined from Equation 8.32a) can then be used to obtain the CIQ point estimate for turning set point *ij* 'on' using the expression:

$$\Omega_{ij}=10^{-[\lambda'_0(\Omega)+\lambda'_{ij}(\Omega)]/10} \tag{8.33}$$

The total loss for set point *ij* acting alone for a single-attribute problem is given by:

$$\Lambda_{ij}=\Omega_{ij}+C_{ij}+\{F_{ij}+[M_{ij}/Y_{RS}]\}/D_0 \tag{8.34}$$

The coefficients C_{ij}, F_{ij}, and $M_{ij}\,/\,Y_{RS}$ represent the variable cost, fixed cost, and prorated investment generated by set point *ij*. When several set points noted as the combination (*ij*) contribute to reducing the single-attribute loss function, Equation 8.33 becomes:

$$\Omega_{(ij)}=10^{-[\lambda'_0(\Omega)+\sum_{(ij)}\lambda'_{ij}(\Omega)]/10} \tag{8.35}$$

8.5 Orthogonal arrays

The choice of experimental design is influenced by the number of independent factors (set points) that are to be evaluated and by the desire to efficiently arrive at an acceptable amount of experimental error for the least effort expended in terms of the number of trials and replications that need to be run. Orthogonal arrays are very popular for experimental design because they are very efficient. Phadke [4] has given a comprehensive discussion of the use of orthogonal arrays within the framework of Taguchi's methodology. There is also an excellent compact booklet on orthogonal arrays by Taguchi and Konishi [5].

The notation used to describe orthogonal arrays is of the form $L_{Rows}[(levels_1)^{\alpha}, (levels_2)^{\beta}, ...]$ in which the subscript *Rows* represents the total number of rows (trials), $(levels_1)^{\alpha}$ represents the number of levels for a number of control factors (columns) equal to α and $(levels_2)^{\beta}$ represents (if present) the number of levels for β additional control factors (additional columns).

The $L_8(2^7)$ array is shown in standard Taguchi notation in Figure 8.9. The two levels for each of the control factors are listed as 1 and 2. Orthogonal arrays have the special property that if all rows containing the occurrences of any main effect set point variable *i* at a level *j* are examined, all of the other main effect variables at each of their levels are present in these rows either once or some whole number of times. For example, all rows containing variable 3 at level 2 also contain the full range of the other factors (Figure 8.10). The similar result for column 4 is shown in Figure 8.11. The lower case letters a, b, and c shown at the bottom of the trials are used to denote the possible interactions between main effects. If these interactions are strong they can confound the results if an additional main effect is evaluated using an interaction column. For example, if factors a and b interact but that is not known at the outset of an experiment and a main effect d is examined in column 3, the resulting measure of the impact of d on the objective function would be in error.

Construction of the $L_9(3^4)$ orthogonal array following the above guidelines is shown in Figure 8.12. The movement from factor to factor, shown as a chain of arrows, represents an experimental trial. For example, the sequence of factors and levels for trial 4 begins with factor 1 at level 2 and then moves to level 1 for factor 2. However, in going to factor 3, level 1 cannot be selected because this level for factor 3 was already selected in trial 1 following factor 2 at level 1. Similarly, in trial 7, level 3 of factor 3 must be selected after level 1 of factor 2 because the other two levels for factor 3 had been selected in trials 1 and 4. Fortunately, the orthogonal arrays most likely to be used in setting up experiments have been included in Taguchi and Konishi's booklet [5].

The φ coefficients expansion for the outcome of a trial with only main effects included is:

$$Y(q) = \varphi_0 + \sum_{ij} \varphi_{ij} X_{ij}(q) \tag{8.36}$$

in which φ_0 is the model baseline for the array being considered. As described more fully in Appendix B, the φ coefficients have the important property:

$$\sum_j \varphi_{ij} = 0. \tag{8.37}$$

If there are only two levels for a given variable i, it follows that:

	Columns						
Trial	1	2	3	4	5	6	7
1	1	1	1	1	1	1	1
2	1	1	1	2	2	2	2
3	1	2	2	1	1	2	2
4	1	2	2	2	2	1	1
5	2	1	2	1	2	1	2
6	2	1	2	2	1	2	1
7	2	2	1	1	2	2	1
8	2	2	1	2	1	1	2
Factor interactions	a	b	a×b	c	a×c	b×c	a×b×c

Figure 8.9 The $L_8(2^7)$ orthogonal array in standard notation. (Source: G. Taguchi and S. Konishi (1987) *Taguchi Methods Orthogonal Arrays and Linear Graphs*, © American Supplier Institute, Allen Park, MI. Reproduced by permission under License No. 961201.)

Trial	Columns 1	2	3	4	5	6	7
1	1	1	1	1	1	1	1
2	1	1	1	2	2	2	2
3	1	2	2	1	1	2	2
4	1	2	2	2	2	1	1
5	2	1	2	1	2	1	2
6	2	1	2	2	1	2	1
7	2	2	1	1	2	2	1
8	2	2	1	2	1	1	2
Factor interactions	a	b	axb	c	axc	bxc	axbxc

Figure 8.10 Selected rows of $L_8(2^7)$ orthogonal array showing that with level 2 .. factor 3 fixed, the other factors cycle twice through their full range. (Source: G. Taguchi and S. Konishi (1987) *Taguchi Methods Orthogonal Arrays and Linear Graphs*, © copyright, American Supplier Institute, Allen Park, MI. Reproduced by permission under License No. 961201.)

$$\varphi_{i0} = -\varphi_{i1} \tag{8.38}$$

and for three levels the corresponding relationship is:

$$\varphi_{i0} = -\varphi_{i1} - \varphi_{i2}. \tag{8.39}$$

The $[X]_\varphi$ $L_8(2^7)$ orthogonal array (Appendix B) is multiplied in Figure 8.13 against the column array of the φ coefficients to yield the outcome array.

From the orthogonality relationship, $\varphi_{i0} = -\varphi_{i1} - \varphi_{i2}$, it follows by inspection that the expressions for the φ coefficients for the $L_8(2^7)$ array are given by (using $Y_q = Y(q)$ shorthand notation):

Trial	Columns 1	2	3	4	5	6	7
1	1	1	1	1	1	1	1
2	1	1	1	2	2	2	2
3	1	2	2	1	1	2	2
4	1	2	2	2	2	1	1
5	2	1	2	1	2	1	2
6	2	1	2	2	1	2	1
7	2	2	1	1	2	2	1
8	2	2	1	2	1	1	2
Full factorial interactions	a	b	axb	c	axc	bxc	axbxc

Figure 8.11 Selected rows of $L_8(2^7)$ orthogonal array showing that with level 2 of factor 4 fixed, the other factors cycle twice through their full range. (Source: G. Taguchi and S. Konishi (1987) *Taguchi Methods Orthogonal Arrays and Linear Graphs*, © American Supplier Institute, Allen Park, MI. Reproduced by permission under License No. 961201.)

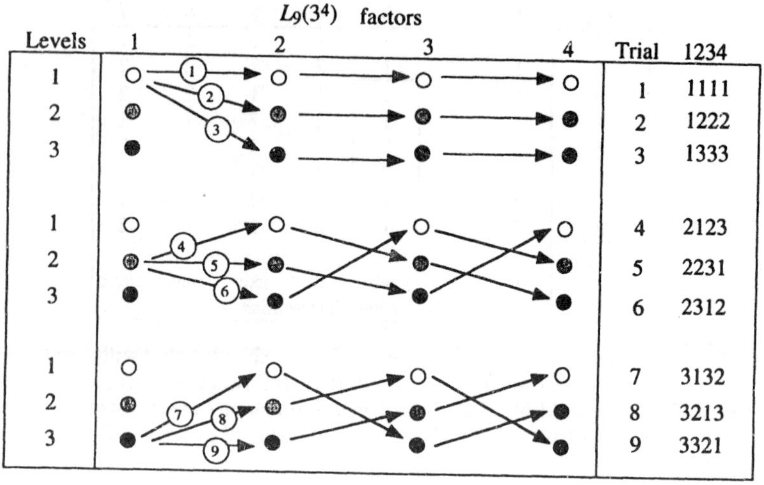

Figure 8.12 Steps in constructing the $L_9(3^4)$ orthogonal array so that each main effect factor goes through its complete cycle when any other main effect factor is held fixed.

$$\varphi_0 = (1/8)\sum_{q=1,8} Y_q$$

$$-\varphi_{10} = \varphi_{11} = (Y_1 + Y_2 + Y_3 + Y_4 + -4\varphi_0)/4$$

$$-\varphi_{20} = \varphi_{21} = (Y_1 + Y_2 + Y_5 + Y_6 + -4\varphi_0)/4$$

$$-\varphi_{30} = \varphi_{31} = (Y_1 + Y_2 + Y_7 + Y_8 + -4\varphi_0)/4$$

$$-\varphi_{40} = \varphi_{41} = (Y_1 + Y_3 + Y_5 + Y_7 + -4\varphi_0)/4$$

$$-\varphi_{50} = \varphi_{51} = (Y_1 + Y_3 + Y_6 + Y_8 + -4\varphi_0)/4$$

$$-\varphi_{60} = \varphi_{61} = (Y_1 + Y_4 + Y_5 + Y_8 + -4\varphi_0)/4$$

$$-\varphi_{70} = \varphi_{71} = (Y_1 + Y_4 + Y_6 + Y_7 + -4\varphi_0)/4$$

The λ coefficients can be determined instead of the φ coefficients for this problem by using the $[X]_\lambda$ representation as described in Appendix B. The λ coefficients should be used when determining whether a change in one or more factors or factor levels will significantly improve, as discussed in detail in Chapter 9, the CIQ, value, or profitability of a specific baseline product relative to one of the experimental trials. The φ coefficients should be used when the objective of the experiment is to simply investigate several factors to see how each influences the CIQ, value, or profits relative to no specific trial, the baseline being the average for all the trials. It is particularly important not to overlook the usually non-zero baseline level coefficient for each set point i given by φ_{i0} especially when more than two levels are involved. The X_{ij} elements in the $[X]_\varphi$ orthogonal array are orthogonal in that the dot product of a pair of columns is zero:

Trial	0	11	21	31	41	51	61	71			
1	1	-1	-1	-1	-1	-1	-1	-1	φ_0		$Y(1)$
2	1	-1	-1	-1	1	1	1	1	φ_{11}		$Y(2)$
3	1	-1	1	1	-1	-1	1	1	φ_{21}		$Y(3)$
4	1	-1	1	1	1	1	-1	-1	φ_{31}	=	$Y(4)$
5	1	1	-1	1	-1	1	-1	1	φ_{41}		$Y(5)$
6	1	1	-1	1	1	-1	1	-1	φ_{51}		$Y(6)$
7	1	1	1	-1	-1	1	1	-1	φ_{61}		$Y(7)$
8	1	1	1	-1	1	-1	-1	1	φ_{71}		$Y(8)$

Figure 8.13 The matrix operations linking the $L_8(2^7)$ experimental design in the **[X]**$_e$ form and the column of unknown coefficients with the column vector of the measured outcomes for the trials.

$$\sum_{q=1}^{z} X_{ij}(q)X_{kl}(q)=0 \qquad (i \neq k).$$

Use of an orthogonal array can exact the price of additional trials for certain mixes of variables and levels. For example, the combination of one variable at two levels and one variable at three levels requires a minimum of three trials plus a baseline yielding four trials. However, a minimum of six trials is required to achieve orthogonality. When this number is doubled to include one replication, the difference becomes eight versus twelve runs. Nevertheless, orthogonal arrays more than make up for the possible penalty of more than minimum trials by being much more efficient in arriving at a given level of experimental error because of fewer replications per trial. The product of the number of trials times the number of replications needed to arrive at a desired level of accuracy will always be less for the orthogonal array than for an experiment which employs an array which is not orthogonal but contains the minimum number of trials. This does not necessarily mean, however, that the orthogonal array always generates a more efficient experiment overall as it may be much easier to make several additional replications at a given trial than to make additional trials.

8.6 Noise factors and the outer array

In addition to control and scaling factors, Taguchi defined noise factors as uncontrollable variables such as environmental temperatures, humidity, dust, snow, altitude, etc. that influence product performance. Noise factors are also prevalent in machines used by the factory to manufacture the product and in the dimensions and characteristics of parts that are assembled to form the subsystem or system. The more robust the product design, the less sensitive the product's performance is to noise factors.

It may be possible, nevertheless, to temporarily control the noise factors during experiments used to determine the optimum set points. When noise factors are controlled in this manner they are said to constitute an outer array for the experiment, the inner array being the control factors themselves (Figure 8.14). For example, it may be known that an annealing furnace temperature will vary ±100 degrees over a two week period. The standard annealing time for parts may be only 1 hour and the furnace will hold its temperature within ±10 degrees for this period. Thus it is possible to run experimental trials at a reasonably well defined temperature and to look at the extremes of temperature as an outer array to evaluate the sensitivity of different material compositions to temperature changes and thus to select the composition that has the least sensitivity.

The example in Figure 8.14 shows an experimental design for evaluating six main effect variables using a $L_8(2^7)$ orthogonal array with column 71 open which can be used to estimate error. The outer array is constructed for the three noise factors using an $L_4(2^3)$ orthogonal array. Four measurements are made for a given trial with each having the noise set points 'on' as noted in the outer array. For example, the second measurement is taken with both H2 and I2 'on' and with G at its baseline, G1.

Inner array

Main effect factor	Level 1	Level 2
A	A1	A2
B	B1	B2
C	C1	C2
D	D1	D2
E	E1	E2
F	F1	F2

Outer array

Noise factor	Level 1	Level 2
G	G1	G2
H	H1	H2
I	I1	I2

$L_4(2^3)$

$L_8(2^7)$

	Inner array									Outer array			
									G2	-1	-1	1	1
	Base	A2	B2	C2	D2	E2	F2	Error	H2	-1	1	-1	1
Trial	0	11	21	31	41	51	61	71	I2	-1	1	1	-1
1	1	-1	-1	-1	-1	-1	-1	-1		g (1,1)	g (1,2)	g (1,3)	g (1,4)
2	1	-1	-1	-1	1	1	1	1		g (2,1)	g (2,2)	g (2,3)	g (2,4)
3	1	-1	1	1	-1	-1	1	1		g (3,1)	g (3,2)	g (3,3)	g (3,4)
4	1	-1	1	1	1	1	-1	-1		g (4,1)	g (4,2)	g (4,3)	g (4,4)
5	1	1	-1	1	-1	1	-1	1		g (5,1)	g (5,2)	g (5,3)	g (5,4)
6	1	1	-1	1	1	-1	1	-1		g (6,1)	g (6,2)	g (6,3)	g (6,4)
7	1	1	1	-1	-1	1	1	-1		g (7,1)	g (7,2)	g (7,3)	g (7,4)
8	1	1	1	-1	1	-1	-1	1		g (8,1)	g (8,2)	g (8,3)	g (8,4)

Figure 8.14 An $L_8(2^7)$ $[\mathbf{X}]_\varphi$ inner array and an $L_4(2^3)$ outer array formed by three noise factors at two levels each.

8.7 The steps for single-attribute parameter design

The steps for parameter design using the approach described above is as follows:

1. Choose an attribute, g, to refine (improve) based upon a priority set by the largest expected increase in customer satisfaction as measured by the improvement of profits or value minus cost.
2. Proceed to brainstorm with experts regarding possible factors and their range which are believed to have promise for reducing the loss function. Also determine the outer array factors needed to simulate the range of noise expected in production conditions.
3. Reduce the list of possible alternatives to those that can be evaluated within the time and budget available.
4. If it seems not unreasonable, limit the levels of alternatives considered to two and ignore possible interactions initially. With these assumptions, choose orthogonal arrays, if feasible, for the inner and outer arrays. Estimate the time and resources needed to perform the experiments. If sufficient time or resources are not available, reduce again the number of alternatives appropriately. For each set point ij to be evaluated, compute the increased combined costs C_{ij} over baseline.
5. Enter the array chosen into the spreadsheet using either the matrix notation for the $[\mathbf{X}]_\varphi$ or the $[\mathbf{X}]_\lambda$ form as appropriate and check to see that the covariance matrix $[\mathbf{X}^T\mathbf{X}]^{-1}$ or the solution matrix $[\mathbf{X}^T\mathbf{X}]^{-1}[\mathbf{X}^T]$ exists. This ensures that there are a sufficient number of linearly independent trials to solve for the unknown coefficients.
6. Select the number of replications, n, for each trial based upon prior experience. If no prior experience exists, choose $n = 5$.
7. Run the experiment and estimate the significance of the results (Chapter 9).
8. With the significance and the expected reduction in loss function known for each set point, assess what action to take based upon the level of investment and variable cost that will be required to implement the changes in the set points.
9. Check your proposed set point changes by running a confirmation experiment complete with replications.
10. If results predicted are verified by the confirmation experiment, go to the next problem in your list of priorities. If the results are not confirmed, check for blunders in experimental technique and in calculations. If no errors are found, consider possible interactions and add the experiments needed to evaluate them to the original design.

8.8 Production examples of parameter design

Taguchi's approach to experimental design has been widely adopted and used to discover design and process modifications which have led to important improvements of product quality. Several case studies are included in Phadke's [6] excellent book on Taguchi methods. The most extensive compilation is in the annual symposia on Taguchi methods sponsored and published by the American Supplier Institute, Allen Park, MI. In what follows a brief description is given of four published experiments to illustrate the diverse nature of the types of problems which have been resolved using Taguchi's approach. The reader is encouraged to read the original studies for insight into how real problems are set up and solved using Taguchi's methodology.

8.8.1 CVD PROCESS EXAMPLE

Phadke [6] investigated four process parameters – temperature, pressure, settling time, and cleaning method – at three levels each for their influence on surface defects (a single attribute) in films produced by chemical vapor deposition (Figure 8.15). No pairwise or higher interactions were assumed significant and the $L_9(3)^4$ orthogonal array was chosen for the experimental design. The temperature effect was found to be most important resulting in a signal-to-noise ratio improvement of 25 dB from this factor alone from baseline.

8.8.2 DIELECTRIC BOND PROCESS OPTIMIZATION

Pingfang Tsai of General Motors [7] studied the dielectric bond process for manufacturing a door trim pull strap assembly (Figure 8.16). The problem involved improving (1) the bond strength between the featured vinyl and the main vinyl of the handle, (2) the bond strength between the main vinyl and the coated steel

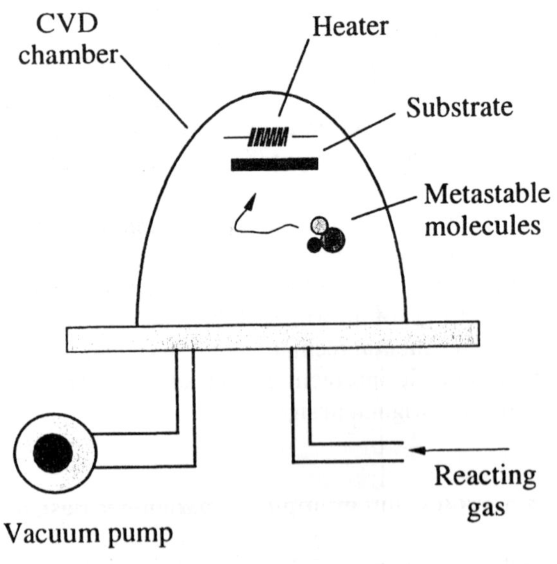

Figure 8.15 Chemical vapor deposition process.

support strap, and (3) bond appearance. This multiattribute optimization problem was solved heuristically by using three different objective functions obtained from the same set of control factors. Tsai investigated four process control factors at three levels each using the $L_9(3^4)$ orthogonal array. Robustness was evaluated by using different color vinyl in the outer array for each trial.

8.8.3 PRINTED CIRCUIT BOARD DRILL BITS

Montmarquet [8], working at ITT, used Taguchi's parameter design methodology to improve the design of 0.013″ diameter bits used in drilling holes in multilayer printed circuit boards. A schematic exploded view of the drilling process is shown in Figure 8.17. The holes, which are called 'vias,' were subsequently plated to provide conduction paths between printed circuits in different layers. Design modifications of the drills were needed as considerable costs were being incurred by their frequent breakage, the printed circuit board being scrapped each time one broke at an average cost of between $200 and $600 for the board. The drills themselves were cheap by comparison. A total of 11 main effects at two levels each and one main effect at four levels were examined using a modified $L_{16}(2^{15})$ design. Also four noise factors were introduced, Montmarquet's experiment being an excellent example of the use of an outer array to introduce noise into an experiment to achieve a robust solution. The percentage of cobalt in the alloy steel was found to be the most important factor in improving drill life.

8.8.4 TRUCK LEAF SPRINGS

Desrochers and Ewing [9] working for Eaton Yale Limited, Suspension Division used Taguchi's experimental design methodology to discover how to best control the processing factors that influenced the free height of a truck spring. The standard NIB1 approach, however, could not be used because the scaling factor represented by the rise of the existing cam over which the springs were formed, the dimension h in Figure 8.18, could not be changed without a large expense. Thus, Desrochers and Ewing were faced with using a NIB2

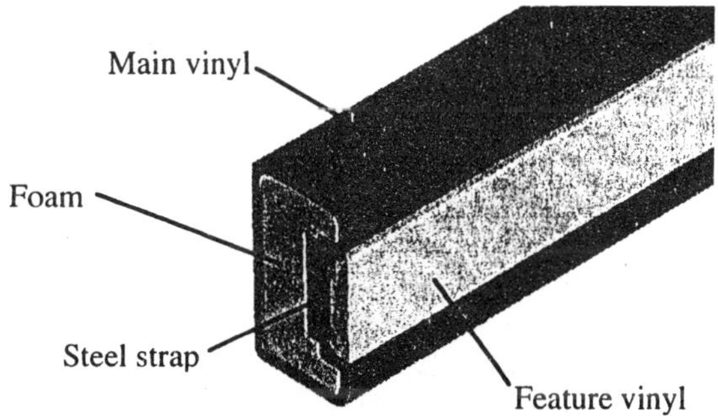

Figure 8.16 Cross section of a door pull strap.

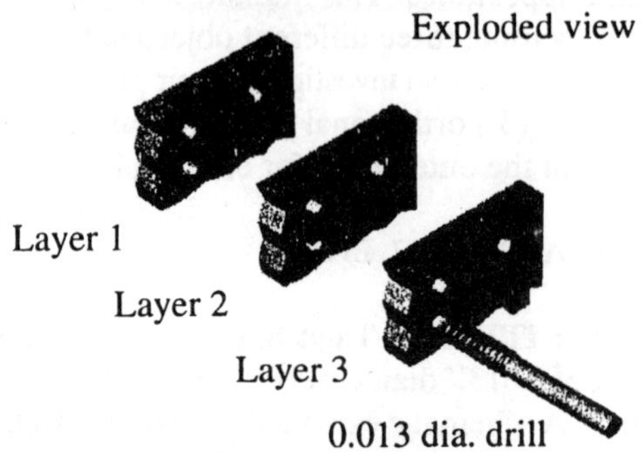

Figure 8.17 Schematic of process for drilling vias for connecting layers in circuit boards.

design of experiment approach to discover the factors which either singly or in combination could be used to simultaneously move the free height of the manufactured springs towards the ideal position and also to reduce, hopefully, the variation in free height between springs. Four main effects and three interactions were investigated using an $L_8(2^7)$ orthogonal array. The temperature of the quenching oil was investigated as a noise factor and it was found to be so important on the free height of the spring that improved controls were introduced in production to minimize the temperature variations. Conversion of the oil quench temperature from a noise factor to a control factor simply changed their design from an L_8 to an L_{16} form.

8.9 A NIB1 case study

To demonstrate the analysis of a parameter design problem using the methods described in this chapter, let us consider a NIB1 problem, for simplicity, having six replications per trial of an attribute g using the $L_4(2^3)$ design shown in Figure 8.19 in the $[\mathbf{X}]_\varphi$ form. The CIQ for this problem is given by the NIB1 expres-

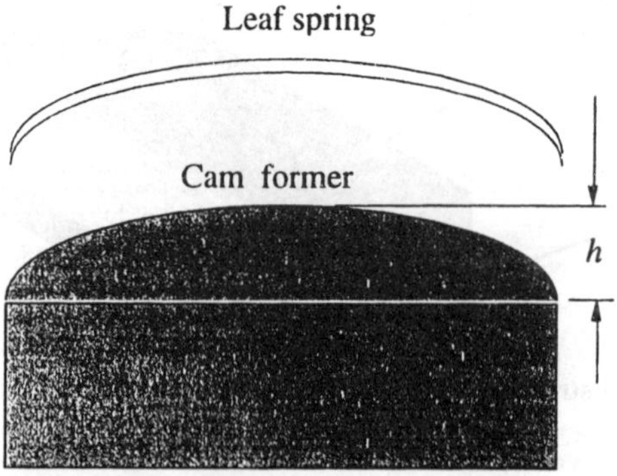

Figure 8.18 Schematic drawing of cam former die for a leaf spring.

sion in Table 8.2. The results for the simulated replications for the attribute g are shown in Figure 8.20 along with the variance and average for each trial. The replications were computer simulated with those for each trial being taken from a normal population whose mean was 40 and whose variance was 5. Consequently no set point (control factor) should have any influence on the outcome. Such a simulated experiment of this nature should not be thought of as a boring non-event, however. It represents what is known as the **null hypothesis** (Appendix A) and is very useful for testing various analysis methods. A full analysis of the experiment is deferred, however, until Chapter 9.

The point estimate for the CIQ for each trial, shown in column Q of Figure 8.20, was computed from the NIB1 expression in Table 8.2 for k_T=2 and g_l=50. The point estimate for the population variance for each trial was computed using Equation 8.30 for ν = 5 df and the point estimate for the population average squared was computed using Equation 8.31. The population variance and mean are independent for a normal distribution so the expected value of their ratio is equal to the ratio of their expected values. Column R shows the transformation of the CIQ to a signal-to-noise ratio. The $\varphi'_{ij}(\Omega)$ coefficients calculated from Equation 8.32a are shown in column T. For this problem, the solution matrix in the general expression, Equation 8.32a, can be replaced by $[\mathbf{X}]_\varphi^{-1}$, the inverse of the design matrix, because the number of unknowns is equal to the number of independent trials. Such an experiment is said to be 'saturated.' All computations were made using a spreadsheet program. The CIQ in cell Q26 was computed using the spreadsheet expression:

```
=$F$2*$F$3^2*(5/(5–2))*O26/(P26^2+(5/(5–2))*O26/6)
```

where \$F\$2 and \$F\$3 represent the cells where k_T and g_l, respectively resided. This expression was dragged down to make the CIQ computations in the remaining rows. The resulting $\varphi'_{ij}(\Omega)$ coefficients were computed using the spreadsheet area expression:

```
=MMULT(MINVERSE($D$10:$G$13),R26:R29).
```

The only departure we have made from the normal Taguchi methodology for a NIB1 experiment at this point is to use Equations 8.30 and 8.31 as point estimates for the population variance and the square of the population mean instead of simply the sample variances and the square of the sample average. Their influence on the CIQ and the resulting loss functions computed from the

	C	D	E	F	G
9	Trial	0	11	21	31
10	1	1	-1	-1	-1
11	2	1	-1	1	1
12	3	1	1	-1	1
13	4	1	1	1	-1

Figure 8.19 The $L_4(2^3)$ experimental design in the $[\mathbf{X}]_\varphi$ form used for the NIB1 case study.

Trial	1	2	3	Replications 4	5	6	Var	Ave	Ω	$-10\times Log(\Omega)$	ij	$\varphi'_{ij}(\Omega)$
1	38.5	42.9	38.7	40.8	40.6	44.4	5.37	40.97	26.65	-14.26	0	-12.25
2	40.3	38.5	38.2	39.0	40.4	40.8	1.18	39.52	6.31	-8.00	11	-1.12
3	42.1	39.6	44.6	41.1	39.2	37.8	5.78	40.72	29.04	-14.63	21	2.20
4	40.0	40.2	38.3	41.2	36.5	40.5	3.02	39.44	16.19	-12.09	31	0.93

Figure 8.20 Computer-simulated replications for the NIB1 case study including the sample variances and averages for the trials.

CIQ are very important when the degrees of freedom are small but their influence on the $\varphi'_{ij}(\Omega)$ coefficients for the signal to noise ratios is only on the baseline term, $\varphi'_0(\Omega)$, because the $\varphi'_{ij}(\Omega)$ are computed as differences between logarithms. As already stated, a point estimate cannot be generated for the population variance or the square of the population mean when the df is less than three which means that the number of replications needs to be four or greater if a point estimate is desired but a confidence range can always be generated when the df is one or more.

Although we have arrived at a set of $\varphi'_{ij}(\Omega)$ coefficients, we have not determined their significance. For example, the coefficient for set point 21 is a positive 2.20 which means that turning set point 21 'on' will reduce the CIQ (improve the value of the product) if we take this favorable finding at face value. But until we demonstrate its significance, we cannot be sure. Perhaps the true effect of set point 21 on the CIQ could be unfavorable (a negative sign), the above finding being just a statistical aberration caused by a small sample size. Of course we know from the normal population used in setting up this hypothetical problem that no set point had an effect on the CIQ, all the findings being no more than statistical error. In a real experiment we would not know this and may be tempted to turn set point 21 'on' at perhaps some additional cost and find out later after many parts were produced that they were no better than the original baseline. Or that they were worse!

8.10 Controversy

In spite of the success and wide usage of Taguchi methods, it is an understatement to say that there is some controversy surrounding the methodology [10, 11]. Some of the issues raised have been simply judgmental, for example, the preference of filling all available opportunities in an orthogonal design with main effects thereby ignoring possible interactions. This is a matter of experience and one would thus be prudent to defer to Taguchi's recommendation on this point. A more fundamental issue involves how to properly estimate the significance of the set point coefficients as determined, for example, from Equation 8.32a or 8.32b for the CIQ signal-to-noise ratio. This issue is addressed in Chapter 9 and in Case Studies 1–4 in which a new approach to determining the significance of a factor on the signal-to-noise ratio is developed and evaluated using random samples from known populations, thereby providing an unambiguous test of the method.

References

1. Taguchi, G. and Wu, Y. (1980), *Introduction to Off-line Quality Control,* Central Japan Quality Association, Nagoya, Japan.
2. Elsayed, E.A. and Chen, A. (1993) Optimal levels of process parameters for products with multiple characteristics. *International Journal of Production Research*, **31**, 1117–32.
3. Taguchi, G. (1993) *Taguchi on Robust Technology Development*, ASME Press, New York, p. 23.
4. Phadke, M. (1989) *Quality Engineering and Robust Design,* Prentice Hall, Englewood Cliffs, NJ, 1989, pp. 41–96.
5. Taguchi, G. and Konishi, S. (1987) *Taguchi Methods Orthogonal Arrays and Linear Graphs*, American Supplier Institute, Allen Park, MI.
6. Phadke, M.S. (1989) *Quality Engineering and Robust Design*, Prentice Hall, Englewood Cliffs, NJ, p. 41.
7. Tsai, P. (1989) Bond process optimization using the Taguchi parameter design approach, in *Seventh Symposium on Taguchi Methods*, American Supplier Institute, Allen Park, MI, pp. 55–72.
8. Montmarquet, F. (1988) Printed circuit drill bit design optimization using Taguchi's methods – .013 diameter bits, in *Sixth Symposium on Taguchi Methods*, American Supplier Institute, Allen Park, MI, pp. 207–16.
9. Desrochers, G. and Ewing, D. (1984) Leaf spring free height analysis, in *Second Symposium on Taguchi Methods*, American Supplier Institute, Allen Park, MI, pp. 38–47.
10. Nair, V.N. (ed.) (1992) Taguchi's parameter design: a panel discussion. *Technometrics*, **34**, 127–61.
11. Logothetis, N. and Wynn, H.P. (1989) *Quality Through Design*, Clarendon Press, Oxford, UK, pp. 87–8.

9 Significance of results

9.1 Introduction

9.1.1 RISK

The introduction of a new product is risky. This is particularly true when a bold departure is planned from what is currently available because it is more difficult to gauge what the customers' responses will be versus a product which will offer several important improvements but is otherwise similar to what is currently available. The other extreme of making no product improvement also carries a very high risk because the product will quickly become outdated in a competitive market. Highly competitive markets also strongly limit extending product development time as a factor for reducing risk. A monopoly can take great care in researching a market before introducing a new product using in-depth surveys and having customers evaluate a large sample of prototypes. But in a highly competitive market, you run the risk of coming in behind your competitors leaving you with reduced share and profits and less resources for financing the development of future products.

9.1.2 RISK MANAGEMENT

Because risk is an ever-present part of the product realization process, intelligent management of risk is vital. Good risk managers also understand that large rewards are often strongly coupled to risk in that products which attempt to move out of the ordinary will generate higher market share if customers agree that the innovations truly generate greater value. The risk management challenge is illustrated in Figure 9.1 in which the two borders enclose the baseline and future product populations. Knowledge of the future product is obtained by extrapolating from the baseline using the results from tests of a limited number of prototypes. What is at risk is the generally very large investment, M, required to develop, manufacture, and distribute the new product. Risk assessment involves measuring the level of confidence, L_C, that the new product will generate an acceptable internal rate of return, I_{RR}, given that you are risking an investment of size M. If the risk-free interest rate given by banks is I_R and the hurdle for improving an investment established by the enterprise is $[1+\eta]\, I_R$, then $I_{RR} \geq I_R\,[1+\eta] \,/\, L_C$. In what follows, we consider the risk associated with the change in a single product attribute for simplicity. Significance testing of a multiattribute problem is considered in Case Study 4.

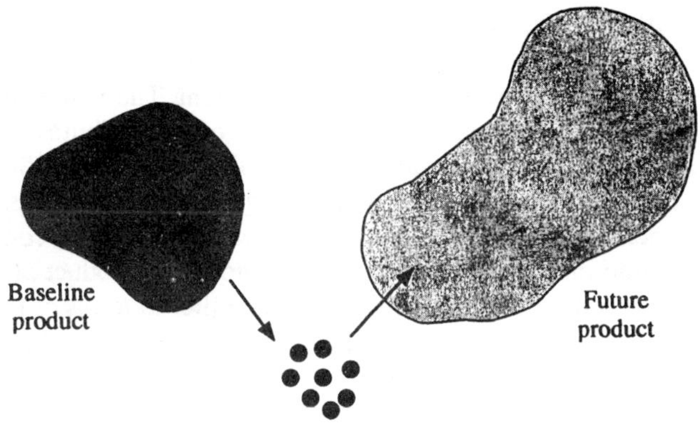

Figure 9.1 Schematic illustration of the process of predicting the behavior of a future product based upon the behavior of a small number of prototypes and the behavior of the product currently in production.

Action should not be taken to develop a new product or process until the risks to profits have been evaluated. For example, if profits for a proposed product change are forecast to increase by at least \$3 000 000 annually over the next two years with 70% confidence, then it is likely that the decision maker would approve the change if no variable cost and no investment increase were required. If, on the other hand, the change required an investment of \$2 000 000, the confidence level may not be high enough for the decision maker to approve, particularly if the forecast also included a 10% chance of a loss.

One also does not have the luxury of playing the law of averages in product development. If an analysis shows that seven out of ten times the new product should be successful, you will get only one try, one roll of the dice, to see if the new product is successful. It is also vital to note that computations of the probability of success are almost always optimistic because not all of the myriad sources of uncertainty will have been included in the analysis. Thus, it is important that the confidence level calculated be high so that the unaccounted for errors should not markedly change the outcome if they do appear.

The first challenge in risk management is to model how value is coupled to the system-level attributes g_i for $i=1, 2, 3, \dots$. The model used here will be the one developed in Chapter 8 which expressed the value of the product as a function of the attributes population average, μ_i, and variance, σ_i^2. Once the model has been selected, the challenge shifts to discovering how to accurately forecast the uncertainty in the population average and variance for the baseline and future products based upon measurements on a limited number of current and prototype products. The final challenge in risk management is to make credible assessments of what the demand and profits will be for the new product. These are the subjects of this chapter.

9.1.3 TYPES OF UNCERTAINTIES

There are two types of uncertainties in risk assessment. The first is the statistical uncertainty in the population means and variances of the attributes for the future product. The second is the strategic uncertainty in its forecast profits. Although these uncertainties are coupled, they represent quite different things. For example, an experiment evaluating two different processing conditions could find that they give the same mean and variance for an important product attribute, one process not being statistically significant from the other in terms of the attribute. However, if one process were considerably less costly than the other, this would cause their forecast profits to be separated by a large amount making the two profit forecasts different with high confidence even though the resulting product attributes were identical. As a result, we will break our analysis of significance into two parts. The first part will be a test of the **statistical significance** that each set point has in changing the mean and variance of the attribute from their baseline levels. The second part will be a test of the **strategic significance** that each set point has in changing the loss function or profits from their baseline levels.

9.1.4 APPROACH TO RISK ASSESSMENT

As described in Chapter 7, the future product can be developed from an existing baseline product using a design of experiment process to determine which of the proposed changes in the components and subsystems are forecast to best improve the product's profitability. Each component or subsystem change being evaluated by the experiment represents a factor specified in terms of the ij set point notation where i refers to the type of factor and j refers to its level, j=0 being the baseline level for factor i. We divide the process of risk assessment into six steps designed to answer the following questions:

1. Are the statistical distributions of the measured attribute for a given trial normal or lognormal or something else?
2. How much is each set point expected to change the variance and average of the attribute?
3. Which set points create changes large enough to be considered as significant in influencing variance?
4. Which set points create changes large enough to be considered as significant in influencing the average?
5. What combination of set points seems best for reducing the CIQ and/or the loss function?
6. Which set points in the best combination are significant in reducing the CIQ and/or loss function?

These six steps are designed to address the issues raised [1, 2, 3] regarding how the significance of the control factors on the S/N ratio should be determined.

9.2 NIB1 example

Consider the NIB1 single-attribute loss function given by Equation 8.29 for a hypothetical product with $k_T = 1$ and $g_I = 1$. The loss function then simplifies to $\Lambda = \Omega = \sigma^2/\mu^2$. The baseline product is noted as 1 and a proposed alternative product, noted as 2, has been made using a different manufacturing process. The measured attributes $g(1, y)$ and $g(2, y)$, respectively, for 1 and 2 for $y=1\rightarrow 10$ replications each are shown in Table 9.1 along with the sample variance and average for the replications. The paired comparison, $L_2(2^1)$, **[X]** array in Table 9.1 represents the simplest design for an experiment, having only two trials and two unknown set points, 0 and 11. More complex designs will be considered after working through this example.

Based upon the results of the experiment, can we assume with high confidence that the loss for the alternative product is less than the loss for the baseline? We need to use the six-step procedure outlined above to answer this question. In step 1 the nature of the two distributions for the basic variable are examined to see if the scatter for each is approximately normally (or log-normally) distributed because this simplifies the analysis if true. The resulting normal probability plots of the replications in Table 9.1 which are shown in Figure 9.2 are satisfactory for assuming that the population is approximately normally distributed. Step 2 involves computing the variance and average for each trial. These are listed in the last two columns in Table 9.1.

9.2.1 STATISTICAL SIGNIFICANCE OF THE TWO VARIANCES

Step 3 involves hypothesizing that the two population variances are equal and computing the F ratio (Appendix A) of the two sample variances which is 2.39 (= 1.273/0.529). Using the spreadsheet function FDIST(2.39,9,9), we find that one

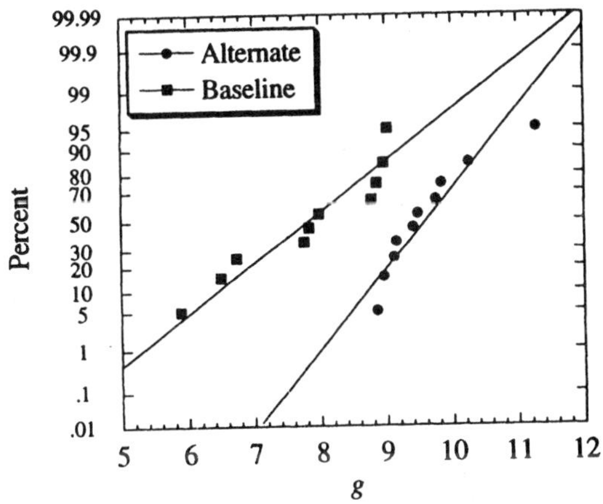

Figure 9.2 Normal probability plot of the replications for trials 1 and 2 in Table 9.1.

Table 9.1 $L_2\ (2^1)$ design for paired comparison experiment including replications and statistics

Trial	*0*	*11*	Replications *1*	*2*	*3*	*4*	*5*	*6*	*7*	*8*	*9*	*10*	*Var*	*Ave*
1	1	0	7.99	8.86	5.89	8.78	7.84	6.74	8.97	7.76	9.03	6.5	1.273	7.836
2	1	1	8.97	8.86	9.48	9.41	9.75	9.15	11.3	10.3	9.12	9.84	0.529	9.612

would be wrong more than 10% of the time in rejecting the hypothesis, noted as $H1_0$, that the two samples had a common population variance. (Using Table A.6b, we see that the F ratio should be greater than 3.179 to reduce the error for accepting hypothesis $H1_0$ to less than 5%.) We conclude that set point 11 does not change variance enough to be considered significant. Although this is far from proof that the two distributions truly had a common population variance, it is prudent to replace the two sample variances with their average or 'pooled' variance $s_P^2 = [1.273 + 0.529]/2 = 0.90$ having $\nu = \nu_P = 18$ (= 2×9) pooled df because it avoids adding a bias to the loss functions which is unwarranted for a high level of confidence in determining if the two loss functions are different.

9.2.2 STATISTICAL SIGNIFICANCE OF THE TWO AVERAGES

Step 4, a determination of the significance of set point 11 on the attribute average, is made using the t distribution given by Equation A.14 to test the hypothesis $H2_0$ that the two population means are the same which for this problem becomes:

$$t = \frac{\bar{g}(2) - \bar{g}(1)}{s_p\sqrt{2/n}}$$

$$= \frac{9.61 - 7.84}{0.95\sqrt{2/10}} = 4.17$$

Using spreadsheet function TDIST(4.17,18,1) = 2.9×10^{-4}, for the single tail of the distribution with t = 4.17 and df = 18, we find the confidence level to be very high that the two population means are not the same; one should be incorrect only three times in ten thousand in assuming that the average for the alternate product is larger than the baseline.

9.2.3 BEST SET POINT

On using the pooled variance and computing the loss function expressions for this problem, we find for step 5 on using Equations 8.30 and 8.31 for the population variance and square of the average, respectively, that the baseline loss is:

$$\Lambda(1) = 1.014/61.5 = 16.5\times10^{-3} \qquad \text{(set point 10 `on')}$$

and that the loss for the alternative is:

$$\Lambda(2) = 1.014/92.5 = 11.0\times10^{-3} \qquad \text{(set point 11 `on')}$$

which results in a favorable difference of -5.5×10^{-3} from baseline.

9.2.4 STRATEGIC SIGNIFICANCE

Although we have determined that the difference between the two averages is significant and that the difference between the two variances is not, we do not know at this juncture the strategic significance for the favorable difference of -5.5×10^{-3} between the two loss functions. Complications arise at this point, step 6, because the loss function is not well characterized statistically except in the limit of large ν (≥ 30). Faced with this difficulty, we turn to computer simulation, the Monte Carlo method. Fortunately, both the variance and average which appear in the expressions for the loss functions follow well-characterized statistics which greatly facilitates the use of computer simulation for significance testing.

When normalized so that its integral from minus to plus infinity is unity, the histogram formed by the points for the random variable μ_S defined as the statistic:

$$\mu_S(q) = \bar{g}(q) + t_S\sqrt{\frac{s_P^2}{n}} \tag{9.1}$$

represents the frequency distribution of possible population means which are consistent with the sample statistics obtained for experimental trial (q) having a pooled sample variance and pooled df given by s_P^2 and ν_P, respectively. Likewise, the histogram formed by the points for the random variable $\sigma_S^2(q)$ defined as the statistic:

$$\sigma_S^2(q) = \frac{\nu_P s_P^2}{\chi_S^2} \tag{9.2}$$

represents the frequency distribution of possible population variances which are consistent with the measured sample statistics obtained for experimental trial (q). The histograms for $\mu_S(q)$ and $\sigma_S^2(q)$ can be generated by selecting points at random from the t and chi-squared distributions and substituting for t_S and χ_S^2 in Equations 9.1 and 9.2, respectively. It is important to note that the population variances for trials $q = 1$ and $q = 2$ are not assumed to be the same in Equation 9.2 even though we pooled the two sample variances. Instead they are shown as being chi-squared distributed about a mean of s_P^2 with ν_P df which supports our claim earlier that the pooling of sample variances does not force the associated population variances to be equal.

Random selection of points for the Monte Carlo computations can be made in several ways. The most direct way is to first generate two lists of random numbers between 0 and 1. One list is then set equal to the values $\wp_C$ for the cumulative distribution for the t statistic and the other list is set equal to the values $\wp_C$ for the cumulative distribution for the χ^2 statistic. The respective values for t and χ^2 corresponding to each of the cumulative distribution points selected can then be determined using tables or functions for the two distributions.

For illustrative purposes, a graphical process is shown in Figure 9.3 for selecting points from a t distribution having 3 df. (A low value of three was chosen for the df to illustrate the strong departures from normal, straight line behavior which can occur for t and chi-squared distributions.) If we assume that a point chosen at random for the cumulative distribution was 0.95, the resulting value for t is seen to be approximately 2.4 for 3 df, the exact selection being 2.353 as obtained from a chart or spreadsheet. An example of this for the χ^2 distribution for 3 df is shown in Figure 9.4. If we assume that a point chosen at random for the cumulative distribution was 0.7, the resulting value for χ^2 is seen to be approximately 3.5.

Once random numbers for $\wp_C$ have been selected, look-up tables for t and χ^2 as shown in Tables 9.2a and 9.2b, respectively, could also be used to select these values for the Monte Carlo process.

Let's now use Monte Carlo simulation to determine the significance that set point 11 has on changing the loss function from its baseline level based upon the two trial distributions shown in Table 9.1. We begin by simulating the uncertainty in the baseline loss, trial 1. Two lists of random numbers need to be generated for this trial. One will be used in selecting the points from the t distribution and the other will be used to select points from the χ^2 distribution. This process is repeated for the alternative, trial 2. For these simulations, $\nu_p = 18$ and the two trial variances are set equal to the pooled variance, s_p^2.

The two random number lists for trial 1 are shown in columns 2 and 5 of Table 9.3. These are used for simulating the baseline t and χ^2 distributions, respectively. The resulting t selections for 18 df and simulated range of uncertainty in the population mean (Equation 9.1) are listed in columns 3 and 4. The resulting χ^2 and simulated uncertainty in the population variance (Equation 9.2) are shown in columns 6 and 7.

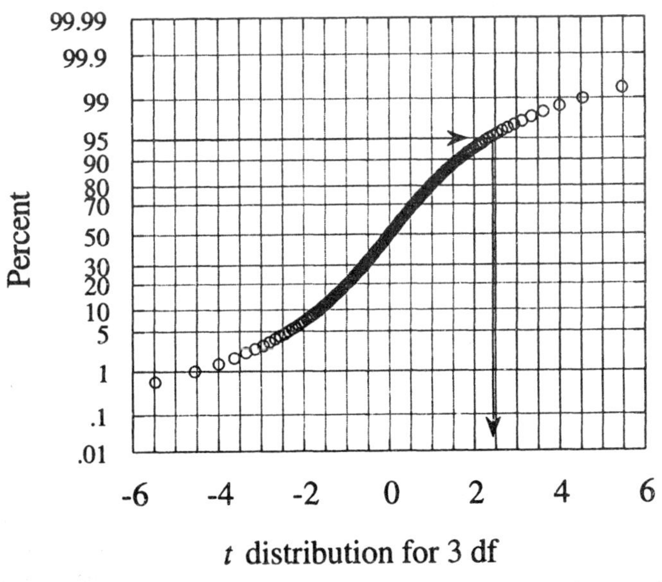

Figure 9.3 Selection of a random number between 0 and 1 for the cumulative probability and the resulting selection of a simulated value for Student's t using the graph.

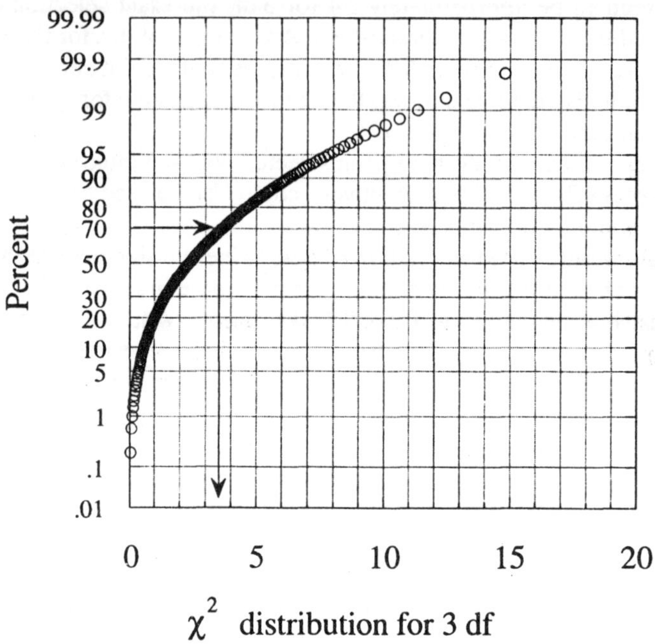

Figure 9.4 Selection of a random number between 0 and 1 for the cumulative probability and the resulting selection of a simulated value for χ^2 using the graph.

Because it is much faster than using either the graphical approach or look-up tables, the values for t and χ^2 were determined for each random number selected using the TINV and CHIINV functions in the spreadsheet. The formula for t is given by (Appendix A):

Table 9.2a Single tail values of t for Student's t distribution

	Cumulative probability									
df	*0.05*	*0.1*	*0.15*	*0.2*	*0.25*	*0.3*	*0.35*	*0.4*	*0.45*	*0.5*
1	−6.314	−3.078	−1.963	−1.376	−1.000	−0.727	−0.510	−0.325	−0.158	0
2	−2.920	−1.886	−1.386	−1.061	−0.816	−0.617	−0.445	−0.289	−0.142	0
3	−2.353	−1.638	−1.250	−0.978	−0.765	−0.584	−0.424	−0.277	−0.137	0
4	−2.132	−1.533	−1.190	−0.941	−0.741	−0.569	−0.414	−0.271	−0.134	0
5	−2.015	−1.476	−1.156	−0.920	−0.727	−0.559	−0.408	−0.267	−0.132	0

	Cumulative probability								
df	*0.55*	*0.6*	*0.65*	*0.7*	*0.75*	*0.8*	*0.85*	*0.9*	*0.95*
1	0.158	0.325	0.510	0.727	1.000	1.376	1.963	3.078	6.314
2	0.142	0.289	0.445	0.617	0.816	1.061	1.386	1.886	2.920
3	0.137	0.277	0.424	0.584	0.765	0.978	1.250	1.638	2.353
4	0.134	0.271	0.414	0.569	0.741	0.941	1.190	1.533	2.132
5	0.132	0.267	0.408	0.559	0.727	0.920	1.156	1.476	2.015

Table 9.2b Values of χ^2 for the chi-squared distribution

df	*Cumulative probability* 0.05	0.1	0.15	0.2	0.25	0.3	0.35	0.4	0.45	0.5
1	0.004	0.016	0.036	0.064	0.102	0.148	0.206	0.275	0.357	0.455
2	0.103	0.211	0.325	0.446	0.575	0.713	0.862	1.022	1.196	1.386
3	0.352	0.584	0.798	1.005	1.213	1.424	1.642	1.869	2.109	2.366
4	0.711	1.064	1.366	1.649	1.923	2.195	2.470	2.753	3.047	3.357
5	1.145	1.610	1.994	2.343	2.675	3.000	3.325	3.656	3.996	4.351

df	*Cumulative probability* 0.55	0.6	0.65	0.7	0.75	0.8	0.85	0.9	0.95
1	0.571	0.708	0.873	1.074	1.323	1.642	2.072	2.706	3.841
2	1.597	1.833	2.100	2.408	2.773	3.219	3.794	4.605	5.991
3	2.643	2.946	3.283	3.665	4.108	4.642	5.317	6.251	7.815
4	3.687	4.045	4.438	4.878	5.385	5.989	6.745	7.779	9.488
5	4.728	5.132	5.573	6.064	6.626	7.289	8.115	9.236	11.070

$$t = \mathrm{IF}(\wp_C > 0.5,1,-1)*$$
$$\mathrm{TINV}(\mathrm{IF}\wp_C > 0.5,2*(1-\wp_C),2*\wp_C),\nu_P)$$

which includes logical IF statements to overcome the fact that the TINV function will not accept a value for $\wp_C$ less than 0.5. The formula for χ^2 is given by:

$$\chi^2 = \mathrm{CHIINV}(1-\wp_C,\nu_P)$$

Column 8 lists the simulated range in uncertainty in the baseline loss function which was computed from the relation $\Lambda_S(q)=\sigma_S^2(q)/\mu_S^2(q)$. This process was repeated for the alternative in Table 9.4.

9.2.5 INCORPORATING COST DIFFERENCES INTO THE DISTRIBUTION

The difference between the two loss functions (alternative loss less baseline loss) in column 8 of Tables 9.3 and 9.4 is shown in Table 9.5. The differences multiplied by 1000 are plotted on normal probability paper (Appendix A) in Figure 9.5. The intersection of the cumulative distribution function with the line for a zero loss difference shows that only 81% confidence can be placed in assuming that the loss for the alternative is less than the loss for the baseline. However, if the alternative represented a reduction in variable cost of $\delta C = 8$ relative to the baseline, that would shift the distribution uniformly and favorably to the left by 8 units as shown. This increases the confidence to over 95% in the alternative's ability to reduce the loss function. If an added investment of M dollars over Y_{RS} years were required for tooling and facilities to implement the alternative, this would shift the distribution uniformly and unfavorably to the right by M/Y_{RS}.

It is important to note that although the error was small, only 3 in 10 000, for assuming that the attribute average for the alternative was greater than that for the baseline, the strategic error in assuming that the loss for the alternative was

Table 9.3 Baseline trial Monte Carlo simulations

1 *Simulation*	*2* RAND()	*3* t	*4* μ	*5* RAND()	*6* χ^2	*7* σ^2	*8* σ^2/μ^2
1	0.688	0.498	7.989	0.809	22.998	0.704	0.0110
2	0.710	0.564	8.009	0.101	10.900	1.486	0.0232
3	0.877	1.199	8.200	0.566	18.330	0.884	0.0131
4	0.695	0.520	7.996	0.850	24.147	0.671	0.0105
5	0.658	0.414	7.964	0.440	16.472	0.984	0.0155
6	0.193	−0.890	7.573	0.715	20.893	0.775	0.0135
7	0.611	0.287	7.926	0.960	29.729	0.545	0.0087
8	0.680	0.476	7.983	0.703	20.650	0.785	0.0123
9	0.328	−0.452	7.704	0.080	10.334	1.568	0.0264
10	0.837	1.011	8.143	0.559	18.231	0.889	0.0134
11	0.368	−0.342	7.737	0.566	18.339	0.883	0.0148
12	0.318	−0.483	7.695	0.408	16.012	1.012	0.0171
13	0.860	1.113	8.174	0.308	14.562	1.113	0.0167
14	0.020	−2.204	7.179	0.966	30.325	0.534	0.0104
15	0.453	−0.121	7.804	0.318	14.700	1.102	0.0181
16	0.588	0.226	7.908	0.764	21.908	0.739	0.0118
17	0.982	2.258	8.517	0.343	15.078	1.074	0.0148
18	0.428	−0.184	7.785	0.492	17.228	0.940	0.0155
19	0.314	−0.493	7.692	0.027	8.374	1.934	0.0327
20	0.757	0.711	8.053	0.026	8.315	1.948	0.0300

less than baseline loss was much higher, being approximately 20% based upon the intercept at the origin (zero) in Figure 9.5 of the distribution without the cost reduction.

Table 9.4 Alternative trial Monte Carlo simulations

1 Simulation	*2* RAND()	*3* t	*4* μ	*5* RAND()	*6* χ^2	*7* σ^2	*8* σ^2/μ^2
1	0.163	−1.011	9.307	0.511	17.496	0.926	0.0107
2	0.673	0.455	9.747	0.128	11.494	1.409	0.0148
3	0.167	−0.993	9.312	0.943	28.325	0.572	0.0066
4	0.809	0.897	9.879	0.744	21.470	0.755	0.0077
5	0.393	−0.276	9.527	0.489	17.181	0.943	0.0104
6	0.684	0.488	9.756	0.912	26.551	0.610	0.0064
7	0.391	−0.281	9.526	0.848	24.093	0.672	0.0074
8	0.061	−1.620	9.124	0.272	14.014	1.159	0.0139
9	0.086	−1.423	9.183	0.988	34.167	0.474	0.0056
10	0.113	−1.253	9.234	0.269	13.977	1.159	0.0139
11	0.347	−0.398	9.490	0.175	12.408	1.306	0.0145
12	0.801	0.867	9.870	0.205	12.946	1.251	0.0128
13	0.219	−0.794	9.372	0.848	24.104	0.672	0.0077
14	0.543	0.110	9.643	0.957	29.494	0.549	0.0059
15	0.812	0.908	9.882	0.231	13.365	1.212	0.0124
16	0.915	1.432	10.040	0.263	13.879	1.167	0.0116
17	0.593	0.238	9.681	0.900	25.983	0.623	0.0067
18	0.156	−1.039	9.298	0.597	18.827	0.860	0.0100
19	0.570	0.178	9.663	0.874	24.957	0.649	0.0070
20	0.341	−0.416	9.485	0.626	19.290	0.840	0.0093

Table 9.5 Simulated alternative loss minus baseline loss

Simulation	*Loss difference*	*Simulation*	*Loss difference*
1	−0.00035	11	−0.00026
2	−0.00833	12	−0.00424
3	−0.00655	13	−0.00900
4	−0.00276	14	−0.00446
5	−0.00512	15	−0.00568
6	−0.00711	16	−0.00025
7	−0.00126	17	−0.00816
8	0.00157	18	−0.00556
9	−0.02079	19	−0.02574
10	0.00019	20	−0.02071

Another benefit of simulation is pedagogical because it illustrates, as seen in Figure 9.5, the scatter in the resulting distribution function. Moreover, the uncertainties in other parameters in the profit or loss function equations – costs, interest rates, etc. – can be handled with relative ease with the simulation methodology if needed. 'What-if' analyses of different cost and investment scenarios can be easily visualized and understood as uniform shifts in the distribution. The downside in using Monte Carlo techniques is that 'canned' programs are generally not available; however, the use of macros in the spreadsheet calculations reduces the time to complete an analysis to a very acceptable level.

9.3 Type I and Type II errors in experimental design

In choosing between an alternative and the baseline, the process is generally biased, as already discussed, to minimize the occurrence of a mistake in which an alternative is selected for development that either offers no improvement over the baseline, or worse, represents a reduction in performance. By having such a bias, however, mistakes will be made of not accepting an alternative which, in fact, would prove to be an improvement if implemented. The first kind of error, which the process is biased to avoid, is called a Type I error. The second, which we often accept, is called a Type II error (Appendix A). The amount of Type I error defines the **significance level** of the experimental design.

9.3.1 DEMONSTRATION OF TYPE I AND TYPE II ERRORS

Consider the results shown in Table 9.6 for an $L_8(2^7)$ experimental design (see Table B.16 for the **[X]** array) having three simulated replications per trial. The λ_{ij} coefficients shown in the last column were determined from the relation:

$$[\lambda]=[\mathbf{X}]^{-1}[\bar{g}]$$

the set point indices ij being shown in the next to last column. The replications shown in Table 9.6 were generated by computer simulation using the expression:

$$g(q,y)=\mu(q)+\zeta$$

where ζ is the standard normal random variable (SNRV) which has a population variance of unity and population mean of zero (Equation A.25). The mean $\mu(q)$ was chosen to be 10 for trials 1, 2, 3, and 4 where set point 11 is 'off' for the $L_8(2^7)$ experimental design and was chosen to be 12 for trials 5, 6, 7, and 8 where set point 11 is 'on.' With these choices for $\mu(q)$, the other six off-baseline coefficients would contribute nothing on average to the outcome if the simulation were repeated a large number of times. The results for these coefficients shown in Table 9.6 are nothing more than experimental error or noise. However, we should find λ_0 and λ_{11} being approximately ten and two, respectively, on averaging a large number of repeated simulations.

Table 9.6 Results for $L_8(2^7)$ experimental design in which only off-baseline set point 11 contributed to the simulated outcomes

Trial	*Replications* 1	2	3	*Var*	*Ave*	*ij*	λ
1	11.93	11.79	9.766	1.465	11.16	0	11.16
2	8.775	11.95	9.658	2.682	10.13	11	1.823
3	8.759	10.13	9.781	0.509	9.557	21	−0.24
4	10.82	10.94	11.55	0.153	11.1	31	−0.08
5	13.06	12.18	11.67	0.492	12.3	41	0.018
6	12.6	12.1	12.72	0.11	12.47	51	0.236
7	11.74	12.97	12.89	0.475	12.53	61	−0.45
8	12.91	12.18	10.69	1.282	11.93	71	−0.84
				Pooled Var = 0.896			

We know from the above simulation model that the samples truly came from normally distributed populations and that no set point has a significant effect on variance. If we did not know this, we could easily show that this was likely to be the case using the procedures to be developed in what follows in this chapter. Consequently, we assume that the trial variances can be represented by a pooled sample variance of 0.896 having 16 df. Note, as stated earlier, this is not the same as saying that the trial variances are equal to each other.

We now want to determine the significance of the λ_{ij} coefficients. This problem is related to our determination earlier of the significance of set point 11 on the average for the experiment shown in Table 9.1. The difference here is that we are dealing with eight trials and seven off-baseline set points, whereas in the paired comparison of Table 9.1 we had only two trials and one off-baseline set point. The added set points increase the opportunities seven-fold for a statistical fluctuation to generate a λ_{ij} coefficient large enough for us to assume it was significant if we based our test of significance on the usual pairwise statistics for t as in Table 9.2a or as given by the spreadsheet function TINV.

We can repeat the computer simulation experiment used to generate the results in Table 9.6 many times to generate distributions for each of the λ_{ij} coefficients. The points shown in Figure 9.6 are for 100 such simulations. The results for the λ_{11} coefficients are plotted as a separate distribution. But the other six coefficients, which are zero on average, were combined into a single cumulative distribution having 600 points total. Using the t statistic (Equation B.27) we can define a critical region limit λ_{CRIT} given by:

$$\lambda_{CRIT} = \frac{t_\alpha s_P}{\sqrt{n}} \sqrt{\mathrm{cov}_{ijij}} \tag{9.3}$$

where $t_\alpha = \mathrm{TINV}(2\alpha, \nu_P)$ is the value of the t statistic for the desired single-tail Type I error level α. If we take $\alpha = 0.01$, we find that $t_\alpha = 2.58$ for 16 df. The term cov_{ijij} is the element on the diagonal of the covariance matrix, $[X^TX]^{-1}$, for set point ij. Substituting 0.896 for the pooled variance, and with $n = 3$ and $cov_{ijij} = 0.5$, we find that $\lambda_{CRIT} = 0.998$.

The critical region line in Figure 9.6 is seen to intersect the line for the combined coefficients at the 99% level, exactly where it should for the pairwise distribution. The number of points in the critical region is equal to the number of Type I errors made during the 100 simulations. The error rate per data point is 1% which is the expected pairwise error rate. But the error rate of importance, the Type I error, is the number of errors per experiment (i.e. per decision) which is 6%.

9.3.2 EXPERIMENT-WISE ERROR

Thus to keep the Type I error rate per decision at 1%, we need to move the single tail boundary of the critical region out to α_{Pair}/x_{OB} where α_{Pair} is the pairwise error rate (1%) and x_{OB} is the number of off-baseline coefficients for the particular experimental design. This procedure for converting the Type I pairwise error to an 'experiment-wise' Type I error is due to Bonferroni [4]. The pairwise errors associated with a specific measurement are readily determined from the appropriate statistical table or spreadsheet statistical function. In testing for sig-

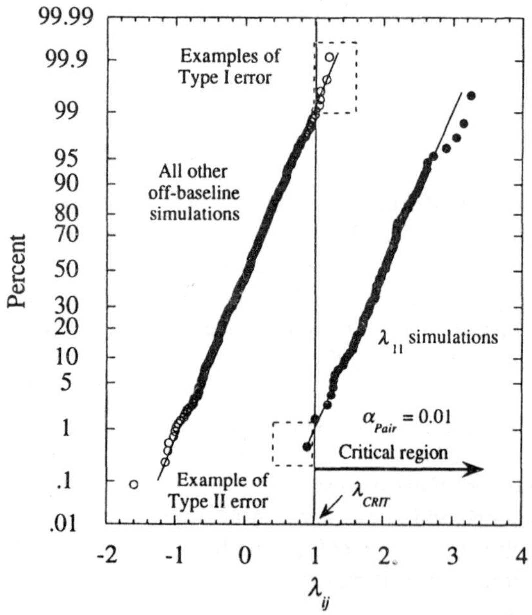

Figure 9.6 Normal probability plots of the seven off-baseline lambda coefficients λ_{ij} determined from 100 simulations of an experiment in which only the coefficient λ_{11} contributed to the outcome. The results for the other six coefficients are plotted together as a single line.

nificance, we will use these to first compute the pairwise error and then multiply it by the number of off-baseline coefficients, x_{OB}, to arrive at the experiment-wise error:

$$\alpha = \alpha_{Pair} x_{OB} \tag{9.4}$$

The relationships between the pairwise and the experiment-wise errors in the presence of 1, 3, 7, and 15 unknown off-baseline coefficients are shown in Table 9.7. The differences can be quite large and should always be accounted for.

A legitimate question at this point is, 'Because the Type I error increases with additional unknowns, why ever consider more than two factors in an experiment?' In answering this question the use of an example is helpful: suppose you wish to evaluate the effects of fifteen separate factors on a specific attribute. You might choose to make fifteen, separate paired comparisons or you might choose to do a combined, single experiment having sixteen trials to evaluate the factors using a one-at-a-time experimental design (Table B.17). The same number of Type I errors (on average) will be made either way if the same critical region is used. However, a combined experiment has two advantages: (1) a more efficient experimental design can be used versus one-at-a-time (an orthogonal array, for example) and (2) an analysis of the sample variances may show that they can be pooled either partially or totally which increases the df. Moreover, by using the combined experiment, you evaluate an agreed-upon possible set of solutions to the problem in concert and can weigh them against each other and by doing so your chances of finding a very good solution to the problem increase. But you still want to have a low Type I error rate per decision, so you make the experiment-wise correction to achieve the level of confidence traditionally used for a decision between a baseline and a single alternative.

9.4 Steps in single-attribute significance testing

The steps in statistical and strategic significance testing explored with the NIB1 problem described in Table 9.1 can be extended to the more general problem of an experimental design having multiple control factors. (Note that multiple

Table 9.7 Relationships between the pairwise and experiment-wise Type I errors

	α			
α_{Pair}	$x_{OB} = 15$	$x_{OB} = 7$	$x_{OB} = 3$	$x_{OB} = 1$
0.001	0.015	0.007	0.003	0.001
0.010	0.150	0.070	0.030	0.010
0.050	0.750	0.350	0.15	0.050
0.100	1.5	0.700	0.300	0.100

control factors or set points should not be confused with multiple attributes.) The same steps for the pair comparison problem appear but the details differ as outlined below for a single attribute problem. An example is considered afterwards to illuminate the process.

Step 1 Test for normal distribution

The statistical tests of significance for $H1_0$ and $H2_0$ are based upon normally distributed variables. Because the basic variables for SIB and LIB problems can never be negative, the logarithms of the basic variables taken here as the natural logs:

$$\ell(q,y) \equiv \begin{cases} \text{SIB: Ln } (g\,(q,y)) \\ \text{LIB: Ln } (\,1/g\,(q,y)) \end{cases} \tag{9.5}$$

are expected to follow the normal distribution. The basic variables $g\,(q,y)$ and $1/g\,(q,y)$ are said to be **lognormally distributed**. The tests for the influence of the set points on variance and the average for SIB and LIB problems are made therefore using the Ln transformed variables defined by Equation 9.5 (Appendix A). But for NIB problems, it is the basic variable itself which is taken as being normally distributed unless known to be otherwise. The check for normality can be made by plotting the replications (or the Ln transformed replications) for each trial on normal probability paper and seeing how well they fit to a straight line. However, when the sample size is small, it may be difficult to assess whether or not the normality condition is satisfied.

Step 2 Compute the set point coefficients for average and variance

For NIB, the averages, $\bar{g}(q)$, and variances, $s_g^2(q)$, are computed for each trial. For SIB and LIB, the average $\bar{\ell}(q)$ and variance $s_\ell^2(q)$ of the Ln transformed variables are computed. Because the variance can never be less than zero, the logarithm of the variance is taken to improve additivity, the variables being primed when their logarithm has been taken. The λ_{ij} and ξ'_{ij} coefficients for the sample averages and −10 log variance, respectively, are then computed from these quantities using the expressions:

$$[\lambda] = \begin{cases} [\mathbf{X}^T\mathbf{X}]^{-1} \quad [\mathbf{X}^T] \quad [\bar{g}] \quad \text{NIB} \\ [\mathbf{X}^T\mathbf{X}]^{-1} \quad [\mathbf{X}^T] \quad [\bar{\ell}] \quad \text{SIB \& LIB} \end{cases} \tag{9.6}$$

$$[\xi'] = \begin{cases} [\mathbf{X}^T\mathbf{X}]^{-1} \quad [\mathbf{X}^T] \quad [-10\log_{10} s_g^2] \quad \text{NIB} \\ [\mathbf{X}^T\mathbf{X}]^{-1} \quad [\mathbf{X}^T] \quad [-10\log_{10} s_\ell^2] \quad \text{SIB \& LIB} \end{cases} \tag{9.7}$$

Step 3 Significance of set points on variance

The pairwise error of the measured off-baseline ξ'_{ij} coefficients for the $H1_0$ null hypothesis that the sample variances had a common population variance is determined by first computing the F_{ij} ratio for each set point given by:

$$F_{ij} \equiv 10^{|G_{ij}|} \tag{9.8}$$

where the parameter whose absolute value is taken in the exponent in Equation 9.8 is given by:

$$G_{ij} = \frac{\xi'_{ij}}{\sqrt{50\,\mathrm{cov}_{ijij}}} \tag{9.9}$$

The pairwise error can be looked up in an F table or calculated from the spreadsheet function FDIST(F_{ij},ν,ν) where ν is the (unpooled) trial degrees of freedom $n - 1$ (assumed constant for all of the trials). The experiment-wise Type I error is then calculated by multiplying the pairwise error by the number of off-baseline coefficients. The derivations of Equations 9.8 and 9.9 are given in Appendix A. For the F_{ij} ratio to be used, the distribution for the ξ'_{ij} coefficients must be symmetric about zero. This will always be true for any $[\mathbf{X}]_\lambda$ array and for two-level, orthogonal $[\mathbf{X}]_\varphi$ arrays. When this condition is not satisfied, the full χ^2 distribution should be evaluated, the results of which are not included here.

When none of the set points are found to be significant for variance, the trial variances can, following normal practice, be averaged to form the pooled variance:

$$s_P^2 = \frac{1}{k}\sum_{q=1}^{k} s^2(q)$$

having $\nu_P = k[n-1]$ df. When one or more set points are found to be significant for variance, a partial pooling of trial variances based upon the resulting pooling array can usually be made to yield the partially pooled trial variances, $s_P^2(q)$, and partially pooled trial df, $\nu_P(q)$, as described in Case Studies 2–4.

Because it desirable to guard against making a change that will increase variance, different Type I error limits can be used for the critical regions denoting favorable ($\xi'_{ij} > 0$) and unfavorable ($\xi'_{ij} < 0$) variance impact. For example, the critical region for a set point taken as being favorable on variance could be set at a Type I experiment-wise error of 1%, whereas the Type I experiment-wise error level for a set point found unfavorable on variance (increases variance) could be taken as 10% with the pooling array constructed accordingly. This decreases the chances that such potentially unfavorable set points on variance would be used because it raises the hurdle on the amount by which such set points must favorably influence the average to offset their potential downside on variance. (For simplicity, the Type I error levels for favorable and unfavorable set points on variance are taken to be the same in the case studies and illustrations which follow but it would be more conservative not to do this.)

Step 4 Significance of set points on average

In considering the $H2_0$ null hypothesis of a common population average for NIB, SIB, or LIB, the pairwise error for the λ_{ij} coefficients can be computed from the approximate t statistic (Appendix B) when the trial variances can only be partially pooled:

$$t_{ij} = \frac{\lambda_{ij}}{\sqrt{s_{ij}^2 / n}} \tag{9.10a}$$

The effective sample variance, s_{ij}^2, and effective df, ν_{ij}, are given by Equations B.29 and B.30, respectively, the trial variances in the expressions being replaced by $s_P^2(q)$. The pairwise error is computed using the function TDIST($|t_{ij}|, \nu_{ij}$,1). The experiment-wise error is calculated by multiplying the pairwise error by the number of off-baseline coefficients. When none of the set points are found to be significant for influencing variance and the assumption of a common population variance is then made, Equation 9.10a becomes a rigorous t statistic given by:

$$t_{ij} = \frac{\lambda_{ij}\sqrt{n}}{\sqrt{s_P^2 \operatorname{cov}_{ijij}}} \tag{9.10b}$$

for $\nu_P = k[n-1]$ pooled df, the pairwise error now being given by TDIST($|t_{ij}|, \nu_P$,1) which is converted into the experiment-wise error in the same manner as above.

Step 5 Determine the best combination of set points

The set point combinations found significant and favorable for variance and/or average now need to be evaluated to see which should, on average, generate a reduction in loss with high confidence. For the NIB2 problem the best combination of set points can only be found by considering all possible combinations; an example of this situation is covered in Case Study 2. For SIB, NIB1, and LIB, the set points which are favorable in reducing the loss with high confidence when considered individually will always represent the best combination of set points for reducing the loss function.

Step 6 Test the strategic significance of the best set point combination

The best set points determined from Step 5 were derived from their average responses. Their strategic significance must be determined next by finding the degree of confidence that can be placed in each set point ij to generate a positive S/N ratio, $\lambda'_{ij}(\Omega)$. This is found using computer simulation to generate the $\lambda'_{ij}(\Omega)$ scatter distribution using the relation:

$$[\lambda'(\Omega)] = [\mathbf{X}^{\mathrm{T}}\mathbf{X}]^{-1}[\mathbf{X}^{\mathrm{T}}][-10\log(\Omega_S(q))] \tag{9.11}$$

The simulated $\Omega_S(q)$ in Equation 9.11 for NIB1 problems is determined from the NIB1 expression in Table 8.2 using simulated population averages and variances given by:

$$\mu_S(q) = \bar{g}(q) + t_S\sqrt{\frac{s^2(q)}{n}} \tag{9.12}$$

$$\sigma_S^2(q) = \frac{\nu s^2(q)}{\chi_S^2} \tag{9.13}$$

Points for simulating the averages and variances in the above expressions are randomly selected from the t and χ^2 distributions using Equations A.22 and A.23, respectively. The trial variances and df in the above expressions and in Equations A.22 and A.23 should be replaced by s_P^2 and $\nu_P(q)$, respectively, if they can be partially pooled or by s_P^2 and $\nu_P = k[n - 1]$ if they can be fully pooled.

For SIB and LIB, the average $\bar{\ell}(q)$ and variance s_ℓ^2 of the Ln transformed variables are used in Equations 9.12 and 9.13 to simulate the range expected for the population mean $\mu_l(q)$ and population variance $\sigma_\ell^2(q)$, respectively. These statistics are then transformed back to compute the simulated CIQ using the expression from Appendix A given by:

$$\Omega_S(q) = k \exp(2\mu_\ell(q) + 2\sigma_\ell^2(q)) \tag{9.14}$$

One hundred or more computer simulations should be made.

The CIQ loss for the best combination of set points and the loss for the production baseline, which, in general, can be different from one of the specific trials, need to be computed for each simulation using the expressions:

$$\Omega_{Best} = 10^{-[\lambda_0'(\Omega) + \sum_{(ij)} \lambda_{(ij)}'(\Omega)]/10} \tag{9.15}$$

where the summation is over the set points found as best and

$$\Omega_{ProdBase} = 10^{-[\lambda_0'(\Omega) + \sum_{(ij)} \lambda_{(PB)}'(\Omega)]/10} \tag{9.16}$$

where the summation is over the set points that define the production baseline. Finally, the difference given by:

$$\Delta\Omega \equiv [\Omega_{Best} - \Omega_{ProdBase}] \tag{9.17}$$

is determined for each of the simulations and the resulting distribution is plotted on normal probability paper. The average of the $\Delta\Omega$ distribution should be negative for the set point combination to be favorable. The value of the cumulative distribution where it intersects the line for $\Delta\Omega = 0$ gives the degree of confidence in assuming that the mean of the loss function for the best combination of set points lies below the mean of the production baseline.

The steps for determining the strategic significance of the ξ_{ij}' variance coefficients and the λ_{ij} average coefficients for the SIB condition are shown in Figures 9.7 and 9.8, respectively. These steps are illustrated in the NIB1 problem which follows and in Case Studies 1–4.

The steps for LIB are the same except that the physical scale transformation from g to $1/g$ must be made first. The steps for NIB1 and many for NIB2 are the

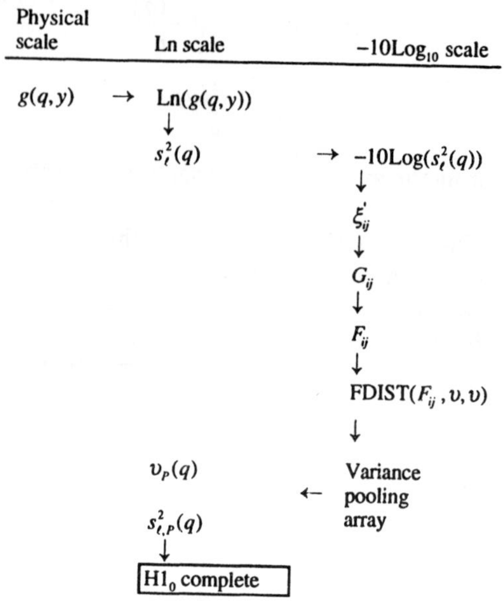

Figure 9.7 The sequence of steps for determining the significance of the ξ'_{ij} SIB variance coefficients.

same as for SIB except that all of the steps that appear under the Ln scale are made instead with the physical scale.

9.5 NIB1 example from Chapter 8

Let's now use the six-step process described above to complete our analysis of the NIB1 problem considered at the end of Chapter 8. Assume that we have

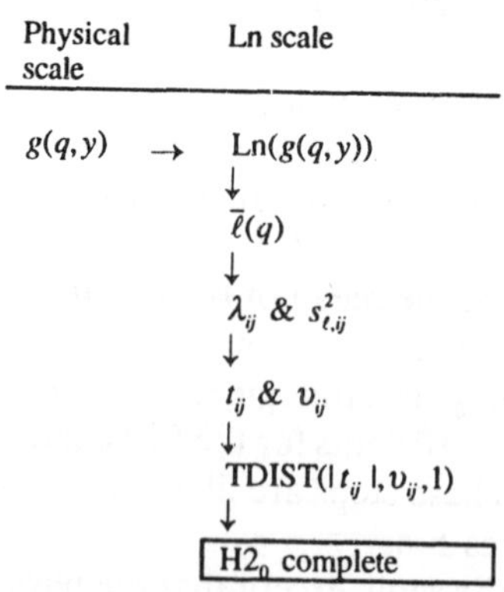

Figure 9.8 The sequence of steps for determining the significance of the λ_{ij} SIB average coefficients.

completed step 1 by plotting the six replications for each trial on normal probability paper and have found that the plots are sufficiently linear to assume that the samples were taken from normally distributed populations. The set point coefficients for averages and variances, step 2, are shown in Figures 9.9 and 9.10, respectively. Figure 9.9 also includes step 3, the testing of the ξ'_{ij} variance coefficients for significance.

No set point is seen to be highly significant for influencing variance based upon the experiment-wise errors. It is important to again emphasize that the use of the F_{ij} ratio test based upon Equation 9.8 is only correct if using an $[\mathbf{X}]_{\varphi}$ orthogonal array when all of the factors have two levels. The spreadsheet expressions used to compute the results appearing in Figure 9.9 are given in Table 9.8.

Because none of the set points were found to have a significant effect on variance, all four of the trial variances were pooled (averaged) to arrive at a pooled variance of 3.84 which was computed in cell O31 of the spreadsheet. Step 3, the testing of the φ_{ij} average coefficients for significance, is also included in Figure 9.10.

Equation 9.10b was used to compute the φ_{ij} coefficients for the sample averages. Each element on the diagonal of the covariance matrix was equal to 0.25 which was computed in cell G16. As was the case of the variance coefficients, no set point is seen to have a significant effect on the average based upon the experiment-wise errors. This, of course, was what was expected as the replications for each trial were selected from populations having the same average and variance. The spreadsheet equations used to compute the results appearing in Figure 9.10 are given in Table 9.9.

Steps 5 and 6 could be skipped for this problem because no set point could have a significant effect on the CIQ as no set point was found to be significant for variance or average. It is important, however, to carry out step 7, to quantify the uncertainty in the CIQ S/N ratio coefficients as displayed by the range of the $\varphi'_{ij}(\Omega_S)$ coefficients given by the simulations. The computations used to generate the simulations are shown in Figure 9.11.

The spreadsheet equations used in Figure 9.11 are given in Table 9.10. A single set of simulations is shown in the last column of Figure 9.11, column AC. Each simulation of these four coefficients was copied and stored to the right of column AC using the 'Copy' and 'Paste Special' commands in the spreadsheet pull down menu. The commands for copying and pasting were recorded as a macro in the spreadsheet so that all that was required to perform 100 simula-

	N	O	P	Q	R	S
36						Exp.
37		ξ'_{ij}	G_{ij}	F_{ij}	Pairwise	-wise
38	ij				error	error
39	0	-5.12	-1.45			
40	11	-1.10	-0.31	2.05	0.23	0.68
41	21	2.35	0.66	4.61	0.06	0.18
42	31	0.94	0.27	1.84	0.26	0.78

Figure 9.9 Calculations used to determine the significance of the ξ'_{ij} variance coefficients.

	U	V	W	X	Y
36					Exp.
37				Pairwise	-wise
38	ij	φ_{ij}	t_{ij}	error	error
39	0	40.163			
40	11	-0.083	-0.208	0.42	1.26
41	21	-0.680	-1.699	0.05	0.16
42	31	-0.042	-0.106	0.46	1.37

Figure 9.10 Calculations used to determine the significance of the φ_{ij} average coefficients.

	R	S	T	U	V	W	X	Y	Z	AB	AC
54											
55	Trial	RAND()	χ^2	σ_S^2	RAND()	t	μ_S	Ω_S	$-10\log\Omega_S$	ij	$\varphi_{ij}(\Omega_S)$
56	1	0.23	15.11	5.08	0.36	-0.36	40.68	15.36	-11.86	0	-12.64
57	2	0.33	16.74	4.59	0.21	-0.82	38.87	15.19	-11.82	11	-0.80
58	3	0.15	13.60	5.65	0.57	0.18	40.86	16.91	-12.28	21	-0.57
59	4	0.01	8.26	9.30	0.80	0.86	40.13	28.87	-14.60	31	0.59

Figure 9.11 Calculations used to develop the Monte Carlo simulations.

tions was to hit key 'h' assigned to the macro 100 times while holding down the two keys of OPTION-COMMAND on the Macintosh® computer. The macro, entitled SimNPaste, is shown in Figure 9.12.

The 100 simulations consisting of 100 columns of four rows each were placed to the right of column AC. These were then copied and pasted into the data spreadsheet of the graphics program, transposed into four columns of 100 rows, and graphed as a normal probability plot. The baseline coefficient simulations were not plotted.

Table 9.8 Spreadsheet equations for results in Figure 9.9

Location	*Spreadsheet equation*	*Type*
O39:O42	=MMULT(MINVERSE($D$10:$G$13),–10*LOG(O26:O29))	Area
P39	=O39/SQRT(50*G16)	Drag
Q40	=10^ABS(P40)	Drag
R40	=FDIST(Q40,5,5)	Drag
S40	=3*R40	Drag

Table 9.9 Spreadsheet equations for results in Figure 9.10

Location	*Spreadsheet equation*	*Type*
V39:V42	=MMULT(MINVERSE($D$10:$G$13),P26:P29)	Area
W40	=V40*SQRT(6)/SQRT($G$16*$O$31)	Drag
X40	=TDIST(ABS(W40),4*5,1)	Drag
Y40	=3*X40	Drag

Table 9.10 Spreadsheet equations for results in Figure 9.11

Location	*Spreadsheet equation*	*Type*
S56	=RAND()	Drag
T56	=CHIINV(1-S56,4*5)	Drag
U56	=(5*4)*$O$31/T56	Drag
V56	=RAND()	Drag
W56	=IF(V56>0.5,1,-1)*TINV(IF(V56>0.5,2*(1-V56),2*V56),4*5)	Drag
X56	=P26+W56*SQRT($O$31/6)	Drag
Y56	=$F$2*$F$3^2*U56/X56^2	Drag
Z56	=-10*LOG(Y56)	Drag
AC56:AC59	=MMULT(MINVERSE($D$10:$G$13),Z56:Z59)	Area

The range found for 100 simulations of the three off-baseline coefficients are shown in Figure 9.13. The uncertainty in the CIQ arises from the uncertainty in the population variance and population average given a pooled sample variance of 3.84 with 20 pooled df and the measured sample averages for $n = 6$. From Figure 9.13, we expect, at better than 99% confidence, that the $\varphi_{ij}'(\Omega)$ coefficients will lie between ±2 dB about zero. This may appear at first to be in conflict with the value of 2.20 in Figure 8.20 for set point $\varphi_{21}'(\Omega)$ which would be significant at an error level of 0.1% if we plotted it in Figure 9.13 and multiplying by three to arrive at the experiment-wise error would only increase it to 0.3%. This paradox is resolved by first noting that the value of 2.20 was computed before examining the impact of the set points on variance and average which clearly showed that no set point should be considered as having a significant effect on the CIQ. The distribution simulated in Figure 9.13 was made after variances were pooled and the df increased from 5 to 20 which increases the slope of the line in Figure 9.13 versus 5 df. If the simulations were computed with 5 df, then set point 21 would not be seen as significant. If one of the set points had been deemed as significant for variance, for example, then its distribution would be offset from the line shown in Figure 9.13 but the resulting line

```
' SimNPaste Macro
' Macro recorded 7/1/95 by harry cook
'
' Keyboard Shortcut: Option+Cmd+h
'
Sub SimNPaste()
   Selection.Copy
   ActiveCell.Offset(0, 2).Range("A1").Select
   Selection.PasteSpecial Paste:=xlValues, Operation:=xlNone, _
      SkipBlanks:=False, Transpose:=False
   Application.CutCopyMode = False
   Selection.Insert Shift:=xlToRight
   ActiveCell.Offset(0, -2).Range("A1:A40").Select
   Calculate
End Sub
```

Figure 9.12 SimNPaste macro.

for comparison would have had a lesser slope as the df would become 10 instead of 20 for this particular problem.

9.6 Unsaturated designs

When there are more trials, k, than the number, w, of unknown coefficients, the experiment is said to be 'unsaturated' and an estimate of the variance of the outcomes and thus the significance of the $\lambda'_{ij}(\Omega)$ or $\varphi'_{ij}(\Omega)$ coefficients can be obtained using the difference squared between the measured outcomes and the least square fits computed from the coefficients. The null hypothesis for the test is that a common population variance and average for the $\lambda'_{ij}(\Omega)$ or $\varphi'_{ij}(\Omega)$ coefficients exist. However, the testing of an additional main effect is given up for each column left unfilled which is wasteful according to followers of Taguchi's methodology. The assumption is also implied that the interactions which appear in the unfilled columns are weak.

If the average general outcome measured for trial q is $Y(q)$ and the least square fit model to this trial is $Y_L(q)$, then the null hypothesis (pooled) pure error of the outcomes is given by:

$$s_P^2 = \frac{1}{k-w}\sum_{q=1}^{k}[Y_L(q) - Y(q)]^2 \tag{9.18}$$

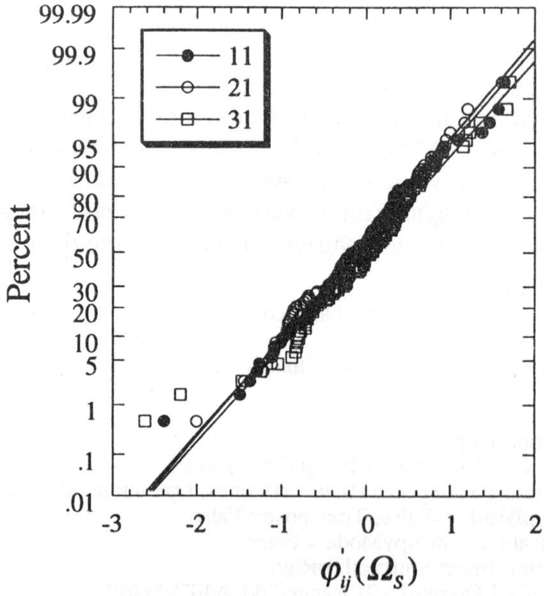

Figure 9.13 Uncertainty in the $\varphi'_{ij}(\Omega)$ coefficients based upon Monte Carlo simulations.

having k – w degrees of freedom where the values for the least square fits are given by:

$$Y_L(q) = \sum_{ij} X_{ij}(q)\lambda_{ij}(Y) \tag{9.19}$$

in which X_{ij} is the element of the $[\mathbf{X}]_\lambda$ array in column ij for trial q. Of course, a similar expression exists for the $\varphi_{ij}(Y)$ coefficients when using the $[\mathbf{X}]_\varphi$ array.

Application of Equation 9.18 to significance testing can be demonstrated using the NIB1 problem already considered above by assuming that no control factor was entered into column 31 in the $[\mathbf{X}]_\phi$ array of Figure 8.19. The resulting solution matrix for the reduced design matrix is shown in Figure 9.14 and is the same as for the full $L_4(2^3)$ except that the row for set point 31 is missing. There are four columns corresponding to the four trials and three rows corresponding to the set points 0, 11, and 21. The general approach to computing a solution matrix using a spreadsheet is to highlight the correct columns and rows and then enter the array expression given by:

```
=MMULT(MINVERSE(MMULT(TRANSPOSE($D$10:$F$13),
     $D$10:$F$13)), TRANSPOSE($D$10:$F$13))
```

The argument for the array is the design matrix. For this problem, it was the array in Figure 8.19 with column 31 dropped. It resided in the area $D$10:$F$13 in the spreadsheet used for the calculations.

The outcomes $Y(q)$ for the trials to be substituted into Equation 9.18 for this specific problem are the S/N ratios for the CIQ given by –10 log(Ω (q)). The computations for the significance of the $\varphi'_{ij}(\Omega)$ coefficients using this approach are shown in Figure 9.15.

The $\varphi'_{ij}(\Omega)$ coefficients are computed in column K. The model outcomes listed in column M were computed from Equation 9.19. The differences squared between the model and actual outcomes (from column Z in Figure 9.11) are given in column N. The variance of 3.46 listed in cell N75 at the bottom of column N represents the variance given by Equation 9.18. This variance is used in Equation 9.10b for $n = 1$ and 1 df to compute the values for t in column P. The pairwise error is then computed in column Q using the TDIST function in the spreadsheet. The experiment-wise error shown in column R is twice the pairwise error as there were only two off-baseline coefficients with set point 31 eliminated. The spreadsheet equations used in Figure 9.15 are given in Table 9.11.

	D	E	F	G	H	I
69			Trials (q)			
70	ij	1	2	3	4	cov_{ijij}
71	0	0.25	0.25	0.25	0.25	0.25
72	11	-0.25	-0.25	0.25	0.25	0.25
73	21	-0.25	0.25	-0.25	0.25	0.25

Figure 9.14 The solution matrix for the reduced $L_4(2^3)$ design matrix (column 31 not used for a main effect).

	J	K	L	M	N	O	P	Q	R
68				Model					Exp.
69					Error			Pairwise	-wise
70	ij	$\varphi'_{ij}(\Omega)$	Trial	$-10\log(\Omega)$	squared	ij	t	error	error
71	0	-12.25	1	-13.33	0.86	0			
72	11	-1.12	2	-8.93	0.86	11	-1.2	0.22	0.44
73	21	2.20	3	-15.56	0.86	21	2.364	0.13	0.25
74			4	-11.16	0.86				
75				Var=	3.46				

Figure 9.15 Evaluation of the significance of the $\varphi'_{ij}(\Omega)$ coefficients using the error determined from the difference squared between the measured S/N ratios for each trial and the S/N ratios computed from the least square fits (Equation 9.18).

Table 9.11 Spreadsheet equations for results in Figure 9.15

Location	*Spreadsheet equation*	*Type*
K71:K73	=MMULT(E71:H73,R26:R29)	Area
M71:M74	=MMULT(D10:F13,K71:K73	Area
N71	=(M71–R26)^2	Drag
P72	=K72*1/SQRT($N$75*172)	Drag
Q72	=TDIST(ABS(P72),1,1)	Drag
R72	=2*Q72	Drag

9.7 Saturated designs

When the design is saturated, a determination of significance can also be made for the $\lambda'_{ij}(\Omega)$ or $\varphi'_{ij}(\Omega)$ coefficients based upon the same null hypothesis as above. For the example considered here, we will formulate the problem in terms of the $\varphi'_{ij}(\Omega)$ coefficients which means that the average of all trials is used as the baseline. A normal distribution can be expected for the null hypothesis because of the central limit theorem and the fact that the $\varphi'_{ij}(\Omega)$ coefficients are determined from a linear combination of average quantities. The normal distribution is reduced to a t distribution because sample measurements must be used for the variance:

$$t_{ij} = \frac{\varphi'_{ij}(\Omega)}{\sqrt{s^2(\varphi'_{ij}(\Omega))}} \tag{9.20}$$

The mean of the $\varphi'_{ij}(\Omega)$ coefficient is not shown in Equation 9.20 because it is zero. The variance in the denominator is given by

$$s^2(\varphi'_{ij}(\Omega)) = \sum_q s^2(-10\log(\Omega(q)))\omega^2_{ij}(q)$$

For the null hypothesis, the population variance $\sigma^2(-10\log(\Omega(q)))$ of the S/N ratio is independent of q and we can thus replace the sample variances on the right-hand side of the above equation with their pooled variance: $s^2(-10\log(\Omega(q)))=s^2_{CP,ij}$ where $s^2_{CP,ij}$ is a complement pooled variance for set point ij. The ij notation distinction must be maintained because the estimate of this pooled sample variance depends upon the set point in question:

$$s^2_{CP,ij} \equiv \left(\frac{1}{w-2}\right)\sum_{\substack{kl \\ \neq ij}} \frac{[\varphi'_{kl}(\Omega)]^2}{\mathrm{cov}_{klkl}} \tag{9.21}$$

where w is the number of unknowns, $w-1$ being the number of off-baseline coefficients. The complement pooled variance is used because the independence between the average and the variance of the normally distributed variable would be violated if the square of the coefficient $\varphi'_{ij}(\Omega)$ being tested in the numerator also appeared in the denominator. On substituting the above result into Equation 9.20, we obtain the t statistic in the desired form:

$$t_{ij} = \frac{\varphi'_{ij}(\Omega)/\sqrt{\mathrm{cov}_{ijij}}}{\sqrt{s^2_{CP,ij}}} \tag{9.22}$$

having $w-2$ df. The pairwise error determined from the t statistic given by Equation 9.22 needs to be multiplied by $w-1$ to obtain the experiment-wise error. The computation of the complement pooled variance and the resulting t statistic are straightforward using the spreadsheet. When, for example, there are only three unknowns (there being only two off-baseline coefficients), it follows that square of the above statistic reduces to a traditional F ratio form given by $s^2_{CP,kl}/s^2_{CP,ij}$, with both the numerator and denominator having 1 df.

This method of significance testing is in keeping with the standard approach used by practitioners of Taguchi's robust design methodology in that it uses the S/N ratios directly without considering separately the significance of the set points on variance and average. However, it differs computationally thereby avoiding the bias [2, 3] (which appears for $w > 4$) and the resulting tendency toward Type I errors when the pooled variance is taken as the average of the sum of the squares of the smallest half or there about (in absolute value) of the $\varphi'_{ij}(\Omega)/\sqrt{\mathrm{cov}_{ijij}}$ coefficients. (This bias is clearly seen in the example below as each of the complement pooled variances is greater than the variance computed from the sum of the four smallest $\varphi'_{ij}(\Omega)/\sqrt{\mathrm{cov}_{ijij}}$ coefficients squared divided by four.)

To illustrate the use of Equation 9.22, consider the evaluation of a NIB1 problem using an $L_8(2^7)$ experimental design represented by an $[\mathbf{X}]_\varphi$ array such as the one shown in Figure 8.13 and saturated with main effects: 11, 21, ..., 71. For simplicity, assume that $k_T = 1$ and $g_I = 1$. The test case chosen was the true null hypothesis for a common normal population having a mean of ten and a variance of unity. A total of five replications were made for each trial, the results being shown in Figure 9.16.

	O	P	Q	R	S	T	U	V	W	X	Y
16				Replications							
17	Trials	1	2	3	4	5		Var	Ave	$E(\mu^2)$	$-10\log(\Omega)$
18	1	9.48	8.96	11.14	9.12	7.91		1.38	9.32	87.47	18.03
19	2	11.24	8.15	9.74	9.99	11.77		2.01	10.18	104.40	17.16
20	3	8.95	10.53	10.41	9.44	9.12		0.54	9.69	94.07	22.39
21	4	9.14	11.46	11.25	10.91	8.90		1.47	10.33	107.30	18.62
22	5	10.33	10.37	9.52	10.53	9.20		0.35	9.99	99.92	24.53
23	6	8.55	9.80	9.04	9.35	8.81		0.24	9.11	83.07	25.48
24	7	10.00	9.05	11.88	11.15	9.29		1.47	10.27	106.13	18.57
25	8	9.76	12.36	10.66	8.76	9.88		1.80	10.29	106.51	17.72

Figure 9.16 Simulation of the $L_8(2^7)$ experiment obeying the null hypothesis. This is used to illustrate tests of significance based upon the direct method for saturated designs. The samples were taken from a common population whose average was ten and whose variance was unity.

The expected value of the square of the mean in cell X18 was computed using the expression:

=W18^2+(4/(4-2))*V18/5

which was dragged down to compute the others in the column. The S/N ratio in cell Y18 was computed using the expression:

=-10*LOG(V18/X18)

which was then dragged down. (The variance used was simply the sample variance as multiplication by 4/(4 – 2) would not change the off-baseline coefficients for the NIB1 problem because they are computed by the taking of logarithms followed by computation of the differences between the S/N ratios.)

The resulting $\varphi'_{ij}(\Omega)$ coefficients are shown in Figure 9.17 along with the computations for arriving at the complement pooled variances and the significance of the coefficients. Each diagonal element of the covariance matrix is 1/8 for the $L_8(2^7)$ experimental design. The complement pooled variance in cell AG19 was computed using the expression:

=(7*$AF$26-AF19)/6

which was then dragged down to compute the others. (Cell AF26 contains the average of all of the terms in column AF from cell AF19 to cell AF25.) The coefficient for set point 31 is seen to yield a 7% pairwise error. When the error is small enough for a coefficient to be deemed significant, then this coefficient can be eliminated from the distribution and the process repeated with the next largest coefficient in absolute value.

If the significance of set point 31 were evaluated according to standard practice using $\varphi'_{31}(\Omega)$ squared in the numerator of the F statistic and the

	AC	AD	AE	AF	AG	AH	AI	AJ
15								Exp.
16			$\frac{\varphi'_{ij}(\Omega)}{\sqrt{\text{cov}_{ijij}}}$	$\left[\frac{\varphi'_{ij}(\Omega)}{\sqrt{\text{cov}_{ijij}}}\right]^2$			Pairwise	-wise
17	ij	$\varphi'_{ij}(\Omega)$			$s^2_{CP,ij}$	t_{ij}	error	error
18	0	20.31						
19	11	1.26	3.57	12.75	10.64	1.09	0.15781	1.10
20	21	-0.99	-2.79	7.79	11.47	-0.82	0.22070	1.54
21	31	2.44	6.91	47.75	4.81	3.15	0.00989	0.07
22	41	-0.57	-1.61	2.59	12.34	-0.46	0.33145	2.32
23	51	-0.59	-1.67	2.80	12.30	-0.48	0.32502	2.28
24	61	0.59	1.66	2.77	12.31	0.47	0.32613	2.28
25	71	0.14	0.39	0.15	12.74	0.11	0.45863	3.21
26			Average=	10.94				

Figure 9.17 Computations used in determining the significance of the $\varphi'_{ij}(\Omega)$ coefficients for the saturated $L_8(2^7)$ experiment obeying the null hypothesis.

pooled variance in the denominator taken as the average of the sum of the squares of the four smallest (in absolute value) of the other $\varphi'_{ij}(\Omega)$ coefficients, as described above, the error found would be 0.87% (or 0.43% for comparison to the single tail t test) and might be taken, incorrectly, as significant, reflecting the bias caused by choosing the four smallest terms in the denominator. A follow-on confirmation experiment, if made, would, hopefully, discover the error. (The F ratio computed using this approach also yields a pairwise error which should be multiplied by three in this case – three possibilities for the numerator – to arrive at an experiment-wise error of 2.6% or 1.3% for comparison to the single-tail t test. This correction is seldom made in practice.) In principle, there is no need for a confirmation experiment if the significance of the set points have been evaluated without a bias. Moreover, if the confirmation experiment is made with the same number of replications as used in the regular experiment, its outcome has no more meaning than the outcomes of the other trials. Thus, when the number of replications are the same, the row denoting the set point configuration of the confirmation trial should simply be added to the [X] array for the original experimental design and its outcome added to the original outcome vector with the coefficients and their significance reevaluated accordingly.

A major advantage of using Equation 9.22 for significance testing is that it can be used when there is only one measurement (one replication) made per trial. (The same is also true for the unsaturated design significance test process described above.) However, it gives little insight into how each set point affects the variance and average independently and, when two or more replications are made per trial, the six-step process described in section 9.4 provides a much more sensitive and rigorous means of determining the significance of the set points on variance, average, and CIQ.

9.8 Concluding remarks regarding the analysis of strategic experiments

The methods described in Chapters 8 and 9 were developed for analyzing the results of 'strategic experiments,' so-called because they are designed to measure the loss of value due to variation and thus demand and profits resulting from the placement of the mean off target and from variation about the mean. They build upon the important concepts and methods first introduced by Taguchi. Modifications of Taguchi's approach were introduced here so that (1) the outcomes of the experiments can be readily used, if desired, to forecast changes in product demand and profitability and (2) the significance of the outcomes can be tested in a manner that minimizes the possibility of making Type I errors as large investments may be made on outcomes deemed as favorable. The modifications are summarized below along with the justification for each:

1. **Make at least four replications per trial if time permits:** an understanding of the sources of variation and the systematic reduction of variance is a major factor in improving the quality of many products. This knowledge is increased as the number of replications (df) is increased.
2. **Determine the significance of the set points on variance:** this gives insight into how each set point influences variance and the resulting pooling array can be used to pool the sample trial variances either partially of fully. The F ratio test described here can be made for the ξ'_{ij} coefficients for any $[\mathbf{X}]_\lambda$ matrix and for two-level $[\mathbf{X}]_\varphi$ matrices.
3. **Determine the significance of the set points on average:** gives insight.
4. **Use the Monte Carlo method to assess significance of set points on the S/N ratio of the CIQ:** the six-step Monte Carlo approach is more sensitive than the two direct methods described above because of its potential for pooling of variance. The approach is also more general with fewer assumptions.
5. **Use the $[\mathbf{X}]_\lambda$ matrix array when the objective is to improve value relative to a baseline:** most new product improvements need to be measured relative to an existing baseline and the $[\mathbf{X}]_\lambda$ array accommodates this. The $[\mathbf{X}]_\varphi$ array should be used, however, when no trial is relevant as a baseline because the average of all the trials becomes the baseline thereby always giving equal or better sensitivity.
6. **When point estimates are needed for the CIQ, the expectations for the population variance and the square of the population average given by Equations 8.30 and 8.31, respectively, should be used in Table 8.2:** a single sample variance and the square of a single sample average give point estimates which are too small for their population counterparts when the df is small.
7. **Unless proven otherwise, the replications for SIB and LIB should be considered as being lognormally distributed:** the basic variable for true SIB and LIB problems can never be negative and thus the variables cannot be normally distributed. The lognormal distribution has the desired property

and thus should be more suitable for describing the statistics for SIB and LIB distributions.

8. **The experiment-wise correction should always be made when assessing the significance of the set points:** as the number of off-baseline set points increase, the chances of making a Type I error for the null hypothesis increases in direct proportion. Bonferroni's technique is the easiest method for making this important correction to the standard pairwise error given by the statistical tables and spreadsheet functions.

References

1. Nair, V.N. (ed.) (1992) Taguchi's parameter design: a panel discussion. *Technometrics*, **34**, 127–61.
2. Logothetis, N. and Wynn, H.P. (1989) *Quality Through Design*, Clarendon Press, Oxford, U.K., pp. 87–8.
3. Box, G. (1988) Signal-to-noise ratios, performance criteria, and transformations. *Technometrics*, **30**, 1–17.
4. Neter, J. and Wasserman, W. (1974) *Applied Linear Statistical Models*, Richard D. Irwin, Inc., Homewood, IL, pp. 146–7.

10 Costs

Costs, in spite of their critical nature, are often poorly understood in relation to the value that they do or do not create. The challenge to the enterprise is to know (1) how costs inside the enterprise add value outside and (2) how to get greater value for each dollar spent. In a competitive market, the downside of the neglect of cost control on the one hand and over-control of costs on the other can be the same, failure of the enterprise. Throwing money at new product development is just as bad as starving it. However, money spent in certain phases of the product development process may generate more value than in other phases. For example, increased spending in the early phases of the design process may generate savings in the finished product that are many times the costs.

10.1 Activity-based costing

Historically, costs have been divided into direct (touch) labor, direct materials, indirect labor, indirect materials and other overheads. The allocation of the fraction of indirect and overhead costs to specific product lines has often been in proportion to direct labor hours required to manufacture the product. This allocation scheme can be inaccurate for many product lines today where direct labor is less than 10–15% of total costs. When poor cost definition is combined with the absence of a clear, direct connection of cost to value generated, the computed profitability of different product lines may well be far from their true nature. The level of inaccuracy can at the extreme lead to the elimination of truly profitable products and the continued production of losers [1].

Activity-based costing is being considered by a variety of firms to foster a better understanding of the value-versus-cost relationship, particularly in regard to making proper investment decisions associated with new technology, flexible manufacturing cells being an important example. The objective is to identify how costs are generated by each activity within the enterprise for the various products being manufactured. Operations discovered which add no direct, tangible value to the product from the customer's viewpoint should be replaced by value adding efforts or eliminated. A conveyer which moves parts from one point to another is a good example of a non-value adding activity. The application of activity-based costing is relatively insensitive to the type of organization. The activity-based strategic accounting concept discussed in Chapter 6, however, is sensitive to the organization structure and is an exploratory idea for providing a performance metric for the system/subsystem organization which can be used to measure the incremental profitability contributed by each unit within that organization.

The flow of product realization over the entire product life cycle needs to be mapped out to fully understand the value and cost opportunities that reside within the diverse product mix of a large enterprise and to identify the specific operating units responsible for generating them. The process flow methodology, which we have already used extensively, is an excellent tool for mapping out the cost structure of activities in the total product realization process flow at any level (Figure 10.1). Costs are assigned to input, output, controls, tools, and waste. An additional source of costs lies inside each production unit in the form of work-in-process (WIP) inventory which ideally should be no greater than the rate of output required to match consumer demand multiplied by the time needed to manufacture and assemble one unit of output. Annual inventory costs are equal to the average amount of WIP multiplied by the cost per unit times the going interest rate plus the costs for the added storage space and handling of the excess. The process flow model is also a useful tool for brainstorming cost reduction and value-adding ideas. It is a more descriptive and complete representation of the actual process than the fishbone (Ishikawa) diagrams often used to guide brainstorming (Figure 10.2).

10.2 Cost drivers

Cost drivers are those factors in design, manufacturing, and distribution that generate costs. For example, within fourteen activity areas at an electronics instruments manufacturing facility, the cost drivers were identified as the number of operations that needed to be performed [2]. The cost driver for part number maintenance was the number of part numbers. The cost driver for masking was the number of points that needed masking and, for final testing, it was the time required for final testing. For hand loading and instrument assembly it was Boothroyd time, which is computed in terms of the number of parts to be assembled and their difficulty in handling and insertion as described in the next section.

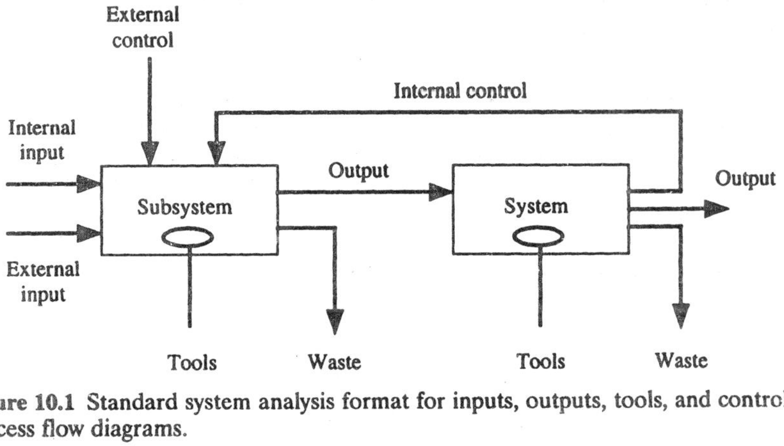

Figure 10.1 Standard system analysis format for inputs, outputs, tools, and controls in process flow diagrams.

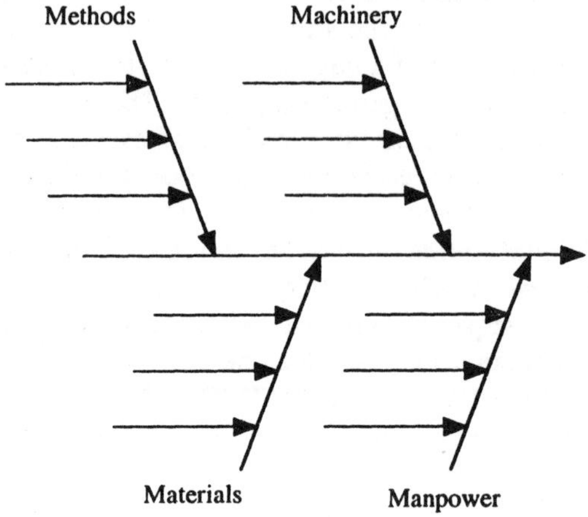

Figure 10.2 Ishikawa's fishbone diagram for brainstorming possible factors.

Griffi, Roth, and Seal [3] identified the cost drivers shown in Table 10.1 for order entry and machine set up. The number of things are cost drivers because they translate into time which translates into money. An insidious bureaucratic cost driver is the number of approval signatures required to implement a purchase order. It represents, according to classical administrative theory, a form of over-control as budgets should be controlled by those responsible for the task. Each approval signature takes time and is a non-value adding effort but many firms are blind to these wastes and, of greater consequence, to the delays created in implementing action. Enterprises that compete poorly generally have lean profits which may be driving them to over-control of costs. This adds waste and time thus making the enterprise even less competitive. Enterprises suffering profit squeezes should examine their control policies to see if they are in such a vicious circle on costs. Any organizational unit that has the ability and responsibility to add value should have the less demanding ability of controlling and reducing its own costs.

Table 10.1 Cost drivers for order entry and machine set up (source: C. Griffi, A.V. Roth, and G.M. Seal (1990) *Competing in World-Class Manufacturing*, Business One Irwin, Homewood, IL, p. 164)

Activity	*Drivers*
Enter orders	1. Number of orders 2. Number of line items
Set up machine	1. Number of setups 2. Complexity (tools, fixtures, materials, machines)

10.3 Boothroyd and Dewhurst design for manual assembly

There are some basic guidelines which have been developed and codified [4] that designers can follow to minimize the costs of manually assembling the components that make up subsystems and systems. Guidelines have also been developed for robotic and automatic assembly [5], the full development of the methodology being contained in the *Design for Assembly Handbook* available from Professors Boothroyd and Dewhurst. A key product or subsystem concept in the Boothroyd and Dewhurst methodology is its design efficiency E_e based upon the minimum number of parts N_m

$$E_e = 3N_m/T_B \tag{10.1}$$

where T_B is the total time in seconds to assemble the actual design (not the minimum number of parts design) evaluated using Boothroyd and Dewhurst's empirical times from the charts given by:

$$T_B = \Sigma_i t_i, \tag{10.2}$$

in which t_i is the handling plus insertion times in seconds to manually assemble component i, the summation being over all parts to be assembled. The factor of three in the equation for E_e is the ideal time in seconds to assemble a single component that is easy to handle and insert. The theoretical minimum number of parts is determined as outlined in Table 10.2.

A given part used in the assembly must satisfy at least one of the three rules for it to be counted as one of the minimum number of parts. Fasteners are often not counted in the minimum number of parts as integral fasteners can be used for many applications.

Manual handling, as shown in Table 10.3, is divided into levels of difficulty (handling times in seconds) based upon whether or not the part is easy to grasp and manipulate, the thickness is greater or less than 2 mm, and the amount of end-to-end (α) and side-to-side (β) rotation angles for symmetry operations. Size is the length in millimeters of the largest orthogonal dimension. The symmetry codes are defined by combinations of the two symmetry angles (Table 10.4). Examples of end-to-end symmetry are shown in Figure 10.3. Side-to-side examples are shown in Figure 10.4.

Table 10.2 Boothroyd and Dewhurst criteria for determining minimum number of parts (source: G. Boothroyd, and P. Dewhurst (1983) Design for assembly: manual assembly. *Machine Design*, **8**, 140–5)

1. The part being inserted into an assembly must be separate because it moves with respect to all other parts already assembled.
2. The part must be made from a different material than or be isolated from all other parts already assembled.
3. The part must be separate from all others in the assembly because necessary assembly or disassembly would be impossible.

Table 10.3 Manual handling times (source: G. Boothroyd and P. Dewhurst (1983) Design for assembly: manual assembly. *Machine Design*, **8**, 140–5. Reprinted with permission from *MACHINE DESIGN*, December 8, 1983. A Penton Publication)

Easy grasp and manipulation

	Thickness > 2 mm			*Thickness ≤ 2 mm*	
Symmetry code	*Size 15+*	*6→15*	*0→6*	*Size 6+*	*0→6*
0	1.13	1.43	1.88	1.69	2.18
1	1.5	1.8	2.25	2.06	2.55
2	1.8	2.1	2.55	2.36	2.85
3	1.95	2.25	2.7	2.51	3.00

Difficult grasp and manipulation

	Thickness > 2 mm			*Thickness ≤ 2 mm*	
Symmetry code	*Size 15+*	*6→15*	*0→6*	*Size 6+*	*0→6*
0	1.84	2.17	2.65	2.45	2.98
1	2.25	2.57	3.06	3.0	3.38
2	2.57	2.9	3.38	3.18	3.7
3	2.73	3.06	3.55	3.34	4.0

Manual insertion times are classified into levels of difficulty based upon whether or not (1) hold down is required to maintain orientation and location, (2) the part is easy to align during assembly, and (3) there is difficulty in reaching the point of assembly with the part and tool. The manual insertion times for the case

Table 10.4 Symmetry codes for angle range (source: G. Boothroyd and P. Dewhurst (1983) Design for assembly: manual assembly. *Machine Design*, **8**, 140–5)

Symmetry code	*Angle range*
0	$\alpha + \beta < 360$
1	$360 \leq \alpha + \beta < 540$
2	$540 \leq \alpha + \beta < 720$
3	$\alpha + \beta = 720$

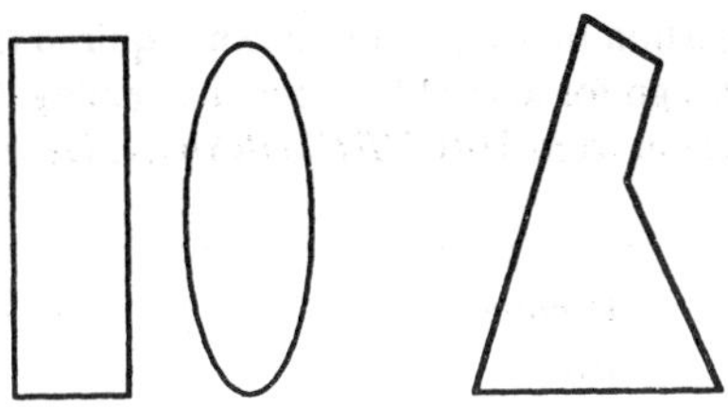

$\alpha = 180$ $\alpha = 360$

Figure 10.3 End-to-end symmetry examples.

when no hold down is required of the previously assembled component while assembling the next are shown in Table 10.5. When hold down is required, the manual insertion times increase by four seconds (roughly doubled) (Table 10.6).

Use of the Boothroyd and Dewhurst methodology may add time to the initial design process but it should pay off with lower costs and faster accomplishments of the later stages of the design process as there should be less of a chance of having to redesign because of assembly difficulties. As designers gain experience with the basic rules of the design for assembly methodology, it should become a natural part of their thought processes and fewer iterations will be required to develop the most efficient design. The key elements are to maximize the symmetry of the individual parts, use as few parts as possible (combine parts), use as few fasteners as possible, assemble all parts along one axis if at all possible, and provide easy access for each part as it is assembled.

Design efficiencies range between 20 and 30% for complex electromechanical assemblies that require extensive wiring and gasketing to as high as 90% for straightforward mechanical assemblies. Other guidelines for minimizing assembly costs [6] are to eliminate assembly adjustments by reducing part geometry variance, providing self-locating features employing generous radii and chamfers, and using component designs that minimize tangling and nesting of parts ahead of assembly. For the fasteners that must be used, standardize on one kind and if that is not possible make sure that they cannot be interchanged resulting in cross threading or safety issues caused by substandard clamping loads.

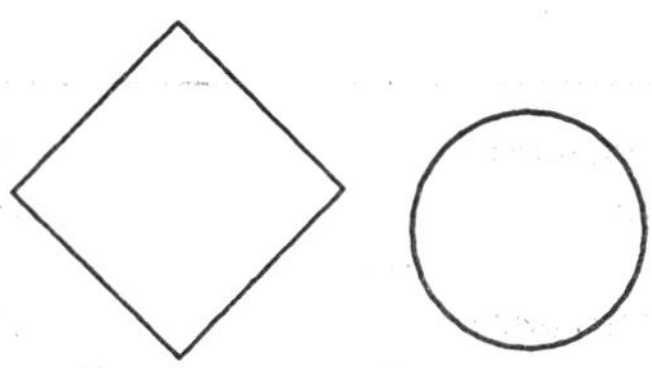

$\beta = 90$ $\beta = 0$

Figure 10.4 Side-to-side symmetry examples.

Table 10.5 Manual insertion times, no hold down required (source: G. Boothroyd and P. Dewhurst (1983) Design for assembly: manual assembly. *Machine Design*, **8**, 140–5. Reprinted with permission from *MACHINE DESIGN*, December 8, 1983. A Penton Publication)

	Alignment easy		*Alignment not easy*	
Reach difficulty	*Resistance* No	*Resistance* Yes	*Resistance* No	*Resistance* Yes
0	1.5	2.5	2.5	3.5
1	4	5	5	6
2	5.5	6.5	6.5	7.5

10.3.1 FIXED-LENGTH SLICER EXAMPLE

A sketch of a fixed-length slicer is shown in Figure 10.5 which has 16 parts, some of which are the same. The analysis of handling times using the symmetry codes of Table 10.4 and the times in Table 10.3 are shown in Table 10.7. The results of the insertion analysis are shown in Table 10.8 along with the total operation times and the minimum number of parts. All parts were assumed to be difficult to align and hold down was assumed to be necessary. The theoretical minimum number of parts was determined using the three criteria described in Table 10.2. The efficiency for the design level shown in Figure 10.5 is 9.2%. A redesign of this device is shown in Figure 10.6 having only three parts.

10.4 Shipping and inventory costs

Inventory, whether it is in finished products that have not been shipped to customers, or raw materials sitting on the dock or somewhere intermediate

Table 10.6 Manual insertion times, hold down required (source: G. Boothroyd and P. Dewhurst (1983) Design for assembly: manual assembly. *Machine Design*, **8**, 140–145. Reprinted with permission from *MACHINE DESIGN*, December 8, 1983. A Penton Publication)

	Alignment easy		*Alignment not easy*	
Reach difficulty	*Resistance* No	*Resistance* Yes	*Resistance* No	*Resistance* Yes
0	5.5	6.5	6.5	7.5
1	8	9	9	10
2	9.5	10.5	10.5	11.5

Table 10.7 Analysis of manual handling times for fixed-length slicer

Part number	*Number of times*	α	β	$\alpha + \beta$	*Symmetry code*	*Easy grasp?*	*Part handling time*
11	1	180	360	540	2	Yes	1.8
10	1	360	360	720	3	Yes	1.95
9	2	360	0	360	1	Yes	3
8	1	360	360	720	3	No (sharp)	2.73
7	1	360	360	720	3	No (sharp)	2.73
6	1	180	0	180	0	Yes	1.13
5A, 5B	2	180	0	180	0	Yes	2.86
4A, 4B	2	180	0	180	0	Yes	2.86
3	1	180	180	360	1	Yes	1.5
2	2	360	0	360	1	Yes	3
1	2	360	0	360	1	Yes	3

between these extremes in the form of WIP, represents costs in the form of carrying charges. There is a direct relationship between the amount of inventory and the distance between sources of output and receivers of input. Consider a component plant that is a distance Δ from the final assembly plant having $N_H/2$ haulers between them carrying parts and $N_H/2$ returning with no parts (Figure 10.7).

Table 10.8 Analysis of manual insertion times for fixed-length slicer

Part number	*Number of times*	*Insertion time**	*Operation time*	*Minimum number of parts*
11	1	6.5	8.3	1
10	1	6.5	8.45	0
9	2	13	16	0
8	1	6.5	9.23	1
7	1	6.5	9.23	0
6	1	6.5	7.63	1
5A, 5B	2	13	15.86	1
4A, 4B	2	13	15.86	0
3	1	6.5	8	0
2	2	13	16	0
1	2	13	16	0
			Totals = 130.56	4

E = 3×4/130 = 9.2%

*All are not easy to align and hold down required.

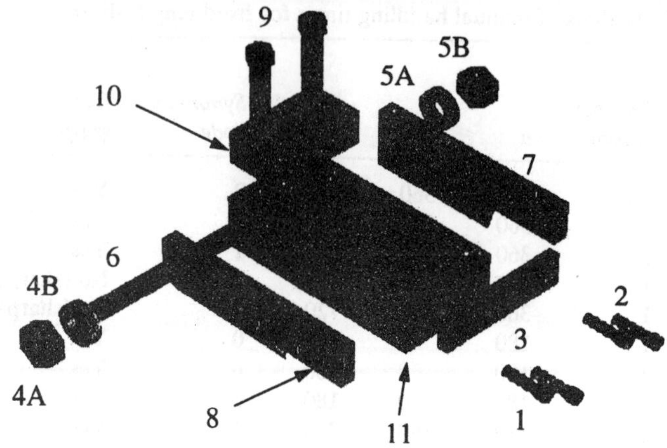

Figure 10.5 Early concept for fixed-length slicer.

The parameters that define the costs of shipping and storage of these parts are given in Table 10.9. The distance between hauler units (d) divided by the velocity of the haulers (v) is equal to the time between the arrival of each hauler (t_H):

$$t_H = d/v.$$

Since the distance between haulers is equal to ($2\Delta/N_H$), the time between units can be written as:

$$t_H = 2\Delta/(vN_H). \qquad (10.3)$$

The average rate at which parts arrive should be balanced against the assembly rate R_A to avoid excess build up of inventory (or too few parts):

$$R_A = B_H/(B_P t_H)$$

It follows from Equation 10.3 that

$$R_A = v\, N_H B_H / (2\, B_P \Delta), \qquad (10.4)$$

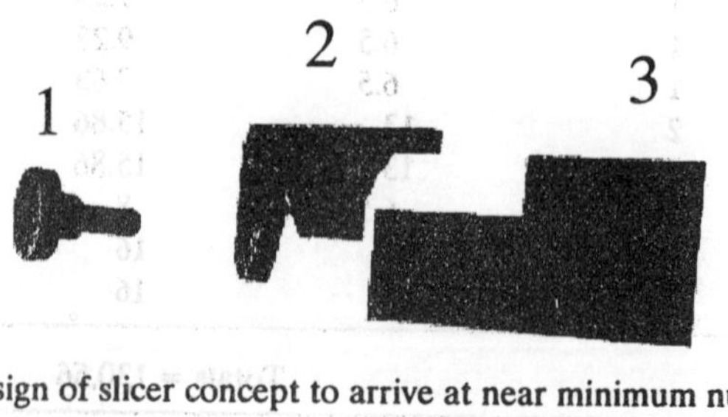

Figure 10.6 Redesign of slicer concept to arrive at near minimum number of parts.

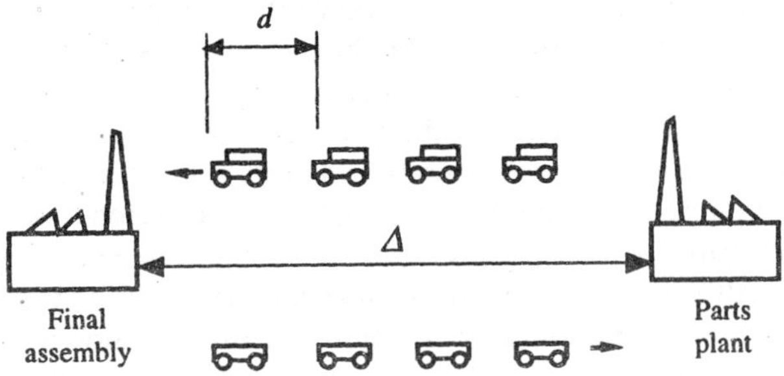

Figure 10.7 Schematic of shipping parts to assembly plant and the return of empty haulers.

and that the number of parts in the pipeline $(N_H B_H / 2B_P)$ is equal to $(R_A \Delta / \nu)$. The annual total costs for shipping and receiving are equal to the sum of the annual inventory costs for materials in the pipeline, in the hauler being filled at the shipping plant, and in the hauler being emptied at the assembly plant plus the operating costs for the haulers and operators. On average the hauler being filled and the hauler being emptied are half-full making the annual inventory costs for them equal to $(IC_P B_H / B_P)$ in which I is the annual interest rate. The inventory costs for the parts in the pipeline are equal to $(IC_P R_A \Delta / \nu)$ and the annual operating costs for the haulers and operators is equal to $N_H[C_H + C_0]$. Thus, the inventory costs including shipping costs are given by:

$$C_{In} = I\,[(C_P B_H / B_P) + (C_P R_A \Delta / \nu)] + N_H (C_H + C_0) \qquad (10.5)$$

Solving for N_H in Equation 9.4 and replacing it by this quantity in Equation 10.5, yields:

$$C_{In} = I\,[(C_P B_H / B_P) + (C_P R_A \Delta / \nu)] + (2\,R_A B_P \Delta / B_H \nu)(C_H + C_0). \qquad (10.6)$$

Table 10.9 Definition of symbols used in analysis of shipping inventory

Symbol	*Definition*
N_H	Number of haulers
B_P	Volume of part
C_P	Cost of part
B_H	Volume of hauler
ν	Velocity of hauler
C_H	Cost of hauler
C_0	Cost of operator
Δ	Distance
C_{In}	Cost of inventory
R_A	Final assembly rate

The total costs of inventory in the supply tube including parts on both docks in Equation 10.6 increases linearly with the distance between the part plant and the final assembly plant as shown schematically in Figure 10.8. The annual operating cost C_H of a hauler should be proportional to its velocity capability raised to some power higher than unity and to its volume raised to a power less than unity. As a result, inventory costs versus hauler velocity should display a minimum (Figure 10.9) resulting from a balance between inventory costs decreasing with speed and operating costs increasing with speed. Total costs should also show a minimum as a function of hauler volume as a result of the balance between transportation costs decreasing with larger hauler volume versus inventory costs for parts on the dock increasing with larger hauler volume (Figure 10.10).

10.5 Just-in-time

The just-in-time (JIT) concept is based on delivery of small lots to the requesting party upstream and as such it is a 'pull' process. From the preceding analysis of inventory costs, the distance between the part plant and final assembly plant should be small for JIT to be cost effective. Improved quality control can also be achieved using JIT as a result of having defective parts contained in small lots. JIT also places a premium on correcting quality problems fast if acceptable throughput is to be achieved.

The same equations used to consider shipments between plants can be used to model shipments between machines within a plant. By moving the machines closer together (reducing Δ), inventory costs can be reduced as well as lot sizes. Toyota understood the importance of JIT to keeping manufacturing costs low and quality high and over time developed the necessary manufacturing environment to implement JIT. Toyota used control cards that were transferred from

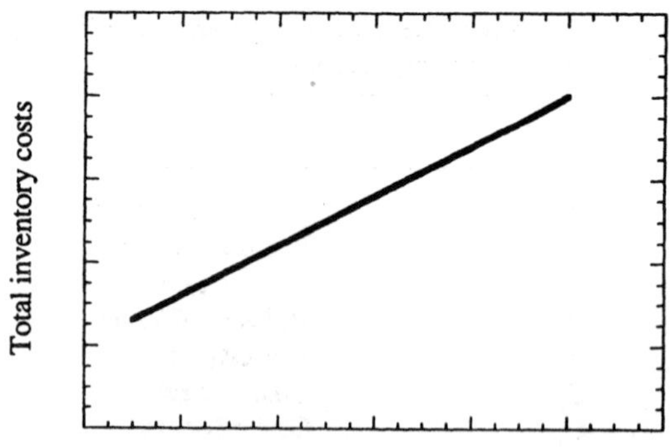

Figure 10.8 Total inventory costs as a function of the distance between final assembler and part manufacturer.

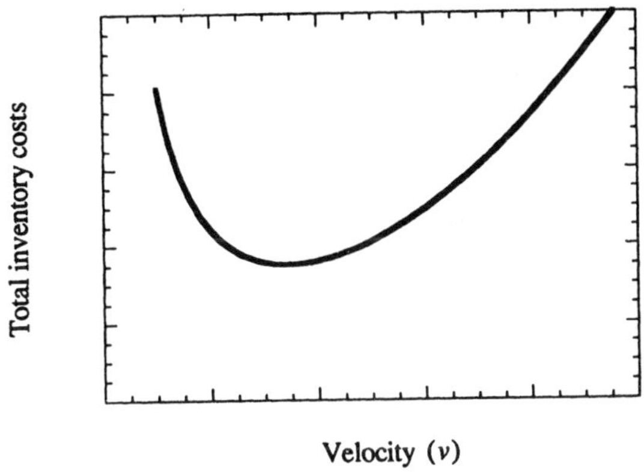

Figure 10.9 Total inventory costs as a function of hauler velocity.

one container to another to signal when a new container of inventory was taken by a machine upstream thereby requesting the machine downstream to begin making parts to refill the inventory. The so-called Kanban (card) process developed by Toyota has been described by Schonberger [7] who was one of the first persons in the U.S. to recognize its impact as a competitive weapon. The basic structure of a simplified JIT pull process is shown in Figure 10.11.

The operator of machine $n + 1$ takes a container of inventory from parts produced by machine n. This signals a need for the operator of machine n to make new parts of the type taken from inventory (the machining line could be making several types of parts). This process cascades downstream because when the operator of machine n takes parts from the inventory made by $n - 1$ of the type needed to meet its own request this signals the operator of machine $n - 1$ to refill the inventory for this type of part. Because the operators will be performing manufacturing operations on different kinds of parts throughout

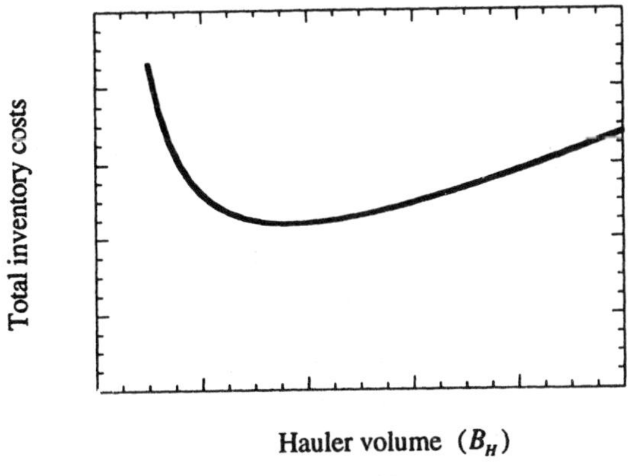

Figure 10.10 Total inventory costs as a function of hauler volume.

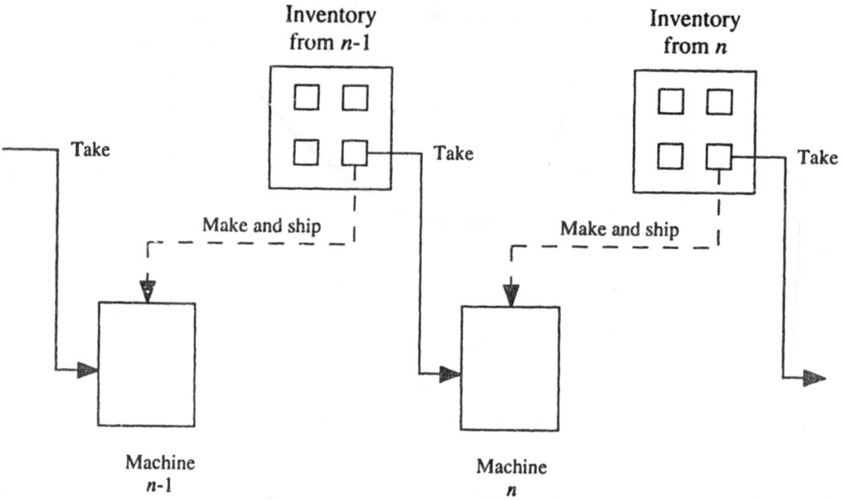

Figure 10.11 Schematic of JIT control of parts on factory floor.

the day, it is necessary that they make quick tool changes and quick setups and that few parts, preferably zero, need to be manufactured before a good part is produced after a new setup.

10.6 Target costing using the S-model

Although the subject of pricing was considered in Chapter 3 in terms of cost and value, **cost-plus**, one of the more common forms of pricing in which price is computed as a percentage mark-up from cost, can lead to products which are improperly priced in relation to their value. The practice of **target costing** is being developed as a replacement for cost-plus to arrive at a more meaningful and competitive price-versus-value relationship for a product [8] and to provide improved control over costs during the development process.

The rule of thumb given by Equation 3.20 is for the price of a future product defined in terms of the price of the product currently being produced and the changes in value and cost. The price of the current product was not questioned. However, the appropriateness of the current price can be evaluated by considering its position in relation to the other products on a value-versus-price trend line as shown schematically in Figure 10.12 (see Figure 3.16 for such a plot for family sedans). Product A can be considered as targeting the most price-sensitive customers for the product segment and product C as targeting the more value-sensitive customers in the segment. Product B is seen as somewhat overpriced vis-à-vis the trend line.

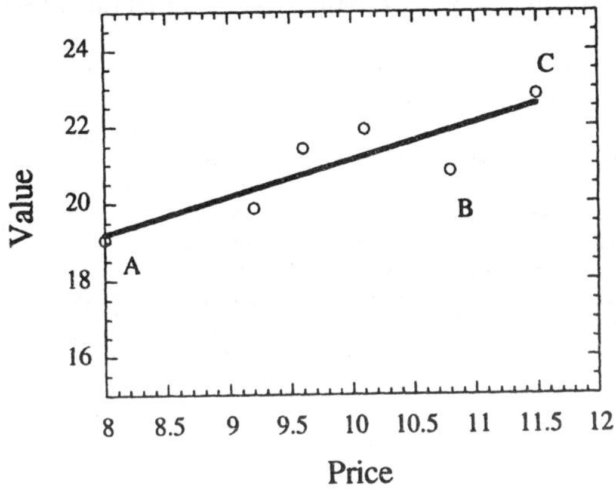

Figure 10.12 Hypothetical value-versus-price trend line for six competitors.

The point chosen along the trend line should balance the positioning of the product line in adjacent segments and reflect the ability of the manufacturer to compete, particularly in terms of brand name value, reliability, and cost. Analysis of family sedans has shown, for example, that the product leader for reliability is often the product that holds the position of highest value and highest price. Thus another manufacturer would be ill-advised to try to take this position away from its current holder without having superior quality. If a point is chosen for a product that is well above the value/price trend, a price war as discussed in Chapter 3 may result.

Once a manufacturer has decided where to place its product on the trend line for the segment of interest, the target for variable cost can be established by equating it to the variable cost which gives the desired internal rate of return, I_{RR}, equal to the interest rate in Equation 3.25 which gives zero for present worth:

$$\sum_{j=0}^{Y_{RS}} \frac{A_i(q,j)}{(1+I_{RR})^j} = 0 \tag{10.7}$$

With the value and price targets taken from the desired point on the value/price trend line and the desired internal rate of return, this relationship can be solved by iterating the left-hand side of Equation 10.7 as a function of variable cost until the cost is found where the summation is approximately zero. The key, of course, to making target costing work as a competitive weapon is to meet the cost and value targets once they have been established.

An example of establishing the target cost using the iterative process with a spreadsheet program is shown in Table 10.10 for a hypothetical new automobile entering a segment with five other competitors. Product development begins in year 0 where the value and price of the new product are targeted at $32 000 and

Table 10.10 Determining target cost based upon internal rate of return analysis using a spreadsheet

1 Year	2 Value	3 Price	4 Investment	5 Comp. (5/6)×(V–P)	6 Annual fixed costs	7 Annual demand	8 Cash flow	9 Discounted cash flow
0	32 000	16 000	300 000 000	14 167	50 000 000	0	-3.50E+08	-3.50E+08
1	32 960	16 480	300 000 000	14 592	51 500 000	0	-3.52E+08	-2.93E+08
2	33 920	16 960	300 000 000	15 017	53 000 000	0	-3.53E+08	-2.45E+08
3	34 880	17 440	0	15 442	250 000 000	110 000	3.66E+08	2.12E+08
4	35 840	17 920	0	15 867	257 500 000	110 000	3.75E+08	1.81E+08
5	36 800	18 400	0	16 292	265 000 000	110 000	3.85E+08	1.55E+08
6	37 760	18 880	0	16 717	272 500 000	110 000	3.94E+08	1.32E+08
7	38 720	19 360	0	17 142	280 000 000	110 000	4.04E+08	1.13E+08
8	39 680	19 840	0	17 567	287 500 000	110 000	4.13E+08	9.61E+07

Present worth = -1.46E+03
Variable cost = 10 863.7

$16 000, respectively. The product is to be introduced, however, in year 3 with the target value and price inflating at 3% per year from zero. The product is discontinued after year 8.

The investment, shown in column 4, is $300 000 000 per year for the first three years and zero thereafter. Column 5 lists the average value minus price expected for the five other competitors multiplied by the ratio of 5/6. This quantity is used in the computation of expected demand based upon Equation 3.8. The coefficient K in Equation 3.8 is taken as equal to 60 units/$ in year zero and discounted by a factor of $1/(1 + 0.03j)$ thereafter. This results in the forecast demand being constant for the new product at 110 000 units from years 3–8 shown in column 7. Annual fixed costs are shown in column 6.

The resulting cash flow is shown in column 8 as computed from Equation 3.14 for a desired internal rate of return equal to 0.20. The variable cost shown in the box at the bottom of column 9 is entered into the expression for cash flow and is changed manually until the present worth reduces to zero within an acceptable error. For this problem, the variable cost of $10 863.70 in year 0 and appreciated annually at 3% results in an error of only –$14 600.

References

1. Berliner, C. and Brimson, J.A. (eds) (1988) *Cost Management for Today's Advanced Manufacturing*, Harvard Business School Press, Boston, MA.
2. Foster, G. and Gupta, M. (1990) Activity accounting: an electronics industry implementation, in *Measures for Manufacturing Excellence*, (ed. R.S. Kaplan) Harvard Business School Press, Boston, MA, pp. 225–68.
3. Griffi, C., Roth, A.V., and Seal, G.M. (1990) *Competing in World-Class Manufacturing*, Business One Irwin, Homewood, IL, p. 164.

4. Boothroyd, G. and Dewhurst, P. (1983) Design for assembly: manual assembly. *Machine Design*, **8**, 140–5.
5. Boothroyd, G. and Dewhurst, P. (1984) *Machine Design*, February 23, 1984, pp. 72–6; *Machine Design*, Jan. 26, 1984, pp. 87–92.
6. Tucks, H. E. (1987) *Designing for Economical Production*, SME, Dearborn, MI, pp. 313–37.
7. Schonberger, R.J. (1982) *Japanese Manufacturing Techniques, Nine Hidden Lessons in Simplicity*, Free Press, New York; (1987) *World Class Manufacturing Casebook, Implementing JIT and TQC*, Free Press, New York.
8. Cooper, R. and Chew, W.B. (1996) Control tomorrow's costs through today's designs. *Harvard Business Review*, **74**, 88–97.

11 Statistical quality control

11.1 End-of-line inspection

Competitive enterprises continue quality improvements after production begins with the focus moving from off-line design of experiment methods to on-line statistical process control. An older form of on-line quality control is 100% inspection at the end of the manufacturing line, the attributes of a product being either acceptable or unacceptable in relation to the target g_T (Figure 11.1).

End-of-line inspection has several shortcomings, however. It is expensive because considerable labor is needed to inspect every part. Moreover, all of the manufacturing costs have been incurred and thus totally wasted for the pieces that fail inspection unless they can be reworked which results in added costs. Furthermore, when inspection is done by people, it is seldom close to being 100% accurate, a sizable fraction (approximately 15%) of the parts made out of specification can get through generating a loss of value to the customer and warranty costs to the manufacturer. This method of quality control also treats markedly different statistical distributions in product attributes as shown in Figure 11.2 as being equivalent from the viewpoint of quality. If the specification shown in Figure 11.2 is for a nominal-is-best situation at $g = g_T$, then the square distribution will generate less customer satisfaction than the peaked distribution. However, the rule of accepting a part if g is between $g_T - \delta$ and $g_T + \delta$ and rejecting it if outside these limits does not differentiate the true quality between the two distributions. Sony observed that the color density of television sets made in Japan and San Diego differed in their statistical distribution and the narrower distribution, not surprisingly, was preferred by customers [1].

11.2 Statistical process control

The use of statistical process control (SPC) methods to monitor the performance of a process reduces the need for 100% inspection and better defines the

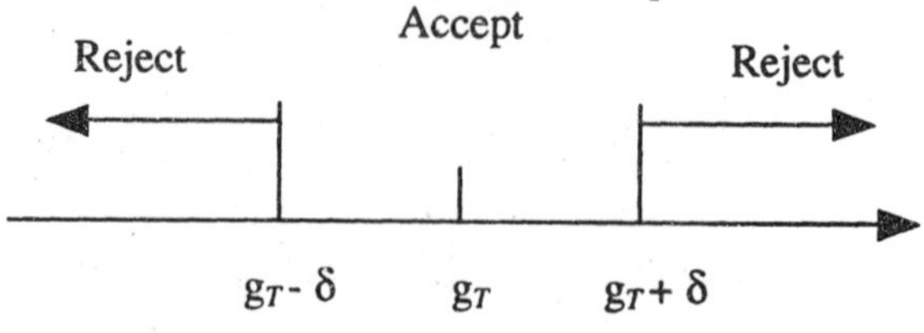

Figure 11.1 Acceptance limits form of quality control.

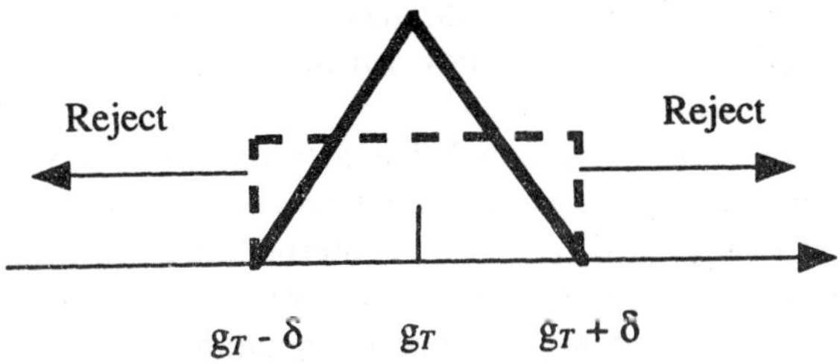

Figure 11.2 Two statistical distributions that are equivalent according to acceptance limits but not according to the customer.

level of quality in the parts produced. Inspection can also be moved from the end of the line to points after each critical operation in the manufacturing sequence thereby detecting problems quickly before additional operations are performed. The information generated can be used to improve the process.

SPC can also be used to extend the DoE process to the production environment with the objective of continuously reducing variance and moving the mean toward the ideal target specification in a cost-effective manner throughout the period that the product is manufactured. The combination of the newer off-line DoE processes for pre-production development and on-line SPC methods for continuous improvement once production has begun results in what we will call statistical quality control (SQC). Implementation of SQC by many manufacturing firms has resulted in marked improvements in product quality over time and a lessening of the use of end-of-line inspection as shown in Figure 11.3.

11.3 Target practice example

Key elements of SPC can be highlighted by considering a straightforward example of two persons shooting a rifle at a bullseye. Person 1 has fired eight shots

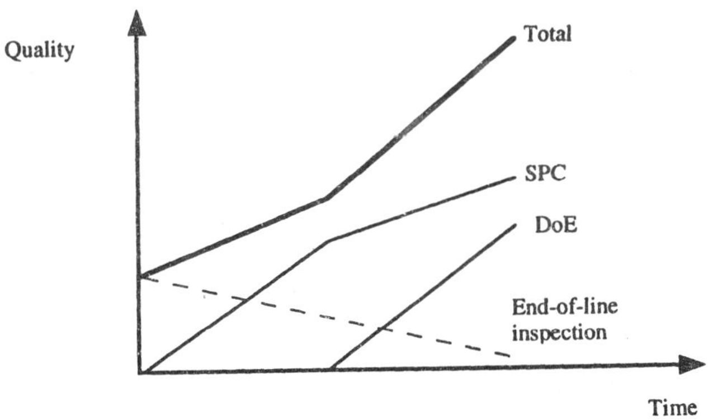

Figure 11.3 Contributions of inspection, SPC, and DoE to improvements in quality over time from the early part of century to the present.

at the target (Figure 11.4) with low variance about the mean but the cluster is off target. Person 2 has fired eight shots which are distributed about the target but have high variance (Figure 11.5).

The systematic error for person 1 was attributable to a special condition involving improper use of the gun sight which could be adjusted and thereby move the cluster for the next trial to be centered about the bullseye. Person 2 had a more difficult challenge requiring a more in-depth examination of the overall process if improvement was to be forthcoming. The process flow chart for the action of firing the rifle at the target is shown in Figure 11.6.

The box T represents the transformation event, the firing of the rifle. The rifle and the person are listed as tools for this box. Input to T were the bullets B for the rifle and food F consumed by the person. The box D represents the delivery system and W represents the wind input to the delivery system. The final symbol represents the target or bullseye. The cause of the difference in accuracy between person 1 and person 2 was not the rifle per se, as it was the same. However, they did not use the same bullets and they ate different food. The eyesight of the two persons also differed as well as their physical strengths.

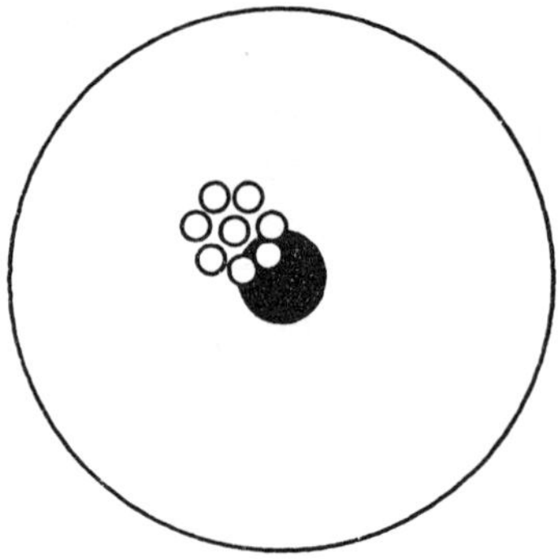

Figure 11.4 Mean off target but low variance, person 1.

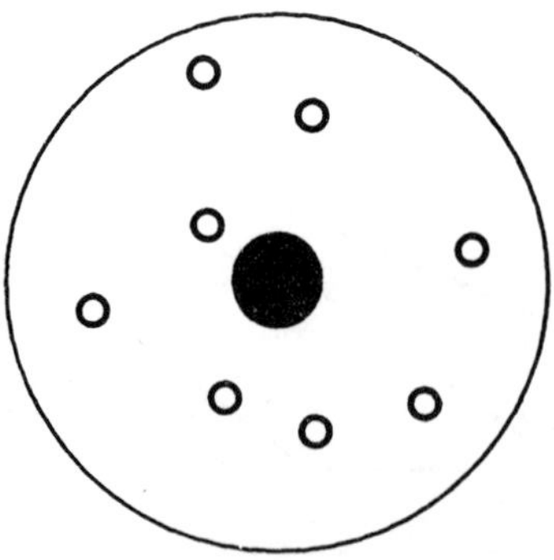

Figure 11.5 Mean on target but high variance, person 2.

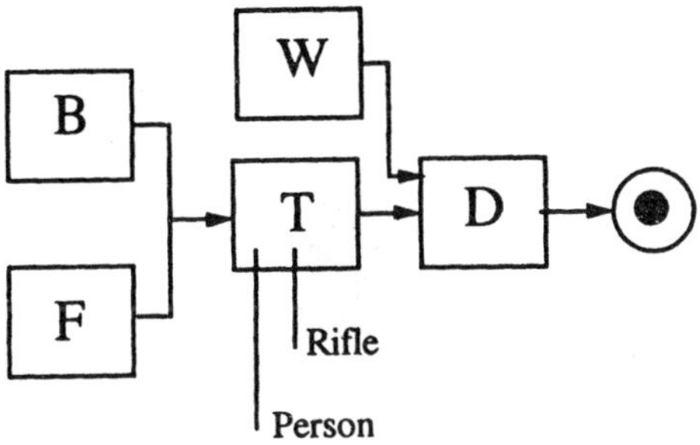

Figure 11.6 Flow chart for firing rifle at bullseye.

Person 2 looked to the results of person 1 as a benchmark and decided to improve by eliminating coffee consumption, requiring bullets that had smaller standard deviation in powder volume, and improving physical strength by weight lifting so as to hold the rifle as steady as person 1. Finally person 2 wore new glasses that corrected astigmatism. The wind was known to blow rather hard and be gusty but this was the same for both persons.

Person 2 was now equal to person 1 in all physical aspects and had the same tools. The experiments were now repeated with the results for person 1 shown in Figure 11.7, the improvement being the proper use of the rifle sight. Person 2, intent on improving, realized that the variance in person 1's cluster was caused by the wind and carefully observed the distance Δg_w between where the last bullet hit and the point of aim. The next shot was fired at $-\Delta g_w$ measured from the bullseye. The results, however, shown in Figure 11.8, were disappointing compared to person 1.

Let's analyze the flaw in person 2's strategy: if the deviation from the bullseye at shot (j) caused by the wind gust is $\Delta g_w(j)$ and the compensation attempted by person 2 is $\Delta g_c(j)$, they are connected by the relationship:

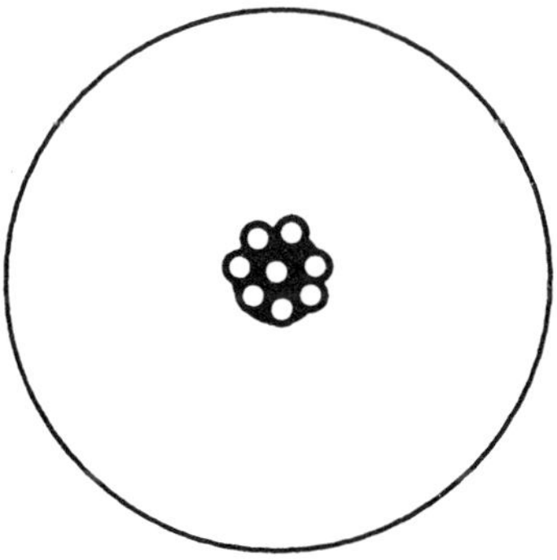

Figure 11.7 Person 1 on target.

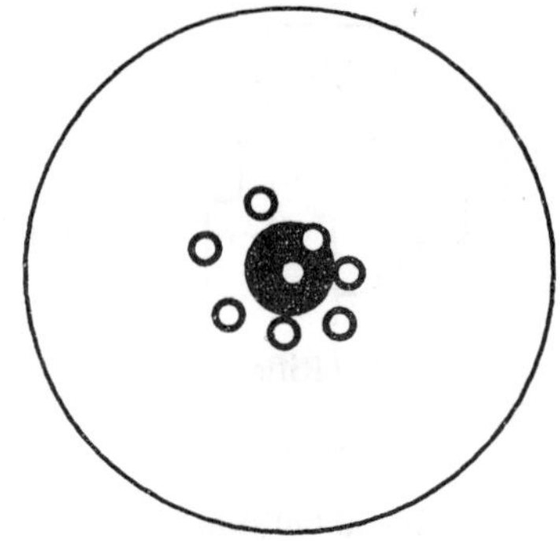

Figure 11.8 Person 2's variance is still too large.

$$\Delta g_c(j+1)=-\Delta g_w(j) \tag{11.1}$$

Thus, the total deviation at (j) is given by

$$\Delta g(j)=\Delta g_w(j)-\Delta g_w(j-1) \tag{11.2}$$

The variance of the deviation after a large number of trials by person 2 would be:

$$s^2 = 2\,s_w^2, \tag{11.3}$$

where s_w is the standard deviation of person 1's shots resulting from the wind alone, there being no correlation between $\Delta g_w(j)$ and $\Delta g_w(j-1)$. The variance in person 2's distribution was greater than person 1's because of the attempt to compensate each time for the error of the last shot which was not possible because the gusts were random.

The moral of this story is that action based upon an improper analysis of statistical data can result in overcorrecting to the detriment of process control. The statistical aspects of the problem must be properly developed and understood if process control is to be improved. Compensation by improving the use of the sight by person 1 was proper because the distribution of shots had been developed and analyzed to yield the right course of action; it was a systematic error that was detected and corrected. Eliminating caffeine and improving strength helped person 2. However, compensation for the wind had no validity from a statistical viewpoint. The error from wind gusts was random; the velocity and direction of the wind gusts could not be predicted from one shot to the next and thus it was impossible to compensate based upon prior events.

The problem of overcompensation is not limited to improperly basing corrective actions on sample sizes of one. Added error will be generated any time that compensation is attempted when the deviation of the sample mean from target is well within the normal statistical behavior expected for the process. Specific rules for judging when to act are based (as discussed below) upon a signal defined by one or more examples of behavior statistically unlikely for a random process.

11.4 Classical long-run SPC

SPC is developed in three stages. The first stage involves a set of experiments which establish the statistical properties of the production process. The second stage consists of routine and ongoing statistical sampling to confirm that the process statistics are being maintained and to detect special causes that cause meaningful deviations. These are deviations deemed highly unlikely when examined against the statistics established for the process at the outset. The third stage involves experimentation to discover ways of reducing product variance or common causes.

Although a process may be under statistical control, it may, nevertheless, be incapable of meeting customer needs. Thus, the first and second stages must document and confirm (1) that the results generated by the production process follow the governing rules of statistical control and (2) that the process is capable of meeting the needs of customers day in and day out. The third stage is an ongoing effort to improve the process so that over time the resulting product meets the rising expectations of the customer in a cost-effective manner.

Stage 1 requires that the process be monitored over a period of time under true production conditions. Monitoring is achieved by taking n samples of the product periodically and measuring important parameters. Usually these parameters are dimensions of the product but they can also be performance factors. The sample size, n, should be no larger than needed to give the accuracy in the mean and standard deviation needed to make intelligent decisions about the properties of interest. This is generally between four and six samples, five often being preferred. Once a sample size n and sampling frequency has been selected, a minimum of $k = 25$ but preferably 50 subgroups, each containing n samples, need to be made to determine the limits for the mean chart.

The means or **X-bar** chart is developed to monitor the behavior of the averages determined from the subgroups. We will consider this chart first although in practice the range or R chart needs to be evaluated before assessing the X-bar chart. The distribution of averages in Figure 11.9 is for 20 subgroups of five samples each taken from a known normal population of mean 10 and variance 2. The resulting standard deviation for the means, $\bar{g}$, is therefore $\sqrt{2/5} = 0.632$ which was used to construct the regions A, B, and C on the means chart. The limit for region C is one standard deviation of the mean, ±0.632, on either side of the dashed line equal to the mean of the means, $\bar{\bar{g}}$. The limits for regions B and A are two (±1.24) and three (±1.89) standard deviations, respectively, on either side of $\bar{\bar{g}}$.

The distribution of points between regions A, B, and C is used in assessing whether or not the process follows the standard rules for statistical control. In addition to examining the behavior of the sample means, the statistical behavior of the standard deviations or ranges for each subgroup of n samples is also used in assessing control.

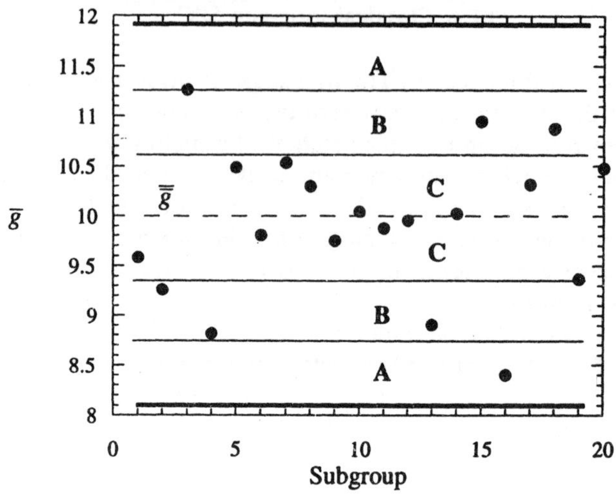

Figure 11.9 Schematic of mean chart showing the scatter from twenty subgroups taken in succession and control regions A, B, and C.

For most real SPC problems, the population statistics will not be known and must therefore be estimated from the sample statistics obtained from the 25 to 50 subgroups. An example of the data required to establish SPC is shown in Table 11.1. A total of $k = 50$ subgroups of $n = 5$ samples each are shown. The four columns on the right show the sample means, $\bar{g}$; variances, s^2; standard deviations, s; and ranges, R (the difference between the maximum and minimum measurements for a sample size n).

Table 11.1 Simulated 50 subgroups for SPC set-up

Sub-Group	*Sample 1*	*2*	*3*	*4*	*5*	*Ave*	*Var*	*Std Dev*	*Range*
1	11.73	12.70	11.24	10.18	11.54	11.48	0.83	0.91	2.52
2	8.70	9.86	8.88	11.38	8.00	9.36	1.71	1.31	3.38
3	11.71	11.94	10.28	7.46	9.41	10.16	3.36	1.83	4.48
4	12.36	8.39	8.51	10.70	9.76	9.94	2.73	1.65	3.98
5	8.73	9.42	8.95	10.41	10.69	9.64	0.76	0.87	1.96
6	9.74	10.52	11.32	9.59	8.59	9.95	1.06	1.03	2.73
7	11.13	9.56	12.54	10.69	11.54	11.09	1.20	1.10	2.98
8	7.71	10.20	9.32	10.01	9.74	9.39	0.99	1.00	2.48
9	8.39	10.65	12.72	9.08	10.48	10.26	2.79	1.67	4.33
10	7.72	8.38	11.09	11.96	7.95	9.42	3.85	1.96	4.24
11	9.32	8.08	11.09	11.02	11.30	10.16	1.98	1.41	3.22
12	9.90	9.27	9.12	8.64	8.27	9.04	0.39	0.62	1.63
13	6.29	9.76	12.15	9.68	8.99	9.38	4.41	2.10	5.87

14	9.24	8.28	11.27	11.54	10.01	10.07	1.87	1.37	3.26
15	9.32	10.44	8.39	11.27	8.04	9.49	1.85	1.36	3.23
16	9.88	9.21	8.97	8.24	8.55	8.97	0.40	0.63	1.63
17	10.99	8.96	9.92	10.57	9.06	9.90	0.80	0.90	2.03
18	10.36	9.09	10.63	9.58	8.15	9.56	1.00	1.00	2.48
19	10.56	9.05	8.93	11.11	11.82	10.30	1.62	1.27	2.89
20	8.00	10.08	6.99	11.29	9.95	9.26	3.00	1.73	4.30
21	9.71	10.53	12.79	11.01	11.05	11.02	1.27	1.13	3.07
22	11.55	9.82	8.99	10.56	7.64	9.71	2.23	1.49	3.91
23	11.00	12.90	9.55	7.83	8.04	9.86	4.52	2.13	5.07
24	10.05	7.73	10.44	7.96	10.14	9.27	1.70	1.30	2.70
25	10.96	7.92	8.66	9.99	10.32	9.57	1.56	1.25	3.04
26	10.41	11.05	10.61	11.62	12.02	11.14	0.46	0.68	1.61
27	11.18	11.66	7.77	9.77	10.60	10.20	2.33	1.53	3.89
28	12.13	9.39	9.39	11.80	8.77	10.30	2.40	1.55	3.36
29	9.74	11.40	8.78	9.40	10.47	9.96	1.02	1.01	2.62
30	9.71	10.90	11.42	12.98	11.21	11.24	1.38	1.18	3.28
31	9.80	10.26	13.27	11.31	10.74	11.08	1.83	1.35	3.48
32	11.05	9.11	13.69	11.93	9.56	11.07	3.43	1.85	4.58
33	10.31	12.03	12.38	8.09	12.38	11.04	3.46	1.86	4.30
34	8.15	8.89	8.23	6.97	9.79	8.41	1.08	1.04	2.82
35	10.79	11.03	6.18	8.74	7.09	8.77	4.69	2.17	4.86
36	12.71	8.03	10.38	11.55	9.75	10.48	3.16	1.78	4.68
37	9.90	10.43	12.52	12.06	9.66	10.91	1.68	1.30	2.86
38	12.86	10.08	8.13	11.10	8.91	10.22	3.47	1.86	4.73
39	9.68	11.04	13.76	11.63	8.05	10.83	4.58	2.14	5.71
40	8.00	9.41	8.99	9.44	10.15	9.20	0.62	0.79	2.15
41	10.12	7.99	9.56	8.47	8.25	8.88	0.84	0.92	2.13
42	11.25	8.26	10.57	10.92	13.29	10.86	3.23	1.80	5.03
43	10.30	10.79	10.56	11.34	10.81	10.76	0.15	0.38	1.04
44	9.66	11.40	12.20	8.21	6.37	9.57	5.59	2.36	5.83
45	12.23	11.05	9.22	9.92	9.17	10.32	1.72	1.31	3.07
46	10.57	13.54	9.65	12.19	7.95	10.78	4.72	2.17	5.59
47	9.11	9.81	10.55	10.72	9.41	9.92	0.49	0.70	1.61
48	12.87	10.91	8.18	13.06	8.45	10.69	5.44	2.33	4.88
49	8.36	9.76	7.83	9.93	12.23	9.62	2.92	1.71	4.39
50	8.83	7.79	10.22	11.17	9.53	9.51	1.67	1.29	3.38
					Averages =	10.04	2.20		3.46

These data were taken from a simulated normal population having a true population mean of 10.00 and a true population variance of 2.00. The mean of the means, $\bar{\bar{g}}$, is 10.04. The pooled sample variance over the 50 subgroups is 2.20 and, as a result, the limits for regions A, B, and C determined from the pooled sample variance would be somewhat more open than their true values. The average of the ranges is noted as $\bar{R}$.

The population standard deviation and the expected value of the sample range, $E(R) = \bar{R}$, for a sample size of n are connected by a theoretical relationship of the form:

$$\sigma = \bar{R}/d_2 \tag{11.4}$$

The factor d_2, shown in Table 11.2, is a function of the sample size, n. As a result of this relationship, the average of the ranges, which is easier to calculate than standard deviation when recording data by hand in a production environment, can be used to set the boundaries of the three regions. When the measurements are entered into a computer database, it is more direct to determine the three regions using the standard deviation calculated by the computer from the pooled variance and not worry about the factor d_2 which is somewhat cumbersome because of its dependence on sample size n.

When the upper and lower limits on the X-bar chart are set at $\pm 3\sigma$, the error would be only 0.0027 (2.7 mistakes per 1000) for assuming that any individual $\bar{g}$ which was measured after the establishment of the SPC chart and which exceeded either the upper or lower limit came from a population having an average different from the population used to establish the limits. However, this same Type I error level does not apply to the collection of 50 points used to establish the limits because the probability of finding a point outside the 3σ limits is 50 times higher for the collection.

In assessing whether or not the distribution of the 50 means follows what is expected for a random variable, it is useful to make a normal probability plot as in Figure 11.10 and look for outliers. The plot should also be approximately linear as a result of the central limit theorem.

The resulting X-bar chart for the 50 subgroups is shown in Figure 11.11. The limits were calculated at ± 1, 2, and 3σ using the standard deviation for the mean computed from the pooled sample variance and equal to $\sqrt{2.20/5}$. Because of the relationship between the population standard deviation and the

Table 11.2 Constant d_2 relating average of range to population standard deviation (source: R.E. DeVor, T. Chang, and J.W. Sutherland (1992) *Statistical Quality Design and Control* © 1992. Adapted by permission of Prentice-Hall, Inc., Upper Saddle River, NJ)

n	d_2	n	d_2
2	1.128	6	2.534
3	1.693	7	2.704
4	2.059	8	2.847
5	2.326	9	2.970
		10	3.078

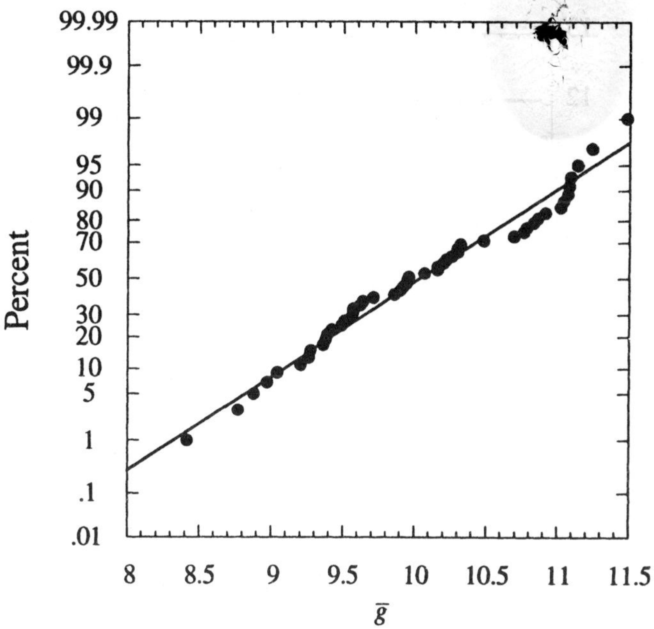

Figure 11.10 Distribution of means from 50 subgroups plotted on normal probability paper.

range given by Equation 6.4, the lines defining the regions A, B and C on the chart of the means can also be constructed using the relations [2]:

$$L_A = \bar{g}_m \pm 3\bar{R} / \left[\sqrt{n}d_2\right], \tag{11.5}$$

$$L_B = \bar{g}_m \pm 2\bar{R} / \left[\sqrt{n}d_2\right], \tag{11.6}$$

$$L_C = \bar{g}_m \pm \bar{R} / \left[\sqrt{n}d_2\right]. \tag{11.7}$$

For a process under statistical control, it is necessary but not sufficient that all of the rules in Table 11.3 are obeyed for the chart of the means [3]. Rules 2 and 3 only apply to zones on one side of the mean of the means line. Rules 1, 4, 5, and 6 also apply to the range chart, to be considered shortly, which has only one zone. These rules are arrived at by assuming that the frequency distribution of means is normal:

$$f(g) = \frac{1}{\sqrt{2\pi}} \exp\left[\frac{-(g-\bar{g})^2}{2s_m^2}\right], \tag{11.8}$$

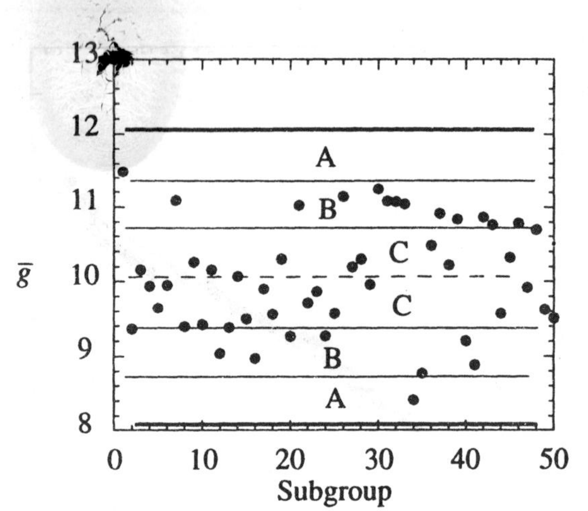

Figure 11.11 SPC X-bar chart limits and regions for sample means developed from the 50 subgroup averages shown.

and that the probability of the behavior described by any one of the events described by one of the rules is roughly equal to the probability, 0.0027, of a sample mean exceeding three standard deviations. For a variable which follows the normal distribution, 68.3% of the observations lie between $\pm s_m$, 95.4% lie between $\pm 2\, s_m$, and 99.7% lie between $\pm 3s_m$.

The range chart is also required to fully assess the rules for statistical process control and should, in fact, be examined before considering the X-bar chart. Lower and upper control limits to be used for future individual range measurements are determined by multiplying $\bar{R}$, the mean of the sample ranges determined from the first 50 subgroups, by the factors D_3 and D_4, respectively, in Table 11.4 which are a function of sample size. Tests 4, 5 and 6 given above for the X-bar chart also apply to the range chart. Intermediate limits as found in the X-bar chart are not used for the range chart. The distribution of ranges in

Table 11.3 Rules defining SPC conditions (source: R.E. DeVor, T. Chang, and J.W. Sutherland (1992) *Statistical Quality Design and Control*, © 1992. Adapted by permission of Prentice-Hall, Inc., Upper Saddle River, NJ)

1.	No points beyond zone A.
2.	No two of three successive points in zone A.
3.	No four out of five points in zone B.
4.	No runs of seven or more points above or below the mean of the means.
5.	No six successive points showing a continuous increase or decrease.
6.	No fourteen successive points should oscillate up and down one after another
7.	No eight points in a row should all fall outside of zone C.
8.	No fifteen successive points should fall in zone C.

the last column of Table 11.1 which were used to establish $\overline{R}$ for the example are shown in Figure 11.12 along with the upper and lower limits used to evaluate future range measurements.

11.5 Process capability

As stated earlier, a process that is under statistical process control may still be ineffective as it may be producing parts that are not acceptable to the customer or not as good or better than competition. If a process is under statistical process control and producing parts that are within tolerance guidelines for customer satisfaction, it is said to be **capable**. The degree to which the process is capable is measured by a coefficient C_{pk} given by:

$$C_{pk} = \frac{USL - \overline{\overline{g}}}{3s}, \tag{11.9}$$

or by

$$C_{pk} = \frac{\overline{\overline{g}} - LSL}{3s}, \tag{11.10}$$

depending on which is less where USL is the upper specification limit and LSL is the lower specification limit. If the standard deviation defined for the sample distribution is s, the standard deviation for the distribution of the sample means (the means from a series of samples) is given by

$$s_m = \frac{s}{\sqrt{n}}, \tag{11.11}$$

where n is the sample size.

Another process capability index, C_p, is defined as the difference between the upper and lower control limits divided by $6s$:

Table 11.4 Range chart limit factors (source: R.E. DeVor, T. Chang, and J.W. Sutherland (1992) *Statistical Quality Design and Control*, © 1992. Adapted by permission of Prentice-Hall, Inc., Upper Saddle River, NJ)

n	D_3	D_4
4	0	2.28
5	0	2.11
6	0	2.00
7	0.08	1.92
8	0.14	1.86
9	0.18	1.82
10	0.22	1.78

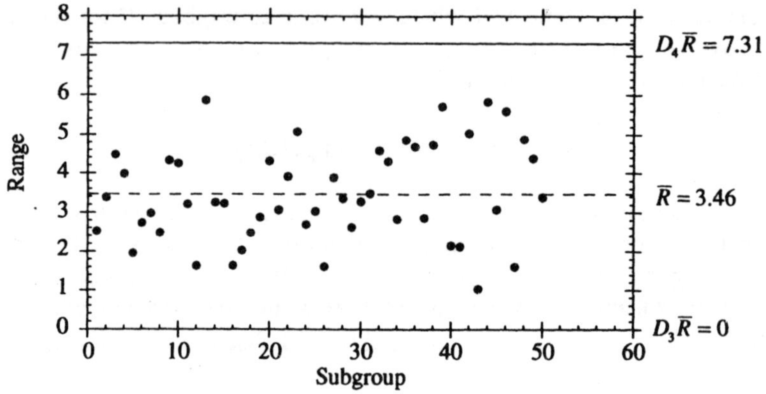

Figure 11.12 SPC range chart for simulated experiment.

$$C_p = \frac{USL - LSL}{6s}. \tag{11.12}$$

Because it is more stringent, the capability index C_{pk} is generally preferred over C_p.

11.6 Statistical tolerancing

Process capability influences the dimensional integrity of assemblies through the 'stack-up' of dimensions. Fortunately, we get a bit of a break when the dimensions are stacked up (added together) if the variations in the dimensions involved are not correlated. In this regard consider a larger dimension, d_{Total}, equal to a sum of N_d smaller dimensions $d(i)$:

$$d_{Total} = \sum_{i=1,N_d} d(i). \tag{11.13}$$

The statistical properties of d_{Total}, its mean and standard deviation, can be computed from the statistical properties of the dimensions that generate it. Because the relation is linear, the mean (average) of d_{Total} is equal to the sum of the means of the individual dimensions, and the variance of d_{Total} if the variances of the individual dimensions are not correlated, is equal to the sum of the individual variances:

$$s_d^2 = \sum_{i=1,N_d} s^2(i). \tag{11.14}$$

If the individual standard deviations are equal, then it follows from the above relation that the standard deviation of the dimensions, given by s_d, is equal to $s\sqrt{N_d}$. What is encouraging about this is that the standard deviation of the sum of the dimensions is less than the sum of the individual standard deviations, sN_d.

A straightforward illustration of this relationship – the 'law of variances' – can be made by using a spreadsheet to simulate the dimensions of the individual $d(i)$ that add up to d_{Total}:

$$d(i) = g\ (\text{random}) \tag{11.15}$$

The random variable g is formed by the average of twelve quasi-random numbers that range uniformly between –0.5 and +0.5 which is then added to 10:

$$g = 10 + \sum_{j=1\to 12} [\text{RAND}() - 0.5]$$

The variance of a single random number that ranges uniformly between 0 and 1 is equal to 1/12 which is equal to the integral of the function x^2 from $x = -0.5 \to +0.5$. As a result of averaging the twelve random numbers used to compute the variable g, the variance of g is unity. The distribution of g itself is not uniform but approximately normal because, using the central limit theorem, it is the sum of twelve randomly distributed quantities.

Listed in Table 11.5 are 100 computations of a dimension d_{Total} each of which is equal to a sum of 20 of the random variables g. This simulates measurement of the stack-up for 100 assemblies of a dimension formed by linearly connecting 20 parts of the same nominal length each having the same, but uncorrelated, standard deviation. Figure 11.13 is a normal probability plot of the 100 numbers shown in Table 11.5.

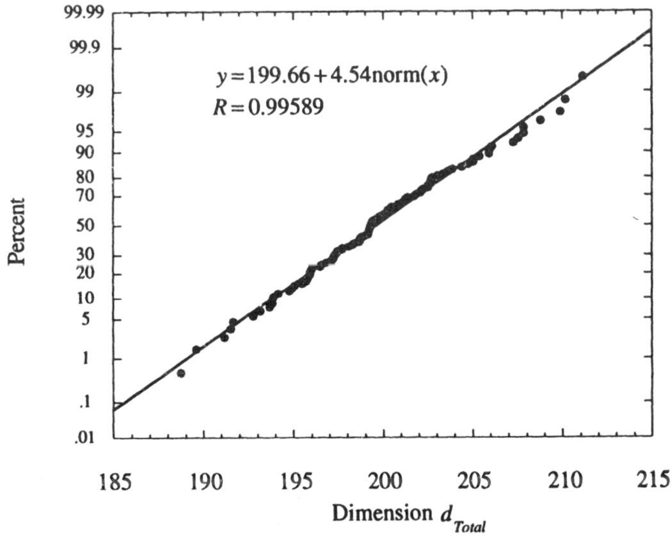

Figure 11.13 Simulated demonstration of the 'law of variances.'

Table 11.5 Overall dimensional stack-up for 20 elements of 10.00 average length having a standard deviation equal to 1.0

200.47	208.78	199.23	200.02	198.7
193.83	202.68	205.39	195.83	202.74
201.73	206.09	194.93	207.82	202.09
211.13	195.77	204.78	197.72	200.8
195.48	189.6	192.78	202.6	207.28
195.04	197.33	198.29	200.24	198.64
198.06	197.23	203.06	201.19	198.77
194.15	195.69	199.16	193.89	197.4
200.17	205.93	205.96	197.25	196.04
203.89	201.25	210.16	199.47	191.53
199.17	197.69	193.17	199.27	200.85
188.71	199.36	207.82	194.78	201.09
199.19	202.67	201.36	199.84	193.83
202.23	202.51	196.49	209.88	201.85
197.08	199.81	199.67	195.91	199.37
191.66	198.7	191.18	197.17	205.05
202.51	200.41	196.81	198.35	202.13
203.37	197.46	204.43	193.68	196.53
199.17	200.25	198.92	195.97	207.52
203.62	199.31	199.31	195.9	205.05

The standard deviation found for the experiment is 4.542. This is in close agreement with the theoretical unbiased standard deviation given by $\sqrt{20}=4.472$ for this problem. The distribution is seen to be very close to normal. This exercise demonstrates the law of variances (Equation 11.14) for the special case where each of the variances are equal. If variances were correlated resulting in all of the standard deviations adding together, the three sigma range of d_{Total} for the above problem, which is equal to 200 ± 13.6, would have become 200 ± 60.

11.7 Short production runs

Short production runs are being made with increasing frequency by manufacturing firms to better target their diverse customer base, to meet just-in-time delivery requirements, and to make the continuing product changes and process improvements needed to stay abreast of highly competitive markets [4]. As a result, there will not be sufficient parts produced at times to generate the 25–50 subgroups needed to establish the classical SPC X-bar and R charts. The statistics for short-run SPC have been studied by Hillier [5], Quesenberry [6], and Rahn [7].

The major issue to be addressed in short production runs as well as long production runs is whether the process is or was capable of meeting customer requirements. It is important to separate this question from the SPC issue of whether the dimensions or properties of the product behaved as random variables during the short production run. Fortunately, we can examine the short run issues using methods discussed in Chapters 8 and 9.

11.7.1 GENERAL CONSIDERATIONS

Consider the case of a short run in which only k subgroups were taken for measurements of a single attribute, the sample size being five for each subgroup. The sampling process can be described in terms of an experimental design written in the form of a $[\mathbf{X}]_\varphi$ matrix, Table 11.6. (See Appendix B for a discussion of design of experiments and related matrix algebra operations.) Each subgroup is interpreted as an experimental trial having five replications. The standard set point notation results in a single factor which has levels $j=1$ through k so we do not require double indices. The variance and average for the five replications for each subgroup are shown in the last two columns.

The $[\mathbf{X}]_\varphi$ form, which defines a baseline equal to the average for all of the outcomes, is used because no trial has any more significance as a baseline than any other trial. The replications shown (which are called samples in SPC) were taken from a simulated normal population having a mean of 10 and a variance of 1.

The general expressions for the elements of the covariance and solution matrices obtained from the $[\mathbf{X}]_\varphi$ matrix for an arbitrary number of subgroups z are shown in Tables 11.7 and 11.8, respectively.

Although the variance coefficients for the 'experiment' can be formally computed using the expression:

$$\xi'(j) = 10\left\{\mathrm{Log}\left(s^2(j)\right) - \overline{\mathrm{Log}(s^2)}\right\} \tag{11.16}$$

Table 11.6 Short production run of seven subgroups interpreted as an experimental design described by a one-at-a-time $[\mathbf{X}]_\varphi$ matrix

	$[\mathbf{X}]_\varphi$											Replications					
Sub group	*0*	*11*	*12*	*13*	*14*	*15*	*16*	. . .	*1*	*2*	*3*	*4*	*5*	*Sub group*	*Var*	*Ave*
1	1	−1	−1	−1	−1	−1	−1	. . .	9.1	10.5	9.7	9.4	13.0	1	2.4	10.3
2	1	1	0	0	0	0	0	. . .	10.1	10.4	9.4	8.5	10.7	2	0.8	9.8
3	1	0	1	0	0	0	0	. . .	11.6	10.2	10.1	10.0	9.9	3	0.5	10.3
4	1	0	0	1	0	0	0	. . .	8.6	9.6	10.9	10.8	8.0	4	1.6	9.6
5	1	0	0	0	1	0	0	. . .	10.0	9.3	9.4	8.9	9.2	5	0.2	9.4
6	1	0	0	0	0	1	0	. . .	10.4	8.6	10.3	8.9	9.4	6	0.7	9.5
7	1	0	0	0	0	0	1	. . .	9.0	10.6	10.3	8.9	10.0	7	0.6	9.8
.	.	.	.	.	.	.	.	. . .	.	.	.	.	.	.	.	.
.	.	.	.	.	.	.	.	. . .	.	.	.	.	.	.	.	.

Table 11.7 Standard form of the covariance matrix for k subgroups

	0	*11*	*12*	*13*	*14*	.	.	.
0	$1/k$	0	0	0	0	.	.	.
11	0	$(k-1)/k$	$-1/k$	$-1/k$	$-1/k$	.	.	.
12	0	$-1/k$	$(k-1)/k$	$-1/k$	$-1/k$	.	.	.
13	0	$-1/k$	$-1/k$	$(k-1)/k$	$-1/k$	.	.	.
14	0	$-1/k$	$-1/k$	$-1/k$	$(k-1)/k$	.	.	.
.	.	.	.	.	.	.	.	.
.	.	.	.	.	.	.	.	.
.	.	.	.	.	.	.	.	.

Table 11.8 Standard form of the solution matrix for k subgroups

	0	*1*	*2*	*3*	*4*	.	.	.
0	$1/k$	$1/k$	$1/k$	$1/k$	$1/k$	.	.	.
11	$-1/k$	$(k-1)/k$	$-1/k$	$-1/k$	$-1/k$	.	.	.
12	$-1/k$	$-1/k$	$(k-1)/k$	$-1/k$	$-1/k$	.	.	.
13	$-1/k$	$-1/k$	$-1/k$	$(k-1)/k$	$-1/k$	.	.	.
14	$-1/k$	$-1/k$	$-1/k$	$-1/k$	$(k-1)/k$	.	.	.
.	.	.	.	.	.	.	.	.
.	.	.	.	.	.	.	.	.
.	.	.	.	.	.	.	.	.

derived from Equation 9.7 for the NIB condition where the dummy index j is the subgroup number, we will not use them because of the complications brought about by their distribution for a purely random process not being symmetric about zero. Instead we will evaluate null hypothesis $H1_0$ for variances using a form of the F ratio as it readily yields a significance test for this one-at-a-time problem.

The nature of the $\varphi(j)$ coefficients for a random process, however, is such that they should be distributed equally about zero, the probability of finding $\varphi = -x$ being the same as the probability of finding $\varphi = x$. The $k-1$ coefficients used to examine whether or not the distribution of the averages from the subgroups can be assumed to have a common population average (null hypothesis $H2_0$) are computed from the NIB expression in Equation 9.6 which reduces to:

$$\varphi(j-1) = \bar{g}(j) - \bar{\bar{g}} \tag{11.17}$$

for this problem for $j > 1$. Instead of the eight rules in Table 11.3 defining the SPC conditions for the X-bar chart, we will have only one short-run rule for the limits on the average coefficients which is the counterpart to rule 1 in Table 11.3. The short-run rule, however, is global in that it considers the averages for all k subgroups, whereas the long-run rule 1 is for a single subgroup taken after the criteria and limits for SPC have been established.

11.7.2 SHORT-RUN VARIANCE LIMITS

Before testing the averages, it is again necessary to first test the hypothesis $H1_0$ of a common population variance. An F test can be readily formed for this purpose using the ratio of the measured variance $s^2(j)$ divided by its complement pooled variance:

$$s^2_{CP}(j) = \left\{-s^2(j) + \sum_i s^2(i)\right\} / [k-1] \tag{11.18}$$

which is equal to the pooled variance from the other subgroups not including j. The above expression is readily programmed into the spreadsheet and can be used to compute each $s^2_{CP}(j)$ alongside a list of measured $s^2(j)$. We compute the variance ratio with $s^2(j)$ having $n-1$ df in the numerator and $s^2_{CP}(j)$ having $[n-1][k-1]$ df in the denominator:

$$F(j) = \frac{s^2(j)}{s^2_{CP}(j)} \tag{11.19}$$

Shown in Table 11.9 are the limits for this ratio for three different experiment-wise errors, a range of subgroups, and $n = 5$ replications. The upper limits were computed from the spreadsheet using the expression:

$$F_{Upper}=\text{FINV}(\alpha/[2*k],4,4*[k-1])$$

where α is the experiment-wise error shown at the top of each column and the lower limits were computed from the expression:

$$F_{Lower}=\text{FINV}(1-\alpha/[2*k],4,4*[k-1])$$

This test for homogeneity of variance resembles Cochran's test [8] but it is different computationally (and simpler when using a spreadsheet) through the use of the complement pooled variance in the denominator and Bonferroni's method for computing the experiment-wise error.

An example of the use of the upper limits on the variance ratio is shown in Figure 11.14 in which the scatter from ten repeats of an experiment are plotted for $n = 5$ and $k = 50$. The samples were taken randomly from a simulated normal population having a variance of unity. No point is seen to exceed either of the two upper limits shown. This is not surprising as the expected number of points out of the 500 total points in 10 experiments that should exceed the 10% upper limit is only 1/2. Use of a 10% error limit means that in 10 SPC experiments of 50 points (subgroups) each, we should, on average, have only one experiment fail the test in that one point, on average, should appear either above the upper limit or below the lower limit (not shown in Figure 11.14).

Table 11.9 Limits on variance ratios and average deviation for short-run SPC

	s^2/s^2_{CP} Experiment-wise Error							$\lvert\varphi\rvert \sqrt{s_P^2/5}$ Experiment-wise Error		
	0.0027		0.05		0.1					
Sub-Groups	Upper	Lower	Upper	Lower	Upper	Lower	Sub-Groups	0.0027	0.05	0.1
3	18.110	0.0136	7.450	0.0613	5.863	0.0890	3	3.387	2.090	1.779
4	12.249	0.0122	6.150	0.0543	5.080	0.0784	4	3.521	2.315	2.016
5	10.183	0.0111	5.638	0.0491	4.773	0.0707	5	3.594	2.455	2.167
6	9.160	0.0102	5.378	0.0451	4.622	0.0648	6	3.642	2.553	2.275
7	8.564	0.0095	5.228	0.0419	4.540	0.0601	7	3.680	2.628	2.358
8	8.175	0.0089	5.136	0.0393	4.494	0.0564	8	3.710	2.688	2.424
9	7.905	0.0085	5.076	0.0372	4.468	0.0532	9	3.734	2.738	2.479
10	7.713	0.0080	5.037	0.0353	4.455	0.0505	10	3.757	2.781	2.526
11	7.567	0.0077	5.012	0.0337	4.450	0.0482	11	3.776	2.818	2.567
12	7.454	0.0074	4.995	0.0323	4.450	0.0462	12	3.794	2.851	2.603
13	7.367	0.0071	4.985	0.0311	4.453	0.0444	13	3.810	2.880	2.635
14	7.298	0.0068	4.980	0.0300	4.459	0.0428	14	3.824	2.906	2.664
15	7.240	0.0066	4.978	0.0290	4.467	0.0413	15	3.839	2.930	2.691
16	7.196	0.0064	4.979	0.0280	4.476	0.0400	16	3.852	2.953	2.715
17	7.156	0.0062	4.981	0.0272	4.486	0.0388	17	3.864	2.973	2.738
18	7.123	0.0061	4.985	0.0265	4.497	0.0377	18	3.875	2.992	2.759
19	7.098	0.0059	4.991	0.0258	4.508	0.0367	19	3.885	3.010	2.778
20	7.076	0.0058	4.997	0.0251	4.519	0.0358	20	3.895	3.026	2.796
21	7.058	0.0056	5.003	0.0245	4.530	0.0349	21	3.905	3.042	2.814
22	7.043	0.0055	5.010	0.0240	4.542	0.0341	22	3.915	3.057	2.830
23	7.029	0.0054	5.018	0.0234	4.553	0.0334	23	3.923	3.071	2.845
24	7.021	0.0053	5.026	0.0229	4.565	0.0327	24	3.932	3.084	2.860

25	7.007	0.0052	5.034	0.0225	4.576	0.0320	25	3.940	3.097	2.873
26	6.999	0.0051	5.043	0.0220	4.587	0.0314	26	3.947	3.109	2.887
27	6.999	0.0050	5.051	0.0216	4.598	0.0308	27	3.955	3.120	2.899
28	6.992	0.0049	5.060	0.0212	4.609	0.0302	28	3.962	3.131	2.911
29	6.985	0.0048	5.068	0.0209	4.620	0.0297	29	3.969	3.142	2.923
30	6.985	0.0047	5.076	0.0205	4.631	0.0292	30	3.976	3.152	2.934
31	6.985	0.0046	5.085	0.0202	4.642	0.0287	31	3.983	3.162	2.944
32	6.978	0.0046	5.094	0.0199	4.652	0.0283	32	3.990	3.172	2.955
33	6.978	0.0045	5.102	0.0196	4.662	0.0278	33	3.996	3.181	2.965
34	6.978	0.0044	5.110	0.0193	4.673	0.0274	34	4.002	3.190	2.974
35	6.978	0.0044	5.119	0.0190	4.683	0.0270	35	4.007	3.198	2.983
36	6.978	0.0043	5.127	0.0187	4.692	0.0266	36	4.013	3.206	2.992
37	6.985	0.0043	5.135	0.0185	4.702	0.0263	37	4.019	3.214	3.001
38	6.985	0.0042	5.144	0.0182	4.712	0.0259	38	4.025	3.222	3.009
39	6.985	0.0041	5.152	0.0180	4.721	0.0256	39	4.029	3.229	3.017
40	6.985	0.0041	5.160	0.0178	4.730	0.0253	40	4.035	3.237	3.025
41	6.985	0.0040	5.168	0.0176	4.739	0.0250	41	4.041	3.244	3.033
42	6.985	0.0040	5.175	0.0174	4.748	0.0247	42	4.044	3.251	3.041
43	6.985	0.0040	5.183	0.0172	4.757	0.0244	43	4.050	3.258	3.048
44	6.992	0.0039	5.191	0.0170	4.766	0.0241	44	4.053	3.264	3.055
45	6.999	0.0039	5.199	0.0168	4.774	0.0238	45	4.059	3.271	3.062
46	6.999	0.0038	5.206	0.0166	4.783	0.0236	46	4.062	3.277	3.068
47	6.999	0.0038	5.214	0.0164	4.791	0.0233	47	4.068	3.283	3.075
48	6.999	0.0038	5.221	0.0162	4.800	0.0231	48	4.071	3.289	3.082
49	7.014	0.0037	5.228	0.0161	4.808	0.0228	49	4.077	3.295	3.088
50	7.014	0.0037	5.236	0.0159	4.816	0.0226	50	4.080	3.300	3.094

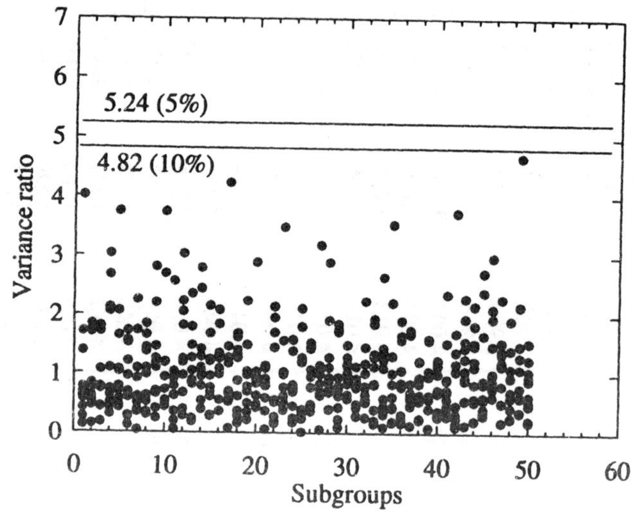

Figure 11.14 The scatter of the variance ratios from ten simulated experiments with $n = 5$ and $k = 50$ for each.

In using limits of the type described here for establishing that a process is random, it is wise not to choose the error too small. For example, use of the three-sigma error of 0.0027 would not be very meaningful because the goal is to detect with some frequency those situations which are not random using the limits. The choice of 10% for the experiment-wise error seems reasonable.

It is also helpful in establishing that a process is random to examine the distribution of the variance ratios using a normal probability plot. The scatter for the ratios computed from samples taken from a simulated normal population for $n = 5$ and $k = 50$ is shown in Figure 11.15. The smooth curve is a plot of the theoretical variance ratio given by *F=FINV*($\wp_C$,4,4*[*k*–1]) which was computed for plotting using 99 equally spaced points over the cumulative distribution $\wp_C$ from 0.01 to 0.99.

11.7.3 SHORT-RUN AVERAGE LIMITS

If the test for assuming a common variance has been met, the conditions imposed on the $k-1$ coefficients, $\varphi(j+1)$ for $j > 1$, for having a common population average over the k subgroups can be evaluated using the pooled variance with the t-distribution. The general short-run SPC limits for the average divided by the square root of the pooled variance over n are shown also in Table 11.9. These were computed from the relationship:

$$\left|\frac{\varphi}{\sqrt{\frac{s_P^2}{n}}}\right| = \sqrt{\frac{k-1}{k}}\,\mathrm{TINV}\left(\frac{\alpha}{k-1},[n-1]k\right) \tag{11.20}$$

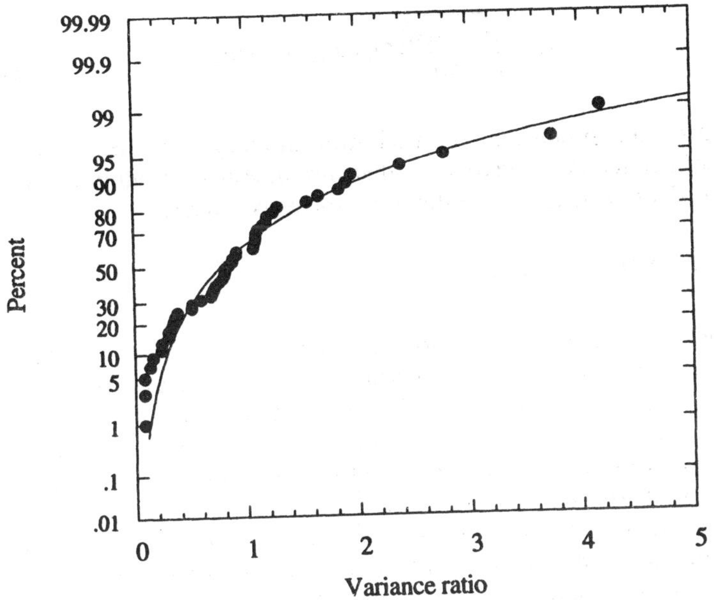

Figure 11.15 The variance ratio scatter from a single simulated experiment plotted on normal probability paper against the theoretical curve for $n = 5$ and $k = 50$.

which follows from Equation 9.10b where α is the total, two-tailed, experiment-wise error.

11.7.4 SHORT-RUN CAPABILITY

The capability of the short-run process to meet customer requirements defined as C_{pkSR} is given by the smaller of:

$$C_{pkSR} = \frac{USL - \bar{\bar{g}}}{\sqrt{s_P^2 \text{TINV}(\alpha,[n-1]k)}} \tag{11.21}$$

or:

$$C_{pkSR} = \frac{\bar{\bar{g}} - LSL}{\sqrt{s_P^2 \text{TINV}(\alpha,[n-1]k)}} \tag{11.22}$$

11.7.5 SHORT-RUN CONTROL

The standard long-run SPC charts are designed to detect a change in processing conditions between a prior set of subgroups of 25 or more and the current subgroup. The equivalent control limit for short-run SPC is given by:

$$\varphi_C = \sqrt{\frac{[k-1]s_P^2}{kn}\mathrm{TINV}(\alpha,[n-1]k)} \tag{11.23}$$

This expression should only be used after an analysis of the variance and average coefficients has validated (1) that the process is random for the subgroup level of interest, k, and (2) that the process is capable.

11.7.6 SHORT-RUN EXAMPLE

An example of the analysis of a short-run problem is shown in Table 11.10 for $n = 5$ and $k = 20$. The simulated measurements were taken from a normal population of mean 10 and variance 0.2. The variance ratios determined for each of the trials are well within the limits shown in Table 11.9. The normal probability plot of the variance ratios shown in Figure 11.16 is seen to closely agree with the theoretical curve. The averages divided by the square root of the quantity equal to the pooled variance of 0.172 divided by 5 also lie well within the limits shown for this statistic in Table 11.9. A normal probability plot of this statistic (Figure 11.17) was made to further confirm that the scatter of the measurements was consistent with that expected for normally distributed random error.

The process capability computed for this example from Equations 11.21 and 11.22 exceeds unity if the upper specification limit, USL, is greater than 11.34 and if the lower specification limit, LSL, is less than 8.77. Stating it once again for emphasis, process capability for short- and long-run SPC is the most fundamental statistical property for the manufacturing process in terms of meeting customer needs. Confidence in using short production runs to achieve low inventories as well as high quality can be gained by demonstrating again and again through short-run SPC evaluations that the process being used is capable. Short production runs also require that set-up times from one job to another be reduced from the times satisfactory for long-run set-ups.

11.8 Continuous improvement (Motorola's Six Sigma Process)

A shortcoming to having just a process capability index as a measure of quality performance is that it suggests that the outcome may be good enough when continuous improvements may be possible that would be cost effective and would improve the net value to society of the product. SPC can provide the structure for ongoing experiments to improve value and reduce costs much like a continuation of the DoE used in the early part of the product realization process. For this approach to be successful, it is necessary that the major variables that affect the process be measured. The results of the trials can be analyzed using the same formalism introduced earlier for DoE.

Table 11.10 A simulated short-run SPC trial

Sub Group	Samples 1	2	3	4	5	Variance s^2	s^2_{CP}	Var Ratio	Average $\bar{g}$	φ	$\lvert\varphi\rvert\sqrt{s_P^2/5}$
1	10.37	10.35	9.69	9.90	9.93	0.090	0.177	0.51	10.05		
2	10.29	9.97	10.31	10.52	10.55	0.054	0.178	0.31	10.33	0.27	1.47
3	9.69	10.52	9.11	9.46	10.14	0.308	0.165	1.87	9.78	−0.27	1.46
4	9.75	9.64	10.73	10.66	9.91	0.266	0.167	1.59	10.14	0.08	0.44
5	9.92	10.41	9.69	10.81	9.75	0.230	0.169	1.36	10.12	0.06	0.33
6	10.32	10.43	10.19	9.93	10.38	0.040	0.179	0.22	10.25	0.20	1.05
7	9.49	10.25	9.83	9.94	10.79	0.243	0.168	1.44	10.06	0.00	0.02
8	9.61	9.62	10.30	10.86	10.24	0.276	0.167	1.66	10.13	0.07	0.38
9	10.55	9.49	10.33	9.79	9.16	0.332	0.164	2.03	9.87	−0.19	1.03
10	10.23	9.72	10.09	10.00	10.28	0.050	0.179	0.28	10.06	0.01	0.04
11	10.53	10.47	9.94	9.87	10.12	0.092	0.176	0.52	10.19	0.13	0.70
12	10.39	9.79	9.71	10.20	10.44	0.116	0.175	0.66	10.11	0.05	0.27
13	9.72	10.42	10.56	9.79	10.36	0.149	0.173	0.86	10.17	0.11	0.61
14	9.89	9.77	10.46	9.78	10.45	0.128	0.175	0.73	10.07	0.01	0.08
15	9.98	9.92	10.20	9.29	10.08	0.127	0.175	0.72	9.89	−0.16	0.87
16	9.95	9.77	10.08	9.79	10.16	0.030	0.180	0.17	9.95	−0.11	0.58
17	10.21	10.51	9.86	10.23	9.78	0.089	0.177	0.51	10.12	0.06	0.33
18	10.76	9.42	9.44	10.39	9.86	0.351	0.163	2.15	9.98	−0.08	0.43
19	9.38	10.48	9.95	10.47	10.42	0.226	0.169	1.34	10.14	0.08	0.44
20	10.27	9.93	9.18	10.06	9.22	0.248	0.168	1.47	9.73	−0.32	1.75
						Average =0.1722			Average =10.06		

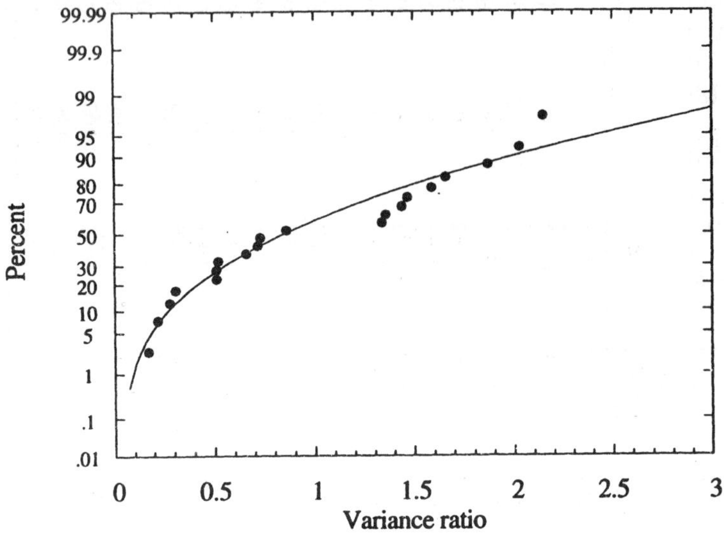

Figure 11.16 The variance ratio scatter from twenty simulated subgroups with five samples for each plotted against the theoretical curve for $n = 5$ and $k = 20$ on normal probability paper.

11.8.1 MOTOROLA'S SIX SIGMA SPC PROGRAM

The Motorola Six Sigma SPC Program [9] is a ten-step management and DoE continuous improvement process having the goal of maintaining all critical processes under statistical control at a level of at least 12σ between the upper

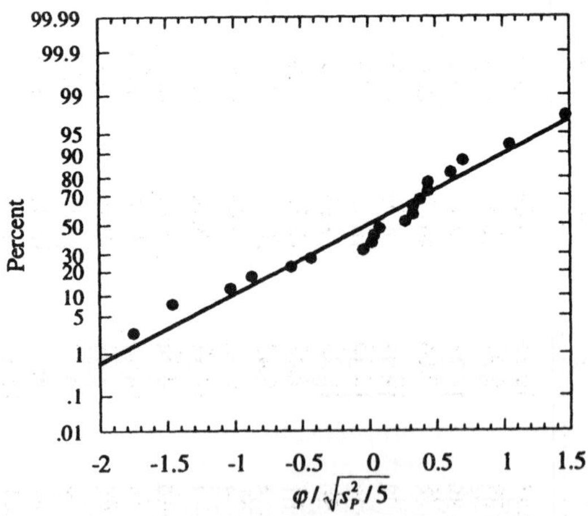

Figure 11.17 Scatter of the normalized average deviation from the grand average for $n = 5$ and $k = 20$.

and lower control limits (the limits being located at $\pm 6\sigma$ and thus the name for the program) and/or a C_{pk} of greater than 1.5 based upon an allowed mean shift of 1.5σ. The ten steps specified by Motorola in their guide are as follows (copyright of Motorola, used by permission):

Step 1 Prioritize opportunities for improvement

Quantify any known or perceived opportunities for improvement. Specify the problems in quantifiable terms such as how much, when, where, and how. Indicate which impact the customer, reliability, product quality, and yields. Identify potential cost savings to the customer and Motorola.

Step 2 Select the appropriate team

Select a small group of people with the product/process knowledge, experience, technical discipline, authority, time, and skill in the specific area of concern. Establish and identify the role of the team and each member. Identify a 'Champion' (in addition to the team leader) who can assist the team and can ensure that the team's recommendations are carried out. The team must decide what and how much it can accomplish.

Step 3 Describe the total process

Utilize a process flow diagram to illustrate the possible variations and alternatives of the process. Include all equipment, manpower, methods, tools, piece parts, and measurement instruments in the process description. Identify all of the known input/output relationships. Highlight any of the alternative work procedures and flows.

Step 4 Perform measure system analysis

Determine precision, accuracy, repeatability and reproducibility of each instrument or gauge used in order to ensure that they are capable. Ensure that there is a P/T ratio of 0.10 or less.

Step 5 Identify and describe the potential critical product(s)/process(es)

List and describe all of the potential critical processes obtained from brainstorming sessions, historical data, yield reports, failure analysis reports, and analysis of line fallout and model the potential problems using graphical illustrations.

Step 6 Isolate and verify critical process(es)

Narrow the potential list of problems to the vital few. Identify the input/output relationship which directly affects specific problems. Verify potential causes of process variability and product problems through engineering experiments, scatter diagrams, and multi-vari charts. Ensure that the data is clear and stratified.

Step 7 Perform process capability study

Identify and define the limitations of the processes. Ensure that the processes are capable of achieving their maximum potential. Identify and remove all variation due to special causes. Determine what the realistic specifications are.

Step 8 Implement optimum operating conditions and control methods. specify: target/tolerances, process controls, preventative and corrective action

Implement permanent corrective action for preventing special cause variations. Demonstrate process stability and predictability. Establish ongoing controls for the process based on prevention of special cause variation using statistical process control techniques. Document the decision tree and reaction plan.

Step 9 Monitor process over time

Develop monitors over time for evidence of improvement. Methods, systems, practices, and procedures must be modified to prevent special cause variations in the processes. Define the specific limitations of the processes. Identify further actions required on the processes.

Step 10 Reduce common cause variation

Process limitations must be recognized. It is only through reduction and elimination of common cause variation, and design for manufacturability that the goal of approaching six sigma can be achieved. Once special cause has been eliminated, only common cause remains. Further reduction of these variables requires work on the system in the form of redesign of equipment or processes. At this point, management commitment and drive is required.

References

1. Taguchi, G. (1993) *Taguchi On Robust Technology Development*, American Society of Mechanical Engineers, New York, p. 20.
2. DeVor, R.E., Chang, T., and Sutherland, J.W. (1992) *Statistical Quality Design and Control*, Macmillan, New York, p. 789.
3. DeVor, R.E., Chang, T., and Sutherland, J.W. (1992) *Statistical Quality Design and Control*, Macmillan, New York, pp. 158–64.
4. Vaughn, T.S. (1994) An alternative framework for short-run SPC. *Production and Inventory Management Journal*, 3rd quarter, 48–52.
5. Hillier, F.S. (1969) X- and R- Chart control limits based on a small number of subgroups, *Journal of Quality Technology*, **1**, 17–25.
6. Quesenberry, C.P. (1991) SPC Q charts for start-up processes and short or long runs. *Journal of Quality Technology*, **23**, 213–24.
7. Rahn, G.E. (1995) Classical versus short-run X-bar charts, Proceedings of the 1995 ASME IMECE Joint Symposium on Concurrent Product and Process Engineering.
8. Walpole, R.E. and Myers, R.H. (1993) *Probability and Statistics for Engineers and Scientists*, 5th edn, Macmillan, New York, p. 474.
9. Hoskins, J., Stuart, R., and Taylor, J., *Statistical Process Control, The Motorola Guide to Statistical Process Control for Continuous Improvement Towards Six Sigma Quality*, Motorola Literature Distribution; P.O. Box 20912. Phoenix, AZ 85036.

Case study 1 NIB1 single-attribute thin film

CS1.1 The problem

A (hypothetical) monopoly wants to improve the quality and profitability of a thin film device whose demand curve is shown in Figure CS1.1. Current annual demand is 500 000 units and current price is $65. Value obtained from the price intercept is approximately $110 and variable cost is $20.

The value of the product is approximately parabolic with thickness having its maximum at 20 (arbitrary units) with critical values, g_c, at film thicknesses of 0 and 40. This value function is shown in Figure CS1.2. The desire is to improve quality and reduce costs through seven proposed process changes.

The baseline process has an average loss of value from an ideal of approximately $0.56 per film computed from the curvature of the value-versus-film-thickness relationship shown in Figure CS1.2 and the measured thickness variance for the baseline process determined from statistical process control measurements over the past month.

CS1.2 The experimental design

The design of experiment array chosen is the $L_8(2^7)$ orthogonal array shown in Table CS1.1. Set points 11 and 21 each represent $0.10 cost reductions. The

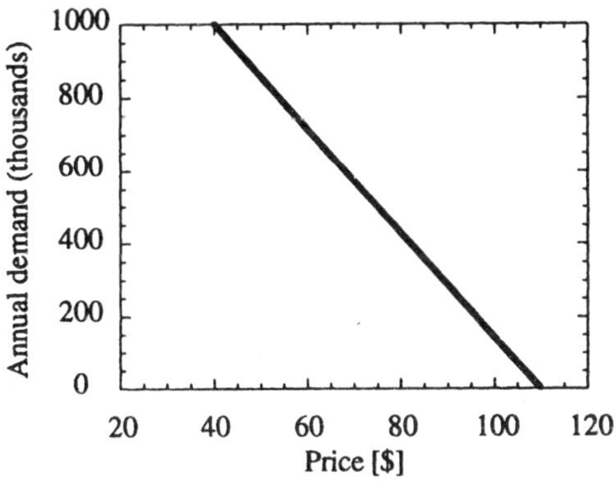

Figure CS1.1 Demand curve versus price for thin film device.

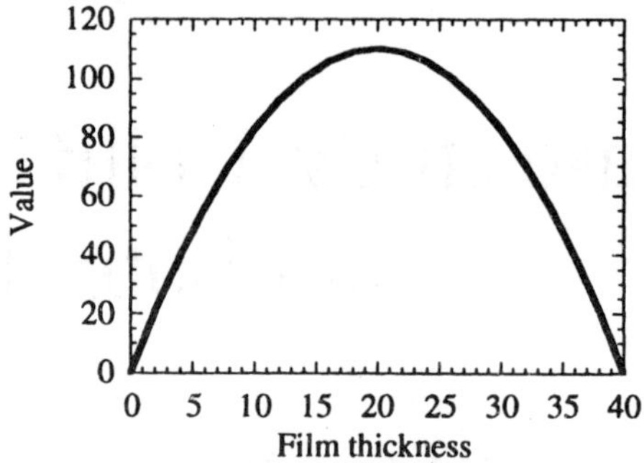

Figure CS1.2 Hypothetical value of thin film device versus film thickness.

remaining proposed process changes were believed from brainstorming sessions to offer the possibility of reducing thickness variance without affecting cost or investment. The time available limited the number of replications to three per trial. The deposition time chosen for the experiments was set below the normal time as some of the new processes being evaluated were expected to improve the rate of deposition, the signal, without, hopefully, appreciably increasing the thickness variance, the noise.

CS1.3 Thickness 'measurements'

Because this case represents a hypothetical problem, the experimental results were generated using computer simulation which allows the results obtained to be compared against the true population statistics. The focus in this case is on how to perform and interpret the calculations for a single-attribute problem having quality loss, variable cost, and profit implications. 'Measured' film thicknesses for the experiments are shown in Table CS1.2. The word 'measured' is placed in quotes throughout the discussion to avoid confusion with the subse-

Table CS1.1 $L_8(2^7)$ design of experiment array for NIB1 problem

Trial	*0*	*11*	*21*	*31*	*41*	*51*	*61*	*71*
1	1	0	0	0	0	0	0	0
2	1	0	0	1	1	1	0	1
3	1	0	1	0	1	1	1	0
4	1	0	1	1	0	0	1	1
5	1	1	0	0	0	1	1	1
6	1	1	0	1	1	0	1	0
7	1	1	1	0	1	0	0	1
8	1	1	1	1	0	1	0	0

quent simulations because the 'measured' thickness $g(q, y)$ for trial q and replication y were themselves drawn from normally distributed populations using the relation:

$$g(q,y)=\mu(q)+\sqrt{\frac{\sigma^2(q)}{2}}\sum_{j=1}^{24}[\text{RAND}()-0.5]_j \qquad \text{(CS1.1)}$$

The results for the $g(q, y)$ are those shown in Table CS1.2. The population means and variances used in Equation CS1.1 are shown in Table CS1.3. These parameters generate a common population variance of unity for all trials and a population mean of 17 when set point 21 is 'on' (trials 3, 4, 7, and 8) and a mean of 14 otherwise.

The statistical properties found for the 'measured' thickness for each trial are shown in Figure CS1.3 and the coefficients computed from them are shown in Figure CS1.4. The ξ'_{ij} coefficients for –10 log of the sample variances and the $\lambda_{ij}(\bar{g})$ coefficients for the sample averages were computed from Equations 9.7 and 9.6, respectively. Taking trial 1 as baseline for the experiment, the average thickness is seen to be 14.321 from cell AP25 in Figure CS1.3, the deviation from the ideal specification of 20 reflecting the reduced deposition time for the experiment versus the normal deposition time for production.

Two CIQ point estimates were computed for the trials in Figure CS1.3. The quantities $\widetilde{\Omega}$ were computed from the NIB1 CIQ expression in Table 8.2 in which the sample variances and sample averages were substituted as point estimates for the population variances and averages. These represent the standard practice for computing the CIQ. The CIQs labeled Ω were computed using the pooled variance shown in cell AQ33 and 16 pooled df in conjunction with Equations 8.29 and 8.30, respectively, for point estimates of the population variances and averages. The justification for pooling is developed in what follows.

Table CS1.2 'Measured' film thickness

	Replications		
Trial	*1*	*2*	*3*
1	14.885	13.882	14.197
2	13.913	12.744	14.241
3	14.603	15.342	16.867
4	17.515	16.952	18.621
5	12.613	15.803	12.990
6	15.716	14.907	12.709
7	16.673	16.399	16.943
8	16.430	17.114	17.330

Table CS1.3 Population means and variances

Trial	μ	σ^2
1	14	1
2	14	1
3	17	1
4	17	1
5	14	1
6	14	1
7	17	1
8	17	1

The $\lambda'_{ij}(\Omega)$ coefficients shown in the last two columns of Figure CS1.4 were computed using the expression:

$$[\lambda'(\Omega)] = [\mathbf{X}^T\mathbf{X}]^{-1}[\mathbf{X}^T][-10\ \mathrm{Log}(\Omega(q))] \qquad \text{(CS1.2)}$$

the spreadsheet area expression being

```
=MMULT(MINVERSE($N$11:$U$18),–10*LOG(AR25:AR32))
```

for the coefficients in column BD and

```
=MMULT(MINVERSE($N$11:$U$18),–10*LOG(AS25:AS32))
```

for the coefficients in column BE. The other parameters used in computing the CIQ for this problem are listed in Table CS1.4. Baseline value is seen from Figure CS1.2 to be $100, which is the quantity used for V_0 (Table CS1.4).

There are two comparisons to make. The first is between the two columns of CIQs in Figure CS1.3 and the second is between the coefficients determined from them in Figure CS1.4. Both pairs of columns are quite different. The ques-

	AO	AP	AQ	AR	AS
22					Pooled
23					
24	Trial	$\bar{g}$	s^2	$\tilde{\Omega}$	Ω
25	1	14.321	0.263	0.141	0.664
26	2	13.633	0.620	0.367	0.734
27	3	15.604	1.334	0.602	0.560
28	4	17.696	0.721	0.253	0.436
29	5	13.802	3.040	1.753	0.716
30	6	14.444	2.422	1.276	0.654
31	7	16.672	0.074	0.029	0.491
32	8	16.958	0.221	0.084	0.475
33	average =	15.391	1.087		

Figure CS1.3 'Measured' properties of the films.

	AT	AU	AV	AW	AX	AY	AZ	BA	BB	BC	BD	BE
21					F				t			
22					Dist	Exp.			Dist.	Exp.		Pooled
23		ξ'_{ij}	G_{ij}	F_{ij}	pair	-wise	$\lambda_{ij}(\bar{g})$	t_{ij}	pair	-wise	$\lambda'_{ij}(\tilde{\Omega})$	$\lambda'_{i}(\Omega)$
24	ij				error	error			error	error		
25	0	5.80					14.321				8.51	1.78
26	11	0.29	0.058	1.14	0.47	3.27	0.155	0.365	0.360	2.52	0.39	0.09
27	21	4.71	0.942	8.75	0.10	0.72	2.683	6.303	0.000	3.7E-05	6.22	1.51
28	31	-1.21	-0.241	1.74	0.36	2.55	0.583	1.370	0.095	0.66	-0.90	0.30
29	41	-0.16	-0.032	1.08	0.48	3.37	-0.606	-1.424	0.087	0.61	-0.48	-0.32
30	51	-3.03	-0.607	4.04	0.20	1.39	-0.784	-1.842	0.042	0.29	-3.47	-0.44
31	61	-8.56	-1.713	51.61	0.02	0.13	-0.010	-0.022	0.491	3.44	-8.57	-0.01
32	71	0.68	0.136	1.37	0.42	2.96	0.119	0.280	0.392	2.74	0.71	0.03

Figure CS1.4 Computed coefficients.

tion is 'Should the trial variances be pooled before calculating the CIQ?' The recommendation here is to pool variances according to the pooling array construction which is made after testing the set points for their influence on variance as described in Chapter 9. Moreover, for this specific problem with only 2 df for the trial sample variances, true point estimates of the CIQ for the individual trials do not even exist without pooling as discussed in Chapter 8.

CS1.4 Test of $H1_0$

The test of the significance of the individual set points on variance can be made using the methods described in Chapter 9. First the F_{ij} ratios for each of the ξ'_{ij} coefficients were computed using Equation 9.8. The pairwise error shown in column AX of Figure CS1.4 (the error for assuming that the hypothesis $H1_0$ of a common population variance was not obeyed) was computed from the spreadsheet function FDIST(F_{ij}, 2, 2). The two 2s represent the unpooled df for the sample variances for each trial. The pairwise results were then multiplied by seven to obtain the experiment-wise Type I error. None of the coefficients are seen to be significant for an error less than 10% (100 × error). This agrees with the fact that the populations for each trial had exactly the same variance. Because none of the 'measured' ξ'_{ij} coefficients are deemed significant, we replace the trial variances with a sample variance s_P^2 averaged (pooled) over all of the trials resulting in 16 df.

Table CS1.4 Parameters used in computing value loss from ideal

g_I	g_C	g_0	V_0	k_T
20	0.0	14	100	0.275

CS1.5 Test of $H2_0$

Based upon the results shown in Figure CS1.4, all of the trial variances were combined to form a single pooled variance s_P^2 equal to 1.087 having 16 df. The hypothesis, $H2_0$, of a common population average was tested next with the pooled variance using the procedures described in Chapter 9. The pairwise error for the λ_{ij} coefficients was computed from the t statistic for set point ij defined as:

$$t_{ij} = \frac{\lambda_{ij}\sqrt{n}}{\sqrt{\sum_{q=1,z} s^2(q)\omega_{ij}^2(q)}} = \frac{\lambda_{ij}\sqrt{n}}{s_P\sqrt{\sum_{q=1,z} \omega_{ij}^2(q)}} \qquad \text{(CS1.3)}$$

The sum over the $\omega_{ij}^2(q)$ under the square root sign in the denominator is equal to 0.5 (= cov_{ijij}). The error results are shown in columns BB and BC of Figure CS1.4. Only set point 21 is seen as being significant. This is expected because we know that the trial samples for this set point were taken from a population that had a population average of 17 versus 14 for the other trials.

CS1.6 Strategic significance

The procedure described in Chapter 9 was used for testing strategic significance. The simulated results for μ_s and σ_S^2 are shown in Figure CS1.5. The simulated CIQ, Ω_S, were computed using the NIB1 expression from Table 8.2. Finally the simulated λ'_{ij} (Ω_S) coefficients shown in Table CS1.5 were determined for the simulated distribution using:

$$[\lambda'(\Omega_S)] = [X^TX]^{-1}[X^T][-10\ \mathrm{Log}(\Omega_S(q))]$$

The results for 100 such repeated simulations are shown in Figure CS1.6.

The simulation results for set point 21 are seen to be distinct from the others in Figure CS1.6. However, this does not in itself answer the question as to

	AH	AI	AJ	AK	AL	AM	AN	AO	AP	AQ	AR
51											
52	Trial	RAND()	t	μ_S	RAND()	χ^2	σ_S^2	Ω_S	$-10\mathrm{Log}(\Omega_S)$	ij	$\lambda'_{ij}(\Omega_S)$
53	1	0.346	-0.403	14.08	0.944	25.85	0.67	0.37	4.284	0	4.28
54	2	0.497	-0.008	13.63	0.089	9.05	1.92	1.14	-0.559	11	0.20
55	3	0.144	-1.098	14.94	0.128	9.91	1.76	0.86	0.636	21	0.51
56	4	0.804	0.878	18.22	0.136	10.05	1.73	0.57	2.421	31	0.30
57	5	0.554	0.139	13.89	0.057	8.20	2.12	1.21	-0.820	41	-0.67
58	6	0.699	0.533	14.76	0.710	18.60	0.93	0.47	3.268	51	-2.64
59	7	0.306	-0.518	16.36	0.304	12.68	1.37	0.56	2.496	61	-0.84
60	8	0.257	-0.667	16.56	0.315	12.84	1.35	0.54	2.652	71	-1.83

Figure CS1.5 Example of one simulated experiment for testing strategic significance.

Table CS1.5 Simulated $\lambda'_{ij}(\Omega_S)$ coefficients determined from the results shown in Figure CS1.5

ij	$\lambda'_{ij}(\Omega_S)$
0	4.04
11	0.85
21	1.92
31	−0.53
41	−1.54
51	−2.24
61	−0.24
71	−1.47

the significance of set point 21 in reducing Ω. Its intersection at zero is seen to be roughly 7% in Figure CS1.6 but this error must be multiplied by seven to obtain the experiment-wise error which becomes 49%! In other words, when hypothesis $H2_0$ is obeyed, there is roughly one chance in every two experiments of the kind just made to find a set point whose coefficient $\lambda'_{ij}(\Omega)$ is as large as the coefficient found for set point 21. This error is much larger than the experiment-wise error found for $\lambda_{21}(\bar{g})$ of 0.0037% in Figure CS1.4. The reason for this is that the variance affects the NIB1 loss function (Equation 8.29) in a

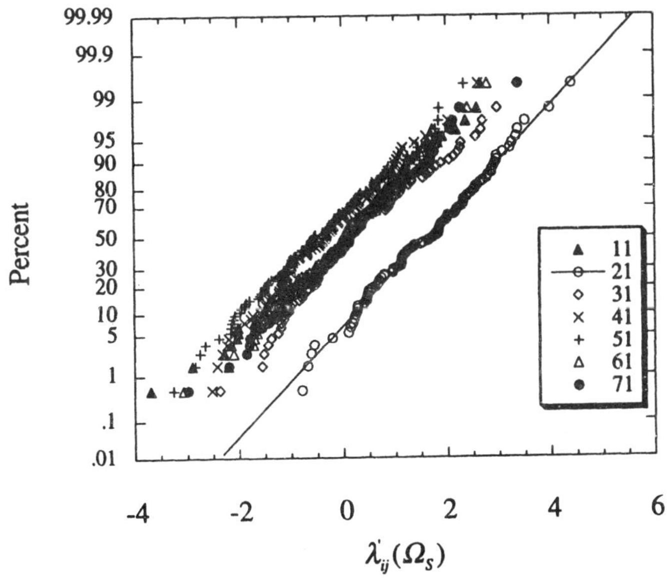

Figure CS1.6 Results of 100 simulations of the $\lambda'_{ij}(\Omega_S)$ coefficients.

strong way in both the numerator and denominator. Although we found that no set point was significant in changing the variance from baseline, there is uncertainty in what the variance is for each set point. The remaining uncertainty in thickness variance with an off-baseline set point 'on' results in the low confidence for set point 21 reducing Ω even though we believe with high confidence that it moves the average film thickness in the right direction.

For completeness, the $\lambda'_{ij}(\Omega_S)$ were converted into the distribution for $\Omega_{21} - \Omega_0$ which is equal to the CIQ loss for set point 21 less the loss for baseline (trial 1). The results shown in Figure CS1.7 before the experiment-wise correction confirm the same level of significance found in Figure CS1.6. Note that, to emphasize a point made earlier, although we pooled variances, we take the population variances $\sigma_S^2(q)$ not as a constant independent of q in simulating the $\lambda'_{ij}(\Omega_S)$ but as a reciprocal chi-squared statistic as determined from a pooled sample variance of 1.087 having 16 pooled df.

CS1.7 Strategic implications

As seen in Figure CS1.7, the median reduction in the cost of inferior quality that set point 21 makes is \$0.20. As stated at the outset of this problem, this set point also yielded a \$0.10 variable cost saving. On combining the value and cost savings shown by the shifted line in Figure CS1.7, we find that the confidence level for this set point in reducing the total loss function is roughly 98% before

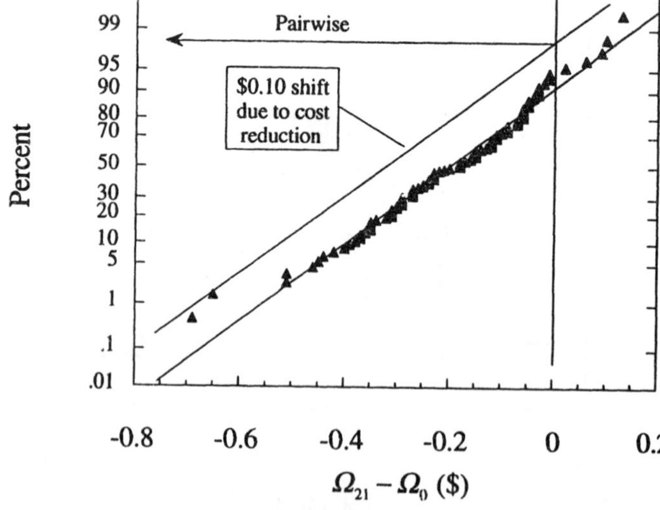

Figure CS1.7 Simulation of the uncertainty in the CIQ for set point 21 relative to the baseline with and without a \$0.10 cost reduction.

making the experiment-wise correction which reduces the confidence level to 86% (= 100 − 7×2).

Should set point 21 be implemented? After all it is a cost reduction and favorable for reducing the cost of inferior quality. The better answer is 'No!' because there is not high confidence in it for reducing the total loss; only 86% of the time should the total loss for 21 be less than the baseline. From inspection of Figure CS1.7, we see that it would be necessary for the cost reduction to be over $0.20 before it could be assumed with reasonable confidence that implementation of set point 21 would reduce the overall loss. By not choosing to implement set point 21 at this juncture, a Type II error would, of course, be made. (In actual practice, this would not be known because the population statistics would not be known.) There is nevertheless considerable evidence from the experiment that set point 21 is favorable but the support is simply not quite strong enough to implement it. What could be done at this juncture is to run additional replications of the baseline and additional replications with set point 21 'on' thereby increasing the df until the confidence level is high enough to make a decision. A question must be answered before undertaking the added work: 'Is the possible added improvement large enough to go to the added expense and effort to run the additional trials?'

Case study 2 NIB2 problem

The NIB2 problem must be treated differently from SIB, NIB1, and LIB because the CIQ is not monotonically increasing or decreasing as a function of the attribute g which makes any attempt to use a linear model for the CIQ S/N ratio futile. NIB1 problems avoid this complication by having a scaling factor for moving the mean to the minimum in the CIQ, if costs are not a function of g, or to the minimum in the loss function, if costs vary with g, leaving any remaining losses to be caused solely by statistical variations about the mean. SIB and LIB problems exhibit monotonic behavior because only one half of the CIQ parabola is used (after the LIB attribute is transformed to its basic variable $1/g$).

However, the linear models for the sample averages and the –10 log of the sample variances in conjunction with the CIQ expression in Table 8.2 are sufficient to analyze all NIB2 problems for finding the set point combination most effective in reducing the CIQ. The overall procedure, which follows, is straightforward, the first four steps being the same as listed in Chapter 9. The procedure is also general in that it could also be used for SIB and LIB problems and it can readily be adapted to problems having value curves composed of two or more connected functions such as the kinked value curve in Figure 5.25. However, for such complex value curves for SIB, LIB, and NIB2, the uncertainty in the attribute variable, g, itself must be explicitly simulated in the expression for the CIQ using the Monte Carlo method (after considerations of pooling of variances have been made) rather than simulating the population average and population variance as done throughout here. Logical statements are then used to connect the attribute g with respect to the proper value function for the domain in which it is found.

Once the pooling array in step 4 has been constructed and the trial variances pooled partially, completely, or not at all according to the significance found for the set points on variance, the coefficients, $\xi'_{ij,P}$, should be computed from –10 times the logarithms of the pooled sample variances instead of the original trial variances, hence the added subscript, P:

$$\xi'_{ij,P} = \sum_{q=1,k} \omega_{ij}(q) - 10\text{Log}(s_P^2(q)).$$

If the design matrix were of the $[\mathbf{X}]_\lambda$ type, the lambda coefficients for the averages would be computed from Equation B.5 with the general outcome variable $[\mathbf{Y}]$ being the column array formed by the sample averages for the trial:

$$\lambda_{ij} = \sum_{q=1,k} \omega_{ij}(q)\bar{g}(q).$$

If the design matrix were of the $[\mathbf{X}]_\varphi$ type, the coefficients for the averages would be computed as:

$$\varphi_{ij} = \sum_{q=1,k} \omega_{ij}(q)\bar{g}(q).$$

Next the **combination matrix,** $[\mathbf{X}]_{Cmb}$, consisting of all of the possible combinations of the main effect variables should be constructed. The interactions under investigation must also be shown in the combination matrix. Each row of this matrix represents a possible combination of set points that needs to be evaluated for its resulting impact on the CIQ.

The –10 log sample variance for each combination (j) needs to be computed from the relation:

$$\xi'_P(j) = \sum_{kl} X_{kl}(q)\xi'_{kl,P}(q)$$

and used to compute the sample variance for combination (j) using the expression:

$$s_P^2(j) = 10^{-\xi'_P(j)/10}$$

Each λ_{kl} (or φ_{kl}) coefficient should be converted into a sample average for the combination:

$$\bar{g}(j) = \sum_{kl} X_{kl}(j)\lambda_{kl}$$

At this point, the Monte Carlo process as described in Chapter 9 should be used to simulate the distribution of possible population averages and variances for each combination using Equations 9.12 and 9.13, respectively. These in turn should be substituted into the NIB2 CIQ expression in Table 8.2 to compute the simulated CIQ for each combination. One of the rows in the combination matrix will represent the production baseline condition and the CIQ for this row should be subtracted from the other CIQs to show the change generated in the CIQ. The resulting CIQ difference distribution for each of the combinations should then be plotted on normal probability paper to determine the significance that each combination has in reducing the CIQ.

With these steps in mind, imagine that we want to reduce customer dissatisfaction with our product caused by the surface roughness of a shaft being off its ideal specification of 53 micro-inches rms. Marketing research showed that consumers were willing to pay \$98 more for products built at the ideal specification versus the baseline product built at 46 rms. This finding results in the coefficient k_T in the NIB2 expression of Table 8.2 being 2 \$/rms^2 (= 98/7^2).

The brainstorming process designed to uncover ways to move the specification to the ideal location identified three main effects at two levels each including their pairwise interactions and fourth main effect at two levels as promising for study. Possible interactions of the fourth main effect with the others were believed to be insignificant. The design array chosen, Figure CS2.1, was an $L_8(2^7)$ orthogonal array with the fourth main effect placed in the last column which would only confound with the triplet interaction between the other three [1]. Trial 3 represented the production baseline. The resulting solution matrix is shown in Figure CS2.2. A total of six replications were made for each trial (Figure CS2.3). The sample variance, –10 log of the sample variance, and the sample average for each trial are also shown in Figure CS2.3.

The six replications for each trial were taken from simulated normal populations using the relation:

$$g(q,y)=\mu(q)+\sqrt{\frac{\sigma^2(q)}{2}}\sum_{j=1}^{24}[\mathrm{RAND}()-0.5]_j\ . \qquad \text{(CS2.1)}$$

The population statistics for each trial are shown in Figure CS2.4. Using these values, only set points 11, 21, and 31 should influence the average (the interactions and set point 41 having no effect on average) and only set points 21 and 41 should influence the log of the variance.

Analysis of the NIB2 problem follows, with certain modifications, the six-step process described in Chapter 9 for SIB, LIB, and NIB1 problems. The ξ'_{ij} variance coefficients shown in column W of Figure CS2.3 were computed using the spreadsheet area expression:

=MMULT(MINVERSE(E5:L12),AD5:AD12).

The F_{ij} ratios in column Y were computed using Equation 9.8 and the pairwise errors in column Z were computed using the spreadsheet function FDIST(F_{ij},ν,ν) where ν is the df, equal to five for this problem before any pooling is made. The experiment-wise error shown in column AA is seven times the pairwise error. This method for determining the significance of the set points on variance is possible with the $[\mathbf{X}]_\varphi$ array in Figure CS2.1 because the absolute value of the elements of its solution matrix in Figure CS2.2 are equal making the ξ'_{ij} symmetric about zero for the hypothesis of a common population variance.

	D	E	F	G	H	I	J	K	L
4	Trial	0	11	21	1121	31	1131	2131	41
5	1	1	-1	-1	1	-1	1	1	-1
6	2	1	-1	-1	1	1	-1	-1	1
7	3	1	-1	1	-1	-1	1	-1	1
8	4	1	-1	1	-1	1	-1	1	-1
9	5	1	1	-1	-1	-1	-1	1	1
10	6	1	1	-1	-1	1	1	-1	-1
11	7	1	1	1	1	-1	-1	-1	-1
12	8	1	1	1	1	1	1	1	1

Figure CS2.1 The design matrix in the $[\mathbf{X}]_\varphi$ form.

	D	E	F	G	H	I	J	K	L
15	*ij*	1	2	3	4	5	6	7	8
16	0	0.125	0.125	0.125	0.125	0.125	0.125	0.125	0.125
17	11	-0.125	-0.125	-0.125	-0.125	0.125	0.125	0.125	0.125
18	21	-0.125	-0.125	0.125	0.125	-0.125	-0.125	0.125	0.125
19	1121	0.125	0.125	-0.125	-0.125	-0.125	-0.125	0.125	0.125
20	31	-0.125	0.125	-0.125	0.125	-0.125	0.125	-0.125	0.125
21	1131	0.125	-0.125	0.125	-0.125	-0.125	0.125	-0.125	0.125
22	2131	0.125	-0.125	-0.125	0.125	0.125	-0.125	-0.125	0.125
23	41	-0.125	0.125	0.125	-0.125	0.125	-0.125	-0.125	0.125

Figure CS2.2 The solution matrix derived from the design matrix in Figure CS2.1.

Set point 21 influenced variance enough to be found as significant in this experiment and set point 41 was taken as significant but borderline. If a large number of replications had been run, both set points would have been found to be 3.7 based upon the population simulations for the trials. The pooling array shown in Figure CS2.6 was constructed based upon these two significant set points. Those trial variances having common set point pairs in rows across columns V and W were pooled. We see by inspection that the trial variances pooled in pairs as follows: trials 1 and 6; trials 2 and 5; trials 3 and 8; and trials 4 and 7, the results being shown in column X of Figure CS2.6. The pooled df is 10 ($= 2\times(6-1)$) for each trial. column AA gives the $\xi'_{ij,P}$ variance coefficients for −10 log of the pooled trial variances. These coefficients will be needed in computing the CIQ for each trial in what follows.

Because we now have good reason to believe that the trial variances did not come from populations having a common variance, we must use the approximate t statistic, Equation B.28, for computing the significance of the set points on the population averages, step 4 in Chapter 9. This requires that we compute the effective sample variance for each set point using Equation B.29 as well as the effective sample df for each set point using Equation B.30. The results of

	V	W	X	Y	Z	AA	AB	AC	AD	AE
3									−10×	
4	Trial	1	2	3	4	5	6	Var	Log(Var)	Ave
5	1	47.52	56.36	50.54	48.11	53.18	47.37	13.19	-11.20	50.51
6	2	44.29	43.45	44.26	43.86	41.67	42.05	1.29	-1.11	43.26
7	3	46.95	46.60	45.58	45.71	46.21	46.09	0.27	5.67	46.19
8	4	41.81	41.22	40.26	41.88	41.50	40.34	0.51	2.95	41.17
9	5	58.31	61.29	60.33	61.58	57.21	61.07	3.20	-5.06	59.96
10	6	50.96	44.40	52.68	50.51	56.28	61.21	32.37	-15.10	52.67
11	7	54.97	54.34	54.77	54.80	56.31	53.41	0.89	0.51	54.77
12	8	51.00	49.86	48.11	50.68	49.33	49.05	1.15	-0.60	49.67

Figure CS2.3 Simulations for the replications for each trial, the trial variances, −10 log of the trial variances, and the trial averages.

	Q	R	S	T
3			-10×	
4	Trial	Ave	Log(Var)	Var
5	1	50	-12.4	17.38
6	2	44	-5.0	3.16
7	3	46	2.4	0.58
8	4	40	-5.0	3.16
9	5	60	-5.0	3.16
10	6	54	-12.4	17.38
11	7	56	-5.0	3.16
12	8	50	2.4	0.58
13	Averages=	50	-5.00	6.07

Figure CS2.4 The population statistics for each trial used in Equation CS2.1 for simulating the replications.

these calculations are listed respectively in columns AC and AF of Figure CS2.7. The effective variances were computed using the spreadsheet area expression given by:

=MMULT(E16:L23^2,X32:X39)

in which the square of the elements in the solution matrix are multiplied by the pooled trial variances in column X of Figure CS2.6. The results in column AE were computed using the area expression given by:

=MMULT(E16:L23^4,X32:X39^2)

in which the fourth power of the elements in the solution matrix are multiplied by the square of the pooled trial variances in column X of Figure CS2.6. The effective df in cell AF32 was then computed from Equation B.30 using the spreadsheet expression:

=Y32*AD32/AE32

which was dragged down to compute the others.

The φ_{ij} coefficients for the averages shown in column AD of Figure CS2.8 were computed using the spreadsheet area expression given by:

=MMULT(MINVERSE(E5:L12),AE5:AE12)

	V	W	X	Y	Z	AA
14						Exp-
15					Pairwise	wise
16	ij	ξ'_{ij}	G_{ij}	F_{ij}	error	error
17	0	-2.99	-1.20	15.77		
18	11	-2.07	-0.83	6.72	0.028	0.198
19	21	5.12	2.05	112.14	0.000	0.000
20	1121	-0.11	-0.04	1.10	0.458	3.205
21	31	-0.47	-0.19	1.55	0.322	2.255
22	1131	-2.32	-0.93	8.44	0.018	0.123
23	2131	-0.48	-0.19	1.56	0.318	2.228
24	41	2.72	1.09	12.21	0.008	0.055

Figure CS2.5 Analysis of the significance of the set points on the sample trial variances.

	U	V	W	X	Y	Z	AA
30							
31	Trial	21	41	s_P^2	ν_P	ij	$\xi'_{ij,P}$
32	1	-1	-1	22.78	10	0	-3.512
33	2	-1	1	2.25	10	11	0.000
34	3	1	1	0.71	10	21	5.035
35	4	1	-1	0.70	10	1121	0.000
36	5	-1	1	2.25	10	31	0.000
37	6	-1	-1	22.78	10	1131	-2.532
38	7	1	-1	0.70	10	2131	0.000
39	8	1	1	0.71	10	41	2.496

Figure CS2.6 The array (columns U, V, and W) used to pool variances based upon the set points found to be significant for variance. The pooled variances are shown in column X and their df in column Y. Column AA lists the variance coefficients computed for –10 log of the pooled variances listed in column X.

The value for t in cell AE18 was computed using the spreadsheet expression

```
=AD18*SQRT(6)/SQRT(AC33)
```

and the pairwise error in cell AF18 was computed using the expression

```
=TDIST(ABS(AE18),AF33,1).
```

The results for these two cells were then dragged down to compute the others.

Because this problem involves four main effects, there are 16 possibilities given by the combination matrix, Figure CS2.9, which need to be evaluated to discover which combination yields the smallest CIQ. The computations leading to the point estimates of the CIQ for each combination are shown in Figure

	AB	AC	AD	AE	AF
28		s_{ij}^2			
29					
30		$= \sum_q \omega_{ij}^2(q) s^2(q)$	s_{ij}^4	$\sum_q \omega_{ij}^4(q) s^4(q)$	ν_{ij}
31	ij				
32	0	0.83	0.68	0.26	26.63
33	11	0.83	0.68	0.26	26.63
34	21	0.83	0.68	0.26	26.63
35	1121	0.83	0.68	0.26	26.63
36	31	0.83	0.68	0.26	26.63
37	1131	0.83	0.68	0.26	26.63
38	2131	0.83	0.68	0.26	26.63
39	41	0.83	0.68	0.26	26.63

Figure CS2.7 Computation of the effective sample variances and effective sample df to be used in evaluating the significance of the set points on the averages based upon the approximate t statistic.

	AC	AD	AE	AF	AG
14					Exp-
15				Pairwise	wise
16	ij	φ_{ij}	t_{ij}	error	error
17	0	49.776			
18	11	4.493	12.11	0.0000	0.0000
19	21	-1.827	-4.92	0.0000	0.0002
20	1121	-0.223	-0.60	0.2768	2.2141
21	31	-3.082	-8.31	0.0000	0.0000
22	1131	-0.014	-0.04	0.4849	3.8789
23	2131	0.553	1.49	0.0742	0.5935
24	41	-0.003	-0.01	0.4964	3.9708

Figure CS2.8 Analysis of the significance of the set points on the sample trial averages.

CS2.10, the spreadsheet expressions used for each of the columns being shown in Table CS2.1. Combination 14 is seen to have the smallest CIQ. Combination 6 represents the baseline condition, the same set point combination as the baseline (trial 3) for the design matrix in Figure CS2.1.

Based upon the population statistics, the minimum CIQ should have been for trial 12, the findings being influenced by experimental error. If the costs were different for each trial, they would need to be added to the CIQ for each trial in Figure CS2.10 and the trial having the minimum loss function should then decide which set point combination was the best of those listed in Figure CS2.9.

	D	E	F	G	H	I	J	K	L
42	Combo	0	11	21	1121	31	1131	2131	41
43	1	1	-1	-1	1	-1	1	1	-1
44	2	1	-1	-1	1	-1	1	1	1
45	3	1	-1	-1	1	1	-1	-1	-1
46	4	1	-1	-1	1	1	-1	-1	1
47	5	1	-1	1	-1	-1	1	-1	-1
48	6	1	-1	1	-1	-1	1	-1	1
49	7	1	-1	1	-1	1	-1	1	-1
50	8	1	-1	1	-1	1	-1	1	1
51	9	1	1	-1	-1	-1	-1	1	-1
52	10	1	1	-1	-1	-1	-1	1	1
53	11	1	1	-1	-1	1	1	-1	-1
54	12	1	1	-1	-1	1	1	-1	1
55	13	1	1	1	1	-1	-1	-1	-1
56	14	1	1	1	1	-1	-1	-1	1
57	15	1	1	1	1	1	1	1	-1
58	16	1	1	1	1	1	1	1	1

Figure CS2.9 The 16 possible combinations of the set points which need to be evaluated for their resulting CIQ relative to the production baseline, combination 6.

	M	N	O	P	Q	R	S
40							
41		$\xi_0' + \sum_{(ij)} \xi_{(ij)}'$	$s_{(ij)}^2$	σ^2	$\varphi_0 + \sum_{(ij)} \varphi_{(ij)}$	$E([g_i - \mu]^2)$	Ω
42	Combo						
43	1	-13.58	22.78	28.47	50.5	12.5	82.0
44	2	-8.58	7.22	9.02	50.5	8.2	34.5
45	3	-8.51	7.10	8.87	43.3	96.6	211.0
46	4	-3.52	2.25	2.81	43.3	95.4	196.4
47	5	-3.50	2.24	2.80	46.2	46.9	99.4
48	6	1.49	0.71	0.89	46.2	46.6	94.9
49	7	1.56	0.70	0.87	41.2	140.2	282.2
50	8	6.55	0.22	0.28	41.2	140.2	281.0
51	9	-8.51	7.10	8.87	60.0	50.6	118.9
52	10	-3.52	2.25	2.81	60.0	49.1	103.9
53	11	-13.58	22.78	28.47	52.7	6.4	69.8
54	12	-8.58	7.22	9.02	52.7	2.1	22.3
55	13	1.56	0.70	0.87	54.8	3.3	8.4
56	14	6.55	0.22	0.28	54.8	3.2	6.9
57	15	-3.50	2.24	2.80	49.7	11.7	28.9
58	16	1.49	0.71	0.89	49.7	11.3	24.3

Figure CS2.10 Computations for the expected value of the CIQ for the 16 combinations listed in Figure CS2.9.

The final step is to use the Monte Carlo process to determine the significance of the set point combinations that yield low values for the CIQ. The procedure, shown in Figure CS2.11, closely follows that given in Chapter 9 except that we determine significance of the low CIQ combinations instead of the significance of the favorable coefficients $\varphi_{ij}'(\Omega_S)$ of the CIQ S/N ratios. The key spreadsheet expression used to generate the results shown in Figure CS2.11 are listed in Table CS2.2.

The last column in Figure CS2.11 lists the combination CIQ minus the CIQ of combination 6 which, as already noted, represented the production baseline for the problem here. The ranges found for combinations 12, 13, and 14 based upon 100 simulations are shown in Figure CS2.12. (The macro described in Chapter 9 was used to automate the cutting and pasting operations involved in the simulations.)

Table CS2.1 The spreadsheet expressions used for the computations in Figure CS2.10

Cell(s)	*Equation*	*Type*
N43:N:58	=MMULT(E43:L58,AA32:AA39)	Area
O43	=10^(-N43/10)	Drag
P43	=($Y$32/($Y$32-2))*O43	Drag
Q43:Q58	=MMULT(E43:L58,AD17:AD24)	Area
R43	=$W$2^2-2*$W$2*Q43+Q43^2+(5/(5-2))*O43/6	Drag
S43	=$W$1*(P43+R43)	Drag

	U	V	W	X	Y	Z	AA	AB	AC
41									
42	Combo	RAND()	χ^2	σ_S^2	RAND()	t	μ_S	Ω_S	$\Omega_S - \Omega_S(6)$
43	1	0.23	6.49	35.08	0.89	1.28	53.01	70.2	-41.0
44	2	0.80	13.45	5.36	0.80	0.89	51.48	15.4	-95.8
45	3	0.43	8.65	8.21	0.43	-0.19	43.07	213.7	102.5
46	4	0.02	3.15	7.14	0.78	0.80	43.76	185.2	74.0
47	5	0.44	8.71	2.57	0.25	-0.71	45.77	109.8	-1.3
48	6	0.86	14.72	0.48	0.05	-1.78	45.58	111.2	0.0
49	7	0.91	16.48	0.42	0.74	0.66	41.39	270.3	159.1
50	8	0.25	6.76	0.33	0.44	-0.15	41.13	282.4	171.2
51	9	0.27	6.96	10.19	1.00	3.22	63.47	239.8	128.6
52	10	0.23	6.55	3.43	0.25	-0.71	59.53	92.1	-19.0
53	11	0.94	17.62	12.93	0.59	0.23	53.12	25.9	-85.3
54	12	0.58	10.19	7.08	0.24	-0.75	51.85	16.8	-94.3
55	13	0.19	6.04	1.16	0.41	-0.23	54.69	8.0	-103.2
56	14	0.53	9.68	0.23	0.21	-0.84	54.60	5.6	-105.6
57	15	0.84	14.35	1.56	0.92	1.56	50.63	14.4	-96.8
58	16	0.30	7.30	0.97	0.48	-0.06	49.65	24.4	-86.8

Figure CS2.11 The computations used to simulate the scatter in the CIQ for each of the 16 combinations. Column AC lists the CIQ for each combination less the CIQ for combination 6.

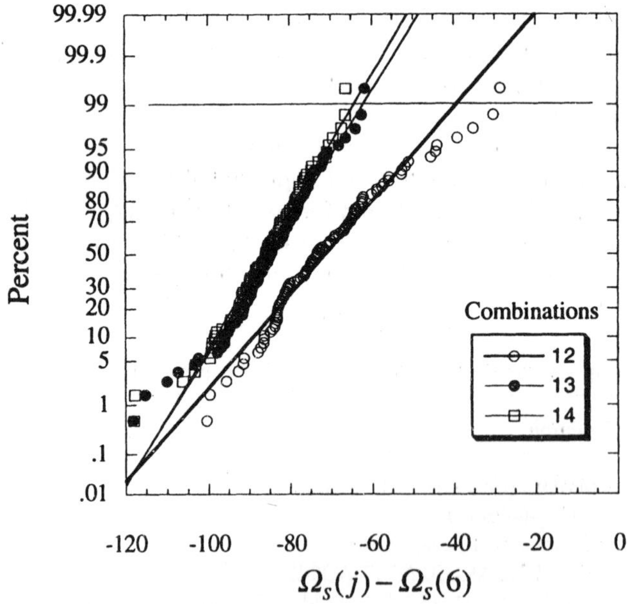

Figure CS2.12 Plot of the CIQ for three low combinations less the CIQ of combination 6 representing the production baseline.

Table CS2.2 The spreadsheet expressions used for the computations in Figure CS2.11

Cell	*Equation*	*Type*
X43	=$Y$32*O43/W43	Drag
AA43	=Q43+Z43*SQRT(O43/6)	Drag
AB43	=$W$1*(X43+($W$2–AA43)^2)	Drag
AC43	=AB43–$AB$48	Drag

The major portion of the distributions lie in the negative region which suggests that the CIQs are expected to be less for these combinations than the baseline. The horizontal line at the 99% pairwise confidence level yields a 93% experiment-wise confidence level (= 100 – (7 × 1)). We see that combination 12 is expected to give a CIQ that is roughly 30 less than baseline at 93% experiment-wise confidence, whereas combinations 13 and 14 are expected to give a CIQ that is 60 less than the baseline at the 93% experiment-wise confidence level. Based upon the actual population statistics for the simulated problem, the reductions should be 94, 78, and 80, respectively, for combinations 12, 13, and 14, the difference between these amounts and the experimental findings being the result of experimental error.

Reference

1. Taguchi, G. and Konishi, S. (1987) *Taguchi Methods Orthogonal Arrays and Linear Graphs*, American Supplier Institute, Allen Park, MI.

Case study 3 LIB problem

This case study illustrates how the techniques developed in Chapter 9 are used to analyze the significance of the factors (set points) in a LIB experiment. Topics covered include the log normal transformation, significance of the set points on variance, significance of the set points on average, and Monte Carlo simulations for determining the significance of the set points on the CIQ. Also illustrated are the use of the effective sample variance (Equation B.29) and the effective sample df (Equation B.30) in conjunction with the approximate t statistic.

We choose a problem having one main effect at four levels, eleven main effects at two levels, and an interaction between main effects 2 and 9. This particular choice of hypothetical variables is similar to the levels and number of variables considered by Montmarquet [1] who showed how the first three columns of the $L_{16}(2^{15})$ orthogonal array could be modified to accommodate the four-level main effect. Figure CS3.1 shows the modified $L_{16}(2^{15})$ design in the $[\mathbf{X}]_\lambda$ form. Figure CS3.2 shows the solution matrix.

It is not necessary to explicitly compute the covariance matrix, $[\mathbf{X}^T\mathbf{X}]^{-1}$, for most problems as all of the information needed to determine the significance of the set points can be readily generated from the solution matrix which for this problem is given simply by the inverse of the design matrix:

=MINVERSE(C13:R28).

	B	C	D	E	F	G	H	I	J	K	L	M	N	O	P	Q	R
12	Trial	0	11	12	13	21	31	41	51	61	71	81	91	10,1	11,1	12,1	2191
13	1	1	0	0	0	0	0	0	0	0	0	0	0	0	0	0	0
14	2	1	0	0	0	0	0	0	0	1	1	1	1	1	1	1	0
15	3	1	0	0	0	1	1	1	1	0	0	0	0	1	1	1	0
16	4	1	0	0	0	1	1	1	1	1	1	1	1	0	0	0	1
17	5	1	1	0	0	0	0	1	1	0	0	1	1	0	0	1	0
18	6	1	1	0	0	0	0	1	1	1	1	0	0	1	1	0	0
19	7	1	1	0	0	1	1	0	0	0	0	1	1	1	1	0	1
20	8	1	1	0	0	1	1	0	0	1	1	0	0	0	0	1	0
21	9	1	0	1	0	0	1	0	1	0	1	0	1	0	1	0	0
22	10	1	0	1	0	0	1	0	1	1	0	1	0	1	0	1	0
23	11	1	0	1	0	1	0	1	0	0	1	0	1	1	0	1	1
24	12	1	0	1	0	1	0	1	0	1	0	1	0	0	1	0	0
25	13	1	0	0	1	0	1	1	0	0	1	1	0	0	1	1	0
26	14	1	0	0	1	0	1	1	0	1	0	0	1	1	0	0	0
27	15	1	0	0	1	1	0	0	1	0	1	1	0	1	0	0	0
28	16	1	0	0	1	1	0	0	1	1	0	0	1	0	1	1	1

Figure CS3.1 The design matrix in the $[\mathbf{X}]_\lambda$ form for an $L_{16}(2^{15})$ orthogonal array modified to accommodate a four-level factor represented in columns D, E and F for set points 11, 12, and 13 and a pairwise interaction between set points 21 and 91 in column R.

	A	B	C	D	E	F	G	H	I	J	K	L	M	N	O
39		1	2	3	4	5	6	7	8	9	10	11	12	13	14
40	0	1.000	0.000	0.000	0.000	0.000	0.000	0.000	0.000	-0.000	0.000	0.000	0.000	0.000	0.000
41	11	-0.250	-0.250	-0.250	-0.250	0.250	0.250	0.250	0.250	0.000	0.000	0.000	0.000	0.000	0.000
42	12	-0.250	-0.250	-0.250	-0.250	0.000	0.000	0.000	0.000	0.250	0.250	0.250	0.250	0.000	0.000
43	13	-0.250	-0.250	-0.250	-0.250	0.000	0.000	0.000	0.000	0.000	0.000	0.000	0.000	0.250	0.250
44	21	-0.250	0.000	0.250	0.000	0.000	-0.250	0.000	0.250	0.000	-0.250	0.000	0.250	-0.250	0.000
45	31	-0.125	-0.125	0.125	0.125	-0.125	-0.125	0.125	0.125	0.125	0.125	-0.125	-0.125	0.125	0.125
46	41	-0.125	-0.125	0.125	0.125	0.125	0.125	-0.125	-0.125	-0.125	-0.125	0.125	0.125	0.125	0.125
47	51	-0.125	-0.125	0.125	0.125	0.125	0.125	-0.125	-0.125	0.125	0.125	-0.125	-0.125	-0.125	-0.125
48	61	-0.125	0.125	-0.125	0.125	-0.125	0.125	-0.125	0.125	-0.125	0.125	-0.125	0.125	-0.125	0.125
49	71	-0.125	0.125	-0.125	0.125	-0.125	0.125	-0.125	0.125	0.125	-0.125	0.125	-0.125	0.125	-0.125
50	81	-0.125	0.125	-0.125	0.125	0.125	-0.125	0.125	-0.125	-0.125	0.125	-0.125	0.125	0.125	-0.125
51	91	-0.250	0.250	0.000	0.000	0.250	-0.250	0.000	0.000	0.250	-0.250	0.000	0.000	-0.250	0.250
52	10,1	-0.125	0.125	0.125	-0.125	-0.125	0.125	0.125	-0.125	-0.125	0.125	0.125	-0.125	-0.125	0.125
53	11,1	-0.125	0.125	0.125	-0.125	-0.125	0.125	0.125	-0.125	0.125	-0.125	-0.125	0.125	0.125	-0.125
54	12,1	-0.125	0.125	0.125	-0.125	0.125	-0.125	-0.125	0.125	-0.125	0.125	0.125	-0.125	0.125	-0.125
55	2191	0.250	-0.250	-0.250	0.250	-0.250	0.250	0.250	-0.250	-0.250	0.250	0.250	-0.250	0.250	-0.250

Figure CS3.2 The solution matrix for the design matrix given in Figure CS3.1.

For example, if the diagonal elements of the covariance matrix were needed, they can be computed by multiplying each row in the solution matrix by its transpose. Computation of the cov_{ijij} elements in this manner is given in Figure CS3.3. The expression entered in cell U40 was

=MMULT(B40:Q40,TRANSPOSE(B40:Q40))

which was then dragged down to generate the remaining elements. The advantage of computing the diagonal elements in this manner is that they are arranged in a single column which is more convenient for their use in subsequent calculations than if they were arranged along a diagonal.

When the design matrix is not square, the solution matrix must be computed using the general least square form given by $[X^TX]^{-1}[X^T]$, and the covari-

	S	T	U
39	*ij*	*ij*	cov_{ijij}
40	0	0	1.00
41	11	11	0.50
42	12	12	0.50
43	13	13	0.50
44	21	21	0.50
45	31	31	0.25
46	41	41	0.25
47	51	51	0.25
48	61	61	0.25
49	71	71	0.25
50	81	81	0.25
51	91	91	0.50
52	10,1	10,1	0.25
53	11,1	11,1	0.25
54	12,1	12,1	0.25
55	2191	2191	1.00

Figure CS3.3 The diagonal elements of the covariance matrix which were computed by summing the squares of the elements across each row of the solution matrix in Figure CS3.2.

ance matrix is often generated as an intermediate step. But even this is not mandatory as the solution matrix can be generated in one step from the design matrix using the general form which for this problem is given by:

```
=MMULT(MINVERSE(MMULT(TRANSPOSE(C13:R28),C13:R28)),
                    TRANSPOSE(C13:R28))
```

Time is saved in keying in this expression by having it as a text macro and adding the equal sign and the argument for the array when used.

The simulated replications for the 16 trials are shown in Figure CS3.4. Imagine that the 'measurements' are for the fatigue life of a spring and that the control factors being investigated relate to design, alloy composition, heat treatment, and surface finish variables. These $g(q, y)$ 'measurements' were simulated in the following manner. First, lognormally transformed simulations for the $y = 1, ..., 5$ replications for each trial were made by selecting points randomly from normal populations having a common variance $\sigma_\ell^2(q) = 1$ for each of the 16 trials using the expression:

$$\mathrm{Ln}(1/g(q,y)) = \mu_\ell(q) + \sqrt{\frac{\sigma_\ell^2(q)}{2}} \sum_{j=1}^{34} [\mathrm{RAND}() - 0.5]_j$$

in which the population averages, $\mu_\ell(q)$, were chosen such that only set points 12 and 41 had an effect. Set point 12 added an amount –1.4 to the baseline of –7.5 when it was 'on' (trials 9 through 12) and set point 41 also added an amount –1.4 when it was 'on' (trials 3, 4, 7, 8, 11, 12, 15, and 16). When both were 'on' (trials 11 and 12) an amount –2.8 was added to the baseline. The simulated $\mathrm{Ln}(1/g(q, y))$ data from which the 'measurements' in Figure CS3.4 were derived are shown in Figure CS3.5 along with their average and variance for each trial.

The computation of the ξ'_{ij} variance coefficients and the test of their significance is shown in Figure CS3.6. The ξ'_{ij} coefficients were computed in column BA using the area expression:

```
=MMULT(B40:Q55,-10*LOG(BZ34:BZ49))
```

which is the spreadsheet representation of Equation 9.7. The F_{ij} ratio (Equation 9.8) in cell BB41 was computed using the expression

```
=10^(ABS(BA41)/(SQRT(50*U59)))
```

which was then dragged down. The pairwise error in cell BC41 was computed using the expression:

```
=FDIST(BB41,4,4)
```

which was then dragged down. The experiment-wise error in the last column of Figure CS3.6 is 15 times the pairwise error. If only the pairwise error were considered, set point 61 would probably be taken as significant for reducing

	CD	CE	CF	CG	CH	CI
33	Trial	1	2	3	4	5
34	1	731	5470	1677	486	2424
35	2	6041	1119	7208	4153	1247
36	3	1499	5311	1701	3129	6086
37	4	2171	11904	6560	26109	27850
38	5	5031	5602	12078	5772	7430
39	6	40201	34372	9175	11023	3641
40	7	710	1714	2684	2043	2086
41	8	6549	3774	437	6669	10714
42	9	3098	31604	6541	4236	33573
43	10	42785	940	3402	24762	8315
44	11	43083	22826	172107	42010	78636
45	12	17824	12067	9597	8185	116636
46	13	12193	5849	8455	8777	4847
47	14	5921	3787	41666	1244	1166
48	15	745	2141	346	6394	2485
49	16	6385	1102	2607	5635	328

Figure CS3.4 The simulated lifetimes for the springs for each trial.

variance since the Type I error would appear to be only be 0.5%. However, when multiplied by the factor of 15 for the experiment-wise correction, the error becomes almost 7%. The lack of any set point significantly influencing variance (less than 1%) based upon the experiment-wise error is in agreement with the simulations having a common population variance for all of the trials.

	BT	BU	BV	BW	BX	BY	BZ	CA
33	Trial	1	2	3	4	5	Var	Ave
34	1	-6.595	-8.607	-7.425	-6.186	-7.793	0.926	-7.321
35	2	-8.706	-7.020	-8.883	-8.332	-7.129	0.776	-8.014
36	3	-7.312	-8.578	-7.439	-8.049	-8.714	0.408	-8.018
37	4	-7.683	-9.385	-8.789	-10.170	-10.235	1.126	-9.252
38	5	-8.523	-8.631	-9.399	-8.661	-8.913	0.123	-8.825
39	6	-10.602	-10.445	-9.124	-9.308	-8.200	0.992	-9.536
40	7	-6.565	-7.447	-7.895	-7.622	-7.643	0.262	-7.435
41	8	-8.787	-8.236	-6.079	-8.805	-9.279	1.592	-8.237
42	9	-8.039	-10.361	-8.786	-8.351	-10.421	1.270	-9.192
43	10	-10.664	-6.846	-8.132	-10.117	-9.026	2.351	-8.957
44	11	-10.671	-10.036	-12.056	-10.646	-11.273	0.583	-10.936
45	12	-9.788	-9.398	-9.169	-9.010	-11.667	1.167	-9.807
46	13	-9.409	-8.674	-9.043	-9.080	-8.486	0.132	-8.938
47	14	-8.686	-8.239	-10.637	-7.126	-7.061	2.129	-8.350
48	15	-6.613	-7.669	-5.846	-8.763	-7.818	1.280	-7.342
49	16	-8.762	-7.005	-7.866	-8.637	-5.794	1.527	-7.613
50						Pooled=	1.040	

Figure CS3.5 The simulated $\mathrm{Ln}(1/g(q,y))$ used to determine the lifetimes in Figure CS3.4. The sample variances and averages for each trial are also shown.

	AZ	BA	BB	BC	BD
37					Exp.-
38				Pairwise	wise
39	ij	ξ'_{ij}	F_{ij}	error	error
40	0	0.334			
41	11	2.029	2.545	0.194	2.907
42	12	-1.973	2.481	0.200	3.003
43	13	-0.550	1.288	0.406	6.090
44	21	-1.333	1.848	0.283	4.250
45	31	-0.170	1.117	0.459	6.880
46	41	2.766	6.059	0.055	0.818
47	51	-0.949	1.855	0.282	4.232
48	61	-4.886	24.093	0.005	0.070
49	71	-0.250	1.177	0.439	6.588
50	81	2.553	5.273	0.068	1.022
51	91	0.101	1.048	0.482	7.237
52	10,1	-0.878	1.771	0.297	4.450
53	11,1	1.868	3.376	0.133	1.992
54	12,1	2.256	4.347	0.092	1.378
55	2191	1.318	1.536	0.344	5.157

Figure CS3.6 Computations used to test the significance of the ξ'_{ij} variance coefficients.

The λ_{ij} coefficients for the average, shown as column BG of Figure CS3.7, were computed according to Equation 9.6 for LIB, the spreadsheet expression for the area being:

=MMULT(B40:Q55,CA34:CA49).

The value of t in cell BH41 was calculated from Equation 9.10a using the spreadsheet expression:

=BG41/SQRT(X41/5),

and the pairwise error in cell BI41 was computed using the expression:

=TDIST(ABS(BH41),AC41,1).

The experiment-wise error was 15 times the pairwise error and we see that set points 12 and 41 are significant which is in agreement with the population statistics used for the trials.

Because none of the set points were found significant for variance, the pooled variance of 1.04 and the pooled df of 64 (= 4×16) would have normally been used in determining the significance of the λ_{ij} in Figure CS3.7. However, this was not done here in order to be able to illustrate the use of Equation B.29 for the effective variance, s^2_{ij}, and Equation B.30 for the effective df, ν_{ij}, which apply when the assumption of a common population variance is not supported or simply doubted and the approximate t distribution, Equation B.28, is used.

	BF	BG	BH	BI	BJ
37					Exp.-
38				Pairwise	wise
39	ij	λ_{ij}	t_{ij}	error	error
40	0	-7.321			
41	11	-0.357	-1.281	0.106	1.596
42	12	-1.571	-4.791	0.000	0.000
43	13	0.091	0.281	0.390	5.857
44	21	0.337	1.013	0.161	2.411
45	31	0.127	0.556	0.290	4.356
46	41	-1.194	-5.236	0.000	0.000
47	51	0.038	0.166	0.434	6.516
48	61	-0.220	-0.964	0.170	2.552
49	71	-0.640	-2.807	0.004	0.055
50	81	0.079	0.347	0.365	5.476
51	91	0.093	0.281	0.391	5.861
52	10,1	0.075	0.327	0.372	5.586
53	11,1	0.084	0.367	0.358	5.365
54	12,1	-0.163	-0.715	0.239	3.585
55	2191	-0.551	-1.207	0.117	1.752

Figure CS3.7 Computations used to test the significance of the λ_{ij} average coefficients.

The computations for s_{ij}^2 and ν_{ij} are made in Figure CS3.8. The four spreadsheet expressions for columns X, Y, AA, and AC are given in sequence by:

```
=MMULT(B40:Q55^2,BZ34:BZ49)
=X40^2
=MMULT(B40:Q55^4,BZ34:BZ49^2)
=(5-1)*Y40/AA40
```

The first and fourth expressions are for areas and the other two are cell expressions which are dragged down. The effective variances computed for this problem are close to the effective variances given by $s_p^2\,\text{cov}_{ijij}$ which is used when all of the variances are pooled but the df are very different.

The last step is to determine the significance of the set points on the CIQ S/N ratio, $\lambda'_{ij}(\Omega)$, using the Monte Carlo simulations. The sequence of computations needed for this are shown in Figure CS3.9. For the simulations, the pooled variance of 1.04 was used along with the pooled df of 64. (Pooling in this manner represents standard practice when hypothesis of a common population variance is not refuted.)

The simulations of the t statistic in column K were made using Equation A.22, the points on the cumulative distribution being given by the random number in column J. The simulated population averages were then computed using Equation 9.12. The simulations for the chi-squared statistic are shown in column Q and the simulated population variances are shown in column R,

	W	X	Y	AA	AC
36		s_{ij}^2			
37					
38		$=\sum_q \omega_{ij}^2(q)s^2(q)$		$\sum_q \omega_{ij}^4(q)s^4(q)$	
39	ij		s_{ij}^4		v_{ij}
40	0	0.926	0.858	0.858	4.0
41	11	0.388	0.150	0.025	23.7
42	12	0.538	0.289	0.046	25.3
43	13	0.519	0.269	0.045	24.2
44	21	0.553	0.306	0.051	23.9
45	31	0.260	0.068	0.006	46.4
46	41	0.260	0.068	0.006	46.4
47	51	0.260	0.068	0.006	46.4
48	61	0.260	0.068	0.006	46.4
49	71	0.260	0.068	0.006	46.4
50	81	0.260	0.068	0.006	46.4
51	91	0.544	0.296	0.055	21.4
52	10,1	0.260	0.068	0.006	46.4
53	11,1	0.260	0.068	0.006	46.4
54	12,1	0.260	0.068	0.006	46.4
55	2191	1.040	1.082	0.093	46.4

Figure CS3.8 Computations used to compute the effective variance and effective df for the set points. These calculations are given only for illustration. They would be needed in using the approximate t statistic to determine the significance of the λ_{ij} coefficients if the hypothesis of a common population variance had not been supported.

	H	I	J	K	L	M	O	P	Q	R	S	T	U	N	W	X
8		Pooled					Pooled									
9	Trial	df	RAND()	t	$\mu_{s,t}$		df	RAND()	χ^2	$\sigma_{s,t}^2$		Ω_s	$-10\,\mathrm{Log}(\Omega_s)$		ij	$\lambda'_{ij}(\Omega_s)$
10	1	64	0.690	0.50	-7.09		64	0.383	60	1.109		6.32E-06	51.99		0	51.99
11	2	64	0.964	1.83	-7.18		64	0.673	69	0.972		4.06E-06	53.92		11	3.14
12	3	64	0.146	-1.06	-8.50		64	0.540	64	1.033		3.25E-07	64.88		12	16.31
13	4	64	0.434	-0.17	-9.33		64	0.774	72	0.922		5.00E-08	73.01		13	-1.44
14	5	64	0.478	-0.06	-8.85		64	0.995	97	0.687		8.11E-08	70.91		21	-3.75
15	6	64	0.629	0.33	-9.38		64	0.499	63	1.052		5.78E-08	72.38		31	0.93
16	7	64	0.302	-0.52	-7.67		64	0.068	48	1.387		3.47E-06	54.60		41	11.47
17	8	64	0.827	0.95	-7.80		64	0.453	62	1.074		1.43E-06	58.46		51	4.08
18	9	64	0.138	-1.10	-9.69		64	0.942	83	0.805		1.91E-08	77.19		61	-1.56
19	10	64	0.251	-0.68	-9.27		64	0.504	63	1.049		7.31E-08	71.36		71	5.40
20	11	64	0.771	0.75	-10.60		64	0.756	71	0.932		4.04E-09	83.94		81	0.77
21	12	64	0.465	-0.09	-9.85		64	0.535	64	1.035		2.22E-08	76.54		91	-2.84
22	13	64	0.196	-0.86	-9.33		64	0.704	70	0.957		5.33E-08	72.74		10,1	-3.46
23	14	64	0.881	1.19	-7.81		64	0.039	46	1.462		3.08E-06	55.11		11,1	1.31
24	15	64	0.490	-0.03	-7.35		64	0.257	56	1.183		4.37E-06	53.60		12,1	2.30
25	16	64	0.393	-0.27	-7.74		64	0.205	55	1.221		2.19E-06	56.59		2191	6.50

Figure CS3.9 Computations used in generating the Monte Carlo simulations of the $\lambda'_{ij}(\Omega_S)$ coefficients shown in Figure CS3.10.

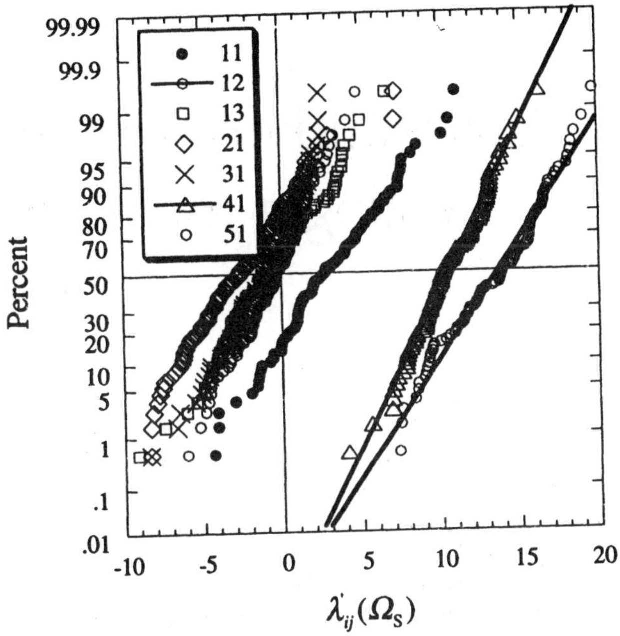

Figure CS3.10 Simulated uncertainty in the $\lambda'_{ij}(\Omega_S)$ coefficients for several set points. The pairwise significance of set points 21 and 41 is given by their cumulative distribution intercepts, which are well below 0.01%, with the vertical line at zero. Their experiment-wise errors are 15 times this and thus below 0.15%.

being computed from Equation 9.13. The CIQ computed from Equation 9.14 is given in column T. Finally, the simulated $\lambda'_{ij}(\Omega)$ coefficients are shown in column X.

A total of 100 simulations were made to determine the significance of set points 12 and 41 on the CIQ S/N ratio. (The macro described in Chapter 9 was used to automate the cutting and pasting operations involved in the simulations.) The distributions found are shown in Figure CS3.10 for set points 12 and 41 and for five of the other set points none of which are seen as significant. The intercepts at the origin for the 12 and 41 distributions appear to be below 0.01% which represents their pairwise error. Their experiment-wise error is therefore below 0.15%.

Reference

1. Montmarquet, F. (1988) Printed circuit drill bit design optimization using Taguchi's methods –.013″ diameter bits, in *Sixth Symposium on Taguchi Methods*, American Supplier Institute, Allen Park, MI, pp. 207–16.

Case study 4 Significance testing a multiattribute problem

This case study is designed to demonstrate a method for evaluating the significance of set points when multiple attributes are involved. Not surprisingly, it follows, in large part, the single-attribute approach discussed in Chapter 9; however, several new issues do arise. One important difference is that a factor in the multiattribute problem can affect one attribute favorably and affect another unfavorably requiring a compromise to be made if it is used, whereas in the single-attribute problem a factor cannot have this type of behavior.

The first step in any multiattribute problem is to get the objective functions for the attributes expressed in a common set of units. For this, we will use the dimensionless value coefficients obtained from Equation 5.6 and written in the form:

$$\nu(g_\varphi) = \frac{V(g_\varphi)}{V_0} \tag{CS4.1}$$

Expressions for these quantities are shown in Table CS4.1 for four types of conditions (SIB, NIB1, NIB2, and LIB) for an arbitrary system attribute, φ, which appears as a subscript. Using the approximation for SIB problems given by:

$$\overline{[V(g_\varphi)]^\gamma} \approx \left[\frac{V_0}{g_{\varphi C}^2 - g_{\varphi 0}^2}\left[g_{\varphi C}^2 - \left[\sigma_\varphi^2 + \mu_\varphi^2\right]\right]\right]^\gamma \tag{CS4.2}$$

and similar approximations for NIB and LIB, exponential weighting, as discussed in Chapter 5, can also be applied to the expressions in Table CS4.1 if required for a particular problem.

The value expression in Table CS4.1 can also be used for single-attribute problems. However, it should be noted that the goal is to maximize the value ratio functions in Table CS4.1, whereas it was to minimize the CIQ functions in Table 8.2. The expressions in the denominators of each of the equations in Table CS4.1 is a constant proportional to the value for trial 1 as determined from the sample measurement, but the expressions in the numerators contain the population statistics for each trial. Because the population statistics must be inferred from the measured sample statistics, they are not known with preci-

Table CS4.1 Value ratio formulas for use in multi-attribute problems with trial 1 as baseline

Condition	$\nu_\varphi(q) = \dfrac{V(g_\varphi)}{V(g_{\varphi 0})}$
SIB	$\dfrac{g_{\varphi C}^2 - [\sigma_\varphi^2(q) + \mu_\varphi^2(q)]}{g_{\varphi C}^2 - [s_\varphi^2(1) + \bar{g}_\varphi^2(1)]}$
NIB1	$\dfrac{[g_{\varphi C} - g_{\varphi I}]^2 - [g_{\varphi I}^2 / \mu_\varphi^2(q)]\sigma_\varphi^2(q)}{[g_{\varphi C} - g_{\varphi I}]^2 - [g_{\varphi I}^2 / \bar{g}_\varphi^2(1)]s_\varphi^2(1)}$
NIB2	$\dfrac{[g_{\varphi C} - g_{\varphi I}]^2 - \left[\sigma_\varphi^2(q) + [\mu_\varphi(q) - g_{\varphi I}]^2\right]}{[g_{\varphi C} - g_{\varphi I}]^2 - \left[s_\varphi^2(1) + [\bar{g}_\varphi(1) - g_{\varphi I}]^2\right]}$
LIB	$\dfrac{1 - g_{\varphi C}^2[\sigma_\varphi^2 + \mu_\varphi^2(q)]}{1 - g_{\varphi C}^2[s_\varphi^2 + g_\varphi^{-1}(1)^2]}$

sion. It is this uncertainty in the population statistics, as in the single-attribute problem, which requires us to determine the significance of a finding before we act upon the outcome of an experiment.

CS4.1 Multiattribute value

This hypothetical case illustrates a multiattribute experiment designed to improve a thin film device whose value to the customer is influenced by its electrical resistance, number of surface defects, and adhesion to the substrate. Because each of these three attributes are possibly subject to modification from changes in any of the set points, three product system specifications, g_1, g_2, g_3, for resistance, number of surface defects, and adhesion, respectively, need to be measured for each experimental trial. The outcome of total value for a given experimental trial (q) is given by:

$$V(q) = V_0\overline{\nu_1(q)\nu_2(q)\nu_3(q)} \tag{CS4.3}$$

The type of loss function suggested for each attribute along with their base, ideal, and critical specifications is shown in Table CS4.2. Resistance is a NIB1 condition having a scaling factor given by a highly accurate adjustment of the width of the film which is generated by changing the geometry of the mask used to define the shape of the film. The number of surface defects is SIB and adhesion is LIB.

Table CS4.2 Base, ideal, and critical specifications for attributes.

System specifications	*Type*	*Notation*	*Base*	*Ideal*	*Critical*
Resistivity	NIB1	g_1	400	475	350
Surface defects	SIB	g_2	4000	0	7000
Adhesion	LIB	g_3	80	∞	30

For these assumptions, the value of the film versus its resistance is a parabola having a maximum at the ideal resistance level shown as 475 (arbitrary units) (Figure CS4.1). The value curves for surface defects and adhesion are shown by Figures CS4.2 and CS4.3, respectively. For the baseline experiment, all of the specifications are at their baseline (trial 1) points, $g_1(1), g_2(1), g_3(1)$. As discussed for the NIB1 thin film problem considered in Case Study 1, the base NIB1 specification refers to the measurement for the baseline trial before scaling factor adjustment. However, the value of the film for Equation CS4.1 is computed as if the adjustment has been made to bring the mean to ideal, the resulting value being less than ideal only as a result of the variance in resistance about the ideal point.

The same four type of set points or factors – temperature, pressure, settling time, and cleaning method – considered by Phadke [1] for the CVD process discussed in Chapter 8 were used for this problem. The $[\mathbf{X}]_\lambda\ L_9(3^4)$ orthogonal experimental design was used and shown here for convenience in Table CS4.3.

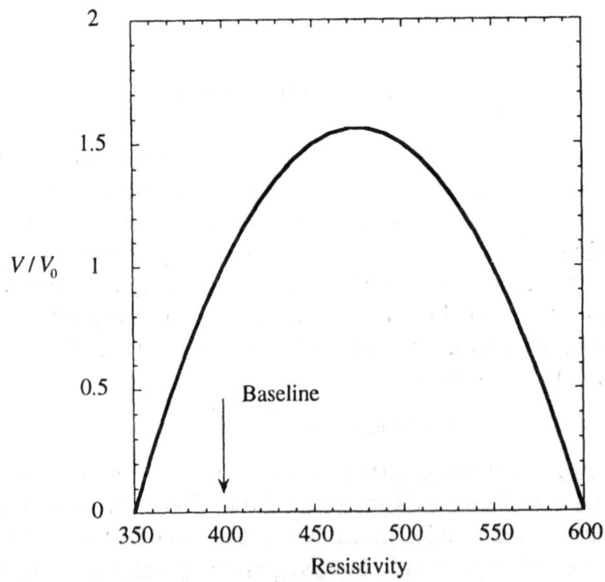

Figure CS4.1 Hypothetical value of film as a function of its electrical resistance.

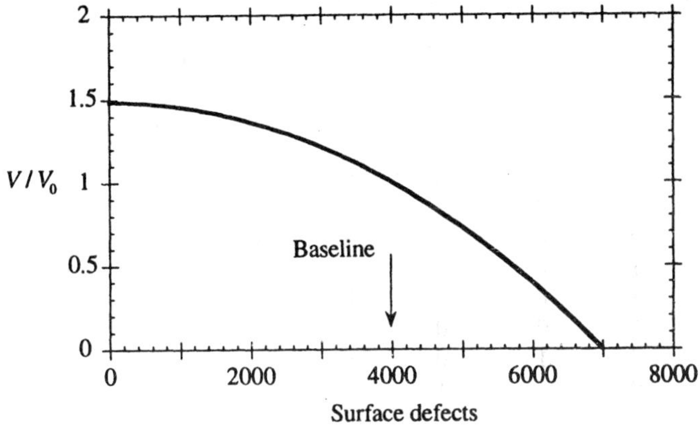

Figure CS4.2 Hypothetical value of film as a function of the number of surface defects.

CS4.2 Sample 'measurements'

The populations were chosen (Table CS4.4) such that only set point 11 affected the mean and variance of resistance; only set point 21 affected the mean and variance of surface defects; and only set point 41 affected the mean and variance of adhesion. Set points 31 and 32 affected none of the attributes.

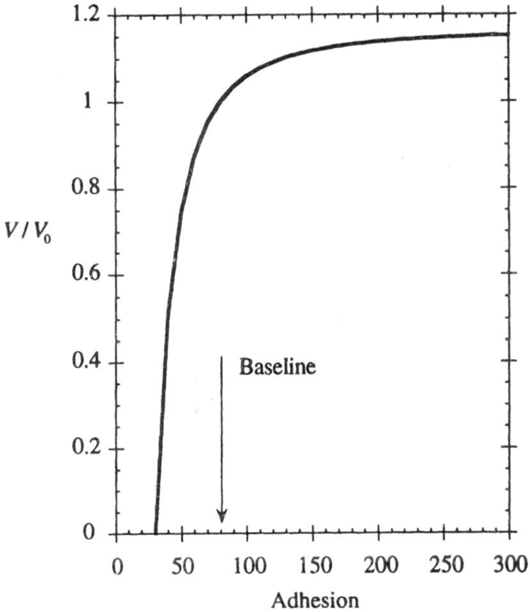

Figure CS4.3 Hypothetical value of film as a function of its adhesion.

Table CS4.3 $L_9(3^4)$ experimental design in $[\mathbf{X}]_\lambda$ form

		Temp		*Pressure*		*Settling time*		*Cleaning method*	
Trial	*0*	*11*	*12*	*21*	*22*	*31*	*32*	*41*	*42*
1	1	0	0	0	0	0	0	0	0
2	1	0	0	1	0	1	0	1	0
3	1	0	0	0	1	0	1	0	1
4	1	1	0	0	0	1	0	0	1
5	1	1	0	1	0	0	1	0	0
6	1	1	0	0	1	0	0	1	0
7	1	0	1	0	0	0	1	1	0
8	1	0	1	1	0	0	0	0	1
9	1	0	1	0	1	1	0	0	0

Lognormal statistics were assumed for surface defects and the reciprocal of adhesion. The basic variables for four replications for each of the three attributes shown in Table CS4.5 were selected from the normal populations for each trial using the random variable:

$$x_\varphi(q) = \mu_\varphi(q) + \sqrt{0.5\sigma_\varphi^2}\sum_{j=1}^{24}[\text{RAND}() - 0.5]_j \qquad \text{(CS4.4)}$$

CS4.3 Coefficients for variances and averages

The resulting sample averages and variances are shown in Table CS4.6 along with the sample variance coefficients, ξ'_{ij}, and sample average coefficients, λ_{ij}, computed from them using Equations 9.7 and 9.6, respectively. The F_{ij} statistic

Table CS4.4 Population means and variances for each trial used to simulate experimental measurements

	Resistivity		*Ln(Surface defects)*		*Ln(1/Adhesion)*	
Trial	μ_1	σ_1^2	μ_2	σ_2^2	μ_3	σ_3^2
1	400	300	8.294	0.05	–4.38	0.08
2	400	300	6.908	0.0025	–5.70	0.004
3	400	300	8.294	0.05	–4.38	0.08
4	550	18	8.294	0.05	–4.38	0.08
5	550	18	6.908	0.0025	–4.38	0.08
6	550	18	8.294	0.05	–5.70	0.004
7	400	300	8.294	0.05	–5.70	0.004
8	400	300	6.908	0.0025	–4.38	0.08
9	400	300	8.294	0.05	–4.38	0.08

Table CS4.5 Simulated 'measurements'

	Resistivity Replications				*Surface Defects Replications*				*Adhesion Replications*			
Trials	*1*	*2*	*3*	*4*	*1*	*2*	*3*	*4*	*1*	*2*	*3*	*4*
1	410	373	387	409	3212	3823	4232	4648	61	122	77	80
2	382	415	390	420	1042	1120	1036	1051	277	298	319	281
3	412	401	384	382	4532	4575	5394	4051	70	90	69	108
4	543	546	549	554	3569	4315	2772	4098	91	85	119	100
5	555	553	543	549	1003	959	1012	1110	54	102	82	52
6	559	552	543	548	3134	4425	3945	3276	298	317	277	272
7	410	376	391	388	4072	5135	4151	4972	323	308	284	295
8	409	438	424	379	1009	964	931	1016	65	51	76	79
9	404	381	400	414	3688	4431	4060	5205	68	79	72	66

was computed from Equation 9.8 and the pairwise error for the ξ'_{ij} coefficients was computed using the spreadsheet function FDIST(F_{ij}, 3, 3). The ξ'_{ij} coefficients found significant based upon the experiment-wise error (eight times the pairwise error) are shown in bold face.

Based upon the significance findings, the trial variances were then pooled following the procedure described in Chapter 9, the three pooling arrays being shown in Table CS4.7a. For resistivity only set point 11 (temperature level 2) was significant and trials 1, 2, 3, 7, 8, and 9 were pooled for 18 df; trials 4, 5, and 6 were pooled for 9 df. For surface defects, only set point 21 (pressure level 2) was significant and trials 1, 3, 4, 6, 7, and 9 were pooled for 18 df; trials 2, 5, and 8 were pooled for 9 df. For adhesion, only set point 41 (cleaning method level 2) was significant and trials 1, 3, 4, 5, 8, and 9 were pooled for 18 df; trials 2, 6, and 7 were pooled for 9 df.

The significance of the λ_{ij} coefficients were then computed using Equation 9.10a. The effective variances s^2_{ij} were computed using Equation B.29 with the sample trial variances replaced by the pooled trial variances $s^2_p(q)$ shown in Table CS4.7a. The effective df ν_{ij} used in computing the pairwise errors were computed from Equation B.30 with the sample trial variances, once again, replaced by their respective pooled trial variances $s^2_p(q)$ shown in Table CS4.7a. The effective variances are shown in Table CS4.7b and can be compared to the pooled trial variances in Table CS4.7a. The averages found to be significant are shown in bold face. The results for both the variance and average coefficients are in agreement with what is expected based upon the known population statistics used in simulating the sample results. The solution matrix and the square of the elements of the solution matrix are shown in Table CS4.8.

Table CS4.6 Averages and variances for the four replications for each trial and tests for common variance and common mean hypotheses

Resistivity

Trials	*Ave*	*Var*	*ij*	ξ'_{ij}	G_{ij}	F_{ij}	*Pair-wise error*	*Exp.-wise error*	*ij*	λ_{ij}	t_{ij}	*Pair-wise error*	*Exp.-wise error*
1	395	322	0	–25.1					0	395			
2	402	362	11	**9.8**	**1.69**	49	0.00	**0.04**	11	**153**	**28**	1.2E–16	**9.4E–16**
3	394	205	12	0.0	0.01	1	0.50	3.97	12	4.3	0.59	2.8E–01	2.26
4	548	21	21	–2.4	–0.41	3	0.23	1.82	21	10.1	1.66	5.7E–02	0.46
5	550	29	22	–0.5	–0.08	1	0.44	3.51	22	3.6	0.59	2.8E–01	2.26
6	550	46	31	2.7	0.47	3	0.20	1.60	31	–2.6	–0.43	3.4E–01	2.69
7	391	191	32	3.1	0.53	3	0.17	1.37	32	–7.4	–1.22	1.2E–01	0.95
8	413	640	41	–0.8	–0.14	1	0.40	3.18	41	–0.5	–0.07	4.7E–01	3.77
9	400	192	42	–0.6	–0.11	1	0.42	3.37	42	3.6	0.59	2.8E–01	2.25

Ln (Surface defects)

Trials	*Ave*	*Var*	*ij*	ξ'_{ij}	G_{ij}	F_{ij}	*Pair-wise error*	*Exp.-wise error*	*ij*	λ_{ij}	t_{ij}	*Pair-wise error*	*Exp.-wise error*
1	8.28	0.025	0	16.02					0	8.28			
2	6.97	0.001	11	–3.09	–0.53	3.42	0.17	1.36	11	–0.12	–2.26	0.018	0.15
3	8.44	0.014	12	–0.17	–0.03	1.07	0.48	3.82	12	0.00	–0.02	0.491	3.93
4	8.20	0.039	21	**10.8**	**1.87**	75	0.00	**0.02**	21	**–1.37**	**–29.8**	0.000	**4E–16**
5	6.93	0.004	22	0.9	0.15	1	0.39	3.13	22	0.04	0.6	0.286	2.29
6	8.20	0.026	31	–0.02	0.00	1.01	0.50	3.98	31	0.05	1.05	0.154	1.23

Trials	Ave	Var	ij	ξ'_{ij}	G_{ij}	F_{ij}	Pair-wise error	Exp.-wise error	ij	λ_{ij}	t_{ij}	Pair-wise error	Exp.-wise error
7	8.42	0.014	32	0.52	0.09	1.23	0.43	3.47	32	0.14	2.67	0.008	0.06
8	6.89	0.002	41	2.05	0.35	2.26	0.26	2.08	41	0.01	0.14	0.445	3.56
9	8.37	0.021	42	1.16	0.20	1.59	0.36	2.86	42	−0.02	−0.34	0.369	2.95

Ln (1/Adhesion)

Trials	Ave	Var	ij	ξ'_{ij}	G_{ij}	F_{ij}	Pair-wise error	Exp.-wise error	ij	λ_{ij}	t_{ij}	Pair-wise error	Exp.-wise error
1	−4.41	0.084	0	10.8					0	−4.41			
2	−5.68	0.004	11	0.38	0.07	1.16	0.45	3.61	11	0.004	0.05	0.480	3.84
3	−4.42	0.046	12	4.51	0.78	6.03	0.09	0.70	12	0.107	1.40	0.090	0.72
4	−4.58	0.022	21	−1.66	−0.29	1.94	0.30	2.40	21	0.192	2.50	0.011	0.09
5	−4.24	0.110	22	2.08	0.36	2.29	0.26	2.05	22	0.116	1.51	0.074	0.60
6	−5.67	0.005	31	5.00	0.87	7.35	0.07	0.54	31	−0.082	−1.07	0.149	1.19
7	−5.71	0.003	32	0.09	0.02	1.04	0.49	3.91	32	−0.027	−0.36	0.362	2.90
8	−4.20	0.039	41	**9.8**	**1.69**	49	0.00	**0.038**	41	**−1.38**	**−20.4**	0.000	**3E–13**
9	−4.27	0.006	42	0.4	0.07	1	0.45	3.580	42	−0.09	−1.0	0.159	1.27

Table CS4.7a Pooling arrays for variances

	Common base	*Resistivity*		*Ln(Surface defects)*		*Ln(1/Adhesion)*	
Trial	*0*	*11*	*Var*	*21*	*Var*	*41*	*Var*
1	1	0	319	0	0.023	0	0.0512
2	1	0	319	1	0.002	1	0.0040
3	1	0	319	0	0.023	0	0.0512
4	1	1	32	0	0.023	0	0.0512
5	1	1	32	1	0.002	0	0.0512
6	1	1	32	0	0.023	1	0.0040
7	1	0	319	0	0.023	1	0.0040
8	1	0	319	1	0.002	0	0.0512
9	1	0	319	0	0.023	0	0.0512

Table CS4.7b Effective degrees of freedom

	ν_{ij}		
ij	*Resistivity*	*Ln(Surface defects)*	*Ln(1/Adhesion)*
0	3	3	3
11	10.8	13.1	12.9
12	18.0	13.1	12.9
21	13.2	10.7	12.9
22	13.2	18.0	12.9
31	13.2	13.1	12.9
32	13.2	13.1	12.9
41	13.2	13.1	10.4
42	13.2	13.1	18.0

CS4.4 The $\lambda_{ij}(\nu_1\nu_2\nu_3)$ value product coefficients

Although we have evaluated the statistical significance of the set points (factors), we now need to verify their strategic significance by computing the distribution of the simulated value product coefficients $\lambda_{ij}(\nu_1\nu_2\nu_3)$ determined from the expression:

$$[\lambda_{ij}(\nu_1\nu_2\nu_3)]=[\mathbf{X}^T\mathbf{X}]^{-1}[\mathbf{X}^T][\nu_1\nu_2\nu_3] \qquad \text{(CS4.5)}$$

Because simulations were needed for three attributes, value coefficients were determined from SNRV computer simulations of the type described in Appendix A to speed up the simulation process. This required that two SNRV 'companion experiments' be run for each attribute. Examples of these are shown in Table CS4.9. The number of SNRV replications in the companion

Table CS4.8 Solution matrix and matrix formed by squaring the elements of the solution matrix

$[X^TX]^{-1}[X^T]$

ij	*q* 1	2	3	4	5	6	7	8	9
0	1	1E–16	1E–16	1E–16	1E–16	1E–16	1E–16	1E–16	1E–16
11	–0.333	–0.333	–0.333	0.333	0.333	0.333	0	0	0
12	–0.333	–0.333	–0.333	0	0	0	0.333	0.333	0.3333
21	–0.333	0.333	0	–0.333	0.333	0	–0.333	0.333	0
22	–0.333	0	0.333	–0.333	0	0.333	–0.333	0	0.3333
31	–0.333	0.333	0	0.333	0	–0.333	0	–0.333	0.3333
32	–0.333	0	0.333	0	0.333	–0.333	0.333	–0.333	0
41	–0.333	0.333	0	0	–0.333	0.333	0.333	0	–0.333
42	–0.333	0	0.333	0.333	–0.333	0	0	0.333	–0.333

$\omega_{ij}^2(q)$

ij	*q* 1	2	3	4	5	6	7	8	9
0	1	1E–32	1E–32	1E–32	1E–32	1E–32	1E–32	1E–32	1E–32
11	0.111	0.111	0.111	0.111	0.111	0.111	0	0	0
12	0.111	0.111	0.111	0	0	0	0.111	0.111	0.1111
21	0.111	0.111	0	0.111	0.111	0	0.111	0.111	0
22	0.111	0	0.111	0.111	0	0.111	0.111	0	0.1111
31	0.111	0.111	0	0.111	0	0.111	0	0.111	0.1111
32	0.111	0	0.111	0	0.111	0.111	0.111	0.111	0
41	0.111	0.111	0	0	0.111	0.111	0.111	0	0.1111
42	0.111	0	0.111	0.111	0.111	0	0	0.111	0.1111

experiment for trial (q) for attribute φ is equal to $\nu_\varphi(q)+1$ and the companion experiment simulations for resistivity shown in Table CS4.9 have nineteen SNRV replications for trials 1, 2, 3, 7, 8, and 9 and ten SNRV replications for trials 4, 5, and 6.

The simulated population statistics were computed from the SNRV statistics using Equations A.27 and A.28 for the variance and average, respectively. The simulated reduced value coefficient ν_1 for resistivity was then computed using the NIB1 expression in Table CS4.1. However, for the surface defects and reciprocal adhesion attributes, the SNRV statistics were first transformed back from the lognormal distributions using Equations A.32 and A.33 before computing the reduced value coefficients ν_2 and ν_3 for surface defects and reciprocal adhesion using the SIB and LIB expressions in Table CS4.1. Typical simulated value products for the three attributes are shown in Table CS4.10.

The computer simulation results are shown in Figure CS4.4. The intercepts at zero determine the significance of the set points on value after being multi-

Table CS4.9 SNRV companion experiments

Trials	*SNRV I Resistivity* 1	2	3	4	5	6	7	8	9	10	11	12	13	14	15	16	17	18	19	*SNRV* Ave.	Var.	*Simulated Resistivity* Ave.	Var.		v_1
1	−2.02	−1.04	0.12	1.42	−0.28	−0.79	0.56	−0.27	−0.37	−0.64	1.37	−1.05	−1.56	−0.64	−1.10	−0.18	0.44	−0.06	1.79	−0.227	1.02	388	510	0	0.980
2	0.33	−1.31	−1.43	−0.74	1.11	−1.43	1.11	−0.84	0.47	−1.96	−0.02	−0.32	−0.84	−0.09	−0.08	1.98	−0.84	0.63	1.82	−0.128	1.25	398	236	11	1.008
3	0.95	−1.42	−0.31	−0.98	−0.85	2.21	0.82	−2.26	1.09	−0.45	−0.92	−0.55	−0.24	−0.53	−0.41	−0.22	0.31	0.05	0.16	−0.188	0.99	389	244	12	1.006
4	0.52	−1.39	0.30	0.29	−0.15	0.65	−1.48	0.21	0.00	−1.46										−0.252	0.73	546	25	21	1.029
5	−0.55	0.35	1.10	−2.13	1.09	−0.33	0.37	−1.93	−0.76	−0.20										−0.298	1.23	548	51	22	1.028
6	0.04	1.67	−1.60	0.85	−0.08	−2.28	1.51	0.87	1.64	1.07										0.371	1.86	553	32	31	1.029
7	−0.21	0.78	1.82	0.79	−0.93	0.33	−0.53	−2.12	1.67	0.49	−2.33	−1.83	−1.30	2.39	0.44	0.64	−1.01	−1.18	2.41	0.016	2.13	391	343	32	0.997
8	−1.34	−1.42	1.18	0.00	−1.59	−0.21	0.61	−1.90	−1.56	−0.56	−0.40	−0.33	−0.33	−0.60	−0.25	0.28	0.38	−0.30	0.49	−0.414	0.70	399	502	41	0.983
9	0.11	1.05	0.80	0.36	0.75	−1.08	−0.18	−0.41	−0.22	−0.57	1.58	0.48	−2.30	−0.47	1.28	0.46	−0.04	0.63	−0.42	0.096	0.81	403	290	42	1.004

	SNRV II Resistivity 1	2	3	4	5	6	7	8	9	10	11	12	13	14	15	16	17	18	19	*SNRV* Var.
1	−0.55	0.18	0.44	0.27	0.11	0.13	0.77	0.54	0.94	−1.23	0.06	1.16	−1.00	0.52	−1.64	0.57	1.22	−0.38	−0.48	0.62
2	0.22	−2.22	1.14	0.92	−2.16	−0.98	0.37	1.94	−0.57	0.34	−0.99	0.10	−1.17	0.60	−0.41	1.96	−0.58	0.34	0.18	1.35
3	−0.80	0.49	1.27	1.51	1.10	−1.38	−1.32	1.07	2.99	1.92	0.86	−0.27	0.28	−0.52	−0.64	−0.05	−0.36	0.67	−0.39	1.31
4	−0.97	−1.11	−0.83	1.26	−0.37	−1.22	−1.61	1.37	−1.39	0.80										1.28
5	0.28	−0.62	0.31	−2.06	0.55	−0.04	0.33	0.14	0.41	0.52										0.63
6	1.20	−0.51	−1.40	−0.04	0.16	0.98	0.23	−1.75	−0.21	−1.36										0.99
7	1.37	0.70	0.39	1.17	−0.02	−0.35	−1.31	0.58	1.30	0.32	−0.99	0.96	−0.40	−0.98	−2.22	−0.53	−0.35	0.61	−0.38	0.93
8	−0.56	−0.33	0.14	0.21	−0.07	−1.00	−1.23	0.54	−0.96	−1.45	1.32	0.56	0.44	−1.32	−0.53	−0.10	−1.33	−0.01	−1.26	0.63
9	−0.29	1.61	−1.23	−0.51	1.69	−1.56	−0.13	1.35	−1.80	0.47	−0.34	0.68	−0.43	−1.03	0.23	−1.19	0.03	−0.14	−1.50	1.10

Trials	*SNRV I Defects* 1	2	3	4	5	6	7	8	9	10	11	12	13	14	15	16	17	18	19	*SNRV* Ave.	Var.	*Lognormal* Ave.	Var.	*Simulated defects* Ave.	Var.		v_2
1	−0.87	0.6	−1.14	−0.23	0.812	0.806	−0.18	0.598	−1.19	1.001	0.96	1.438	−0.27	0.708	−0.13	0.365	1.053	0.943	−0.5	0.251	0.632	8.36	0.024	4306	450817	0	0.92
2	0.976	0.832	0.923	−2.22	−0.92	−1.12	0.383	1.79	0.141	0.287										0.107	1.431	6.97	0.001	1070	1542	11	1.46
3	1.065	−1.06	−0.05	−0.01	1.112	0.521	−0.61	1.199	1.373	0.413	0.392	1.275	1.88	0.633	0.915	−0.93	0.014	−1.33	−0.05	0.355	0.802	8.53	0.019	5127	507887	12	0.68
4	1.847	0.244	0.218	−2.11	0.09	−0.14	−0.13	−0.99	−0.55	−0.33	0.898	0.421	0.776	−1.39	−0.37	−0.21	−0.56	1	0.99	−0.02	0.855	8.19	0.022	3663	303973	21	1.08
5	−0.7	1.05	−0.82	1.141	2.345	−0.37	−0.28	0.036	1.379	0.824										0.461	1.082	6.96	0.004	1056	4488	22	1.46
6	−0.69	1.345	1.162	1.012	0.94	1.067	−0.59	−0.58	0.924	−0.36	0.763	−0.08	0.037	−1	−0.61	−0.39	−0.98	0.138	0.008	0.112	0.622	8.24	0.029	3841	429562	31	1.03
7	0.258	−0.34	−0.32	−0.15	2.463	1.328	−0.07	1.474	−0.76	−0.04	−0.58	1.258	−0.07	2.336	0.971	−1.49	0.395	1.188	0.295	0.429	1.089	8.52	0.025	5097	664846	32	0.68
8	−0.55	−0.4	−0.85	0.232	−2.05	0.374	0.249	−0.86	0.667	1.089										−0.21	0.842	6.87	0.003	964	2607	41	1.47
9	−1.68	−1.33	0.185	1.587	−0.2	2.187	−0.16	1.232	0.194	−1.92	−0.54	1.09	−1.16	1.636	0.507	2.6	0.071	−1.07	1.879	0.27	1.834	8.42	0.018	4564	382891	42	0.85

	SNRV II Defects																			*SNRV*
	1	*2*	*3*	*4*	*5*	*6*	*7*	*8*	*9*	*10*	*11*	*12*	*13*	*14*	*15*	*16*	*17*	*18*	*19*	*Var.*
1	0.796	−0.35	−0.37	0.137	−1.04	−1.51	0.762	2.064	0.117	−1.63	−1.75	−0.91	0.125	0.469	−1.04	0.325	−0.11	0.129	−1.31	0.971
2	0.521	−0.93	−0.6	−2.07	2.531	0.842	−0.49	0.432	0.077	−1.31										1.657
3	−1.99	0.628	0.185	0.076	−0.08	−0.89	−0.21	−0.8	−1.97	1.226	−2.17	1.501	−1.42	0.203	0.779	0.802	0.568	−0.98	0.276	1.219
4	1.116	0.145	0.242	−0.69	0.226	0.077	0.852	−0.22	−2.12	−2.46	0.422	1.393	0.634	−0.08	−0.24	0.874	0.831	1.315	−0.15	1.041
5	0.276	0.892	0.199	−0.46	−0.85	−0.67	0.284	1.483	−0.55	−0.36										0.556
6	1.135	−1.7	0.406	0.79	−1.12	0.352	−0.24	−0.4	0.097	0.973	0.138	−1.61	−0.09	1.492	−1.11	−0.47	−0.6	−0.23	−0.97	0.813
7	−0.36	1.776	−0.08	−0.6	−0.8	0.769	−1.39	−0.25	−0.22	0.091	0.467	1.535	0.446	0.945	−0.59	0.764	−1.56	0.007	1.745	0.923
8	−0.79	0.596	−0.48	0.536	−1.47	−0.92	0.286	−0.58	0.588	1.404										0.797
9	0.481	−0.14	−2.11	0.825	0.863	1.118	0.802	−1.38	−0.15	−1.22	0.766	−1.1	−0.29	1.121	0.454	−1.6	0.304	−2.51	0.118	1.281

	SNRVI 1/Adh																			*SNRV*		*Lognormal*		*Simulated 1/Adh.*			
Trials	*1*	*2*	*3*	*4*	*5*	*6*	*7*	*8*	*9*	*10*	*11*	*12*	*13*	*14*	*15*	*16*	*17*	*18*	*19*	*Ave.*	*Var.*	*Ave.*	*Var.*	*Ave.*	*Var.*		v_3
1	0.358	−0.03	−0.46	0.023	0.504	−0.4	−1.07	1.035	0.242	−0.38	−0.75	2.369	0.7	1.266	−1.68	−1.17	−1.46	0.169	0.684	−0	1.014	−4.41	0.029	0.0123	4.5E−06	0	1.019
2	1.458	0.837	0.322	−1.35	1.245	1.119	2.172	−0	0.571	−1.4										0.497	1.344	−5.67	0.009	0.0035	1.1E−07	11	1.152
3	−0.41	−2.28	−1.13	−1.17	1.3	1.604	1.237	0.909	0.531	−1.28	0.036	−1.16	−0.19	0.238	−0.37	0.171	0.373	1.476	0.306	0.009	1.145	−4.41	0.043	0.0124	6.7E−06	12	0.997
4	−0.63	2.538	0.133	0.03	0.022	0.71	−2.34	0.279	0.503	−0.97	1.473	−0.27	−0.11	−0.49	0.662	0.461	−1.07	1.786	−0.4	0.122	1.193	−4.57	0.063	0.0107	7.4E−06	21	1.037
5	1.559	0.958	−1.44	0.248	0.272	0.367	−0.16	1.78	−0.51	1.282	−0.24	0.91	−0.39	−0.77	−0.54	−0.23	0.242	−1.07	−1.02	0.066	0.829	−4.23	0.033	0.0148	7.4E−06	22	0.928
6	1.864	−0.46	0.025	0.074	−1.52	−0.42	−1.1	−0.7	−0.74	0.232										−0.28	0.86	−5.68	0.003	0.0034	3.5E−08	31	1.152
7	−0.29	−0.37	−1.19	1.267	0.322	0.484	0.674	0.825	1.068	−2.09										0.071	1.128	−5.71	0.005	0.0033	5.6E−08	32	1.153
8	−0.59	1.391	−0.02	−0.42	1.638	−0.41	0.073	1.887	0.816	−0.92	−0.6	−0.02	0.364	−0.78	−1.38	1.003	−0.17	0.002	0.071	0.102	0.791	−4.19	0.074	0.0157	1.9E−05	41	0.886
9	−1.13	0.859	0.048	−1.13	−1.27	−0.62	1.823	1.672	0.28	−0.07	−0.81	−0.5	−0.81	2.418	0.578	0.77	−1.16	−0.89	0.478	0.028	1.233	−4.27	0.052	0.0144	1.1E−05	42	0.935

	SNRV II 1/Adh																			*SNRV*
	1	*2*	*3*	*4*	*5*	*6*	*7*	*8*	*9*	*10*	*11*	*12*	*13*	*14*	*15*	*16*	*17*	*18*	*19*	*Var.*
1	0.488	0.651	1.102	−0.83	−1.9	−0.05	1.659	−0.25	1.989	−1.35	0.472	0.55	2.336	−0.36	0.904	0.648	−0.22	−0.08	−3.09	1.741
2	0.024	−0.86	−0.96	−0.44	0.387	−1.05	0.013	0.184	0.197	1.075										0.459
3	−1.03	−1.05	−0.51	−0.3	−0.23	1.515	−1.05	−0.29	−1.38	0.156	0.038	0.541	2.622	0.086	0.168	−0.2	−1.64	−0.28	1.887	1.194
4	−1.31	−1.52	−0.72	0.582	1.282	−0	−0.3	0.891	0.946	1.021	1.488	−0.51	0.141	1.018	−0.76	1.089	0.469	−0.58	0.279	0.812
5	1.919	−0	−1.07	1.346	−1.37	1.119	1.038	1.631	3.054	−0.51	−0.15	−0.17	−0.82	−0.65	−0.68	−0.48	0.647	1.562	−1.11	1.531
6	−0.47	1.046	0.362	1.371	−0.31	−0.42	0.872	1.404	1.444	−2.2										1.365
7	−0.38	−0.14	0.763	0.541	−1.77	−1.6	0.333	0.582	−0.77	−0.14										0.793
8	0.766	0.45	−0.75	−1.15	1.507	0.72	−0.25	−0.41	−0.38	0.138	−0.08	−0.43	−0.76	0.51	−0.43	0.833	−1.62	−1.57	−0.2	0.689
9	0.396	−0.21	−0.72	−1.53	−1.42	1.881	1.199	1.565	−0.74	−0.14	0.102	0.868	−0.57	0.142	0.016	−0.56	1.595	0.49	0.939	0.977

Table CS4.10 Typical value product simulations for the three attributes and resulting value coefficients for the set points

Trials	*Simulated* $\nu_1\nu_2\nu_3$	*ij*	*Simulated* $\lambda_{ij}(\nu_1\nu_2\nu_3)$
0	0.916	0	0.916
1	1.699	11	0.159
2	0.681	12	−0.145
3	1.151	21	0.508
4	1.396	22	−0.050
5	1.225	31	0.075
6	0.785	32	−0.186
7	1.281	41	0.200
8	0.797	42	0.001

plied by the number of off-baseline coefficients, eight. Although this plot shows that set points 11, 21, and 41 have favorable strategic significance as expected, set point 32 is shown to have a 96% experiment-wise confidence level (= 100 − 8[100 − 99.5]) that it would reduce total value if implemented. Because we know this is not correct from the true population statistics, this high level of confidence is simply the result of statistical fluctuations but in a real experiment we, of course, would not have this insight.

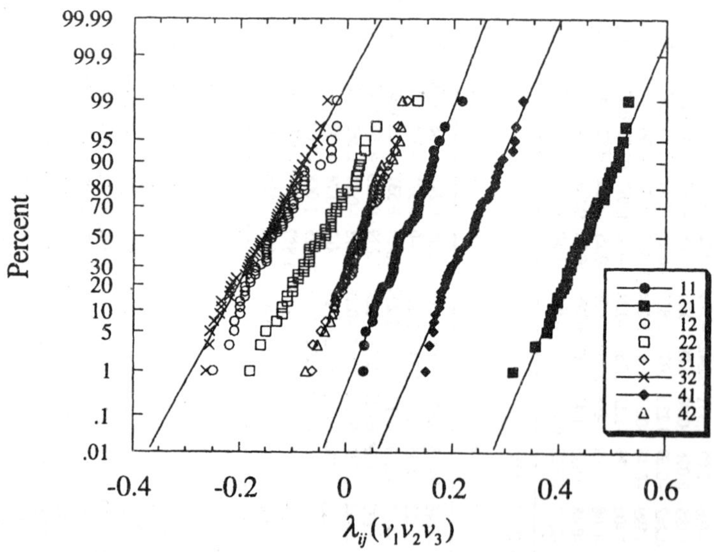

Figure CS4.4 The $\lambda_{ij}(\nu_1\nu_2\nu_3)$ coefficients which measure the contribution of the set points to the total value of the film as determined from their combined influence on the attributes of resistivity, surface defects, and adhesion.

What does this tell us about significance testing? Clearly it emphasizes the fact that to avoid Type I error we must only accept findings that generate very high levels of confidence and accept the fact that this will result in having Type II errors. The use of a 99% experiment-wise confidence level for this problem would have resulted in no Type I errors but one Type II error, set point 11 not being accepted.

However, because we have also made the test for statistical significance, we have added insight into the issue of strategic significance. On examining Table CS4.6, we see that set point 32 was found to increase the SIB Ln surface defects by an amount of 0.14 which led to a level of confidence of 94% for an unfavorable change in value for this attribute if this set point were turned on. For the other two attributes, set point 32 is seen to have little significance. Thus our conclusion for set point 32 should be that it should not be turned on because it is unfavorable but the confidence level is not high that it is unfavorable. We see, on the other hand, that set point 11 exhibits a high level of statistical significance for reducing resistivity variance and increasing its average both of which are favorable for a NIB1 attribute. This set point is also seen to have little significant impact on the other attributes. Thus, we can take the result for set point 11 in Figure CS4.4 to be sufficient to turn this set point on with high confidence that it will increase value.

Reference

1. Phadke, M. (1989) *Quality Engineering and Robust Design,* Prentice Hall, Englewood Cliffs, NJ, 1989, pp. 41–96.

Case study 5 Error analysis for direct value (DV) method

In developing a survey to determine the value of an alternative versus a baseline product, there will be uncertainty in the value determined due to errors in selecting the respondents from the correct customer segment of interest, in the model used to calculate value, and in the statistical error arising from the use of a sample size that is a small fraction of the population of interest. The objective of this case study is to evaluate the error arising from sample size and its consequences.

Figure CS5.1 shows a plot typical of the kind used to determine the neutral price P_N (Chapter 5) from a DV survey. Assume the survey, which is hypothetical (the results being determined using computer simulation), asked respondents to choose between a baseline vehicle made by manufacturer A at a price of \$16 000 and an alternative vehicle made by manufacturer B which was identical in terms of it specifications and style to A. Four paired choices were made by each respondent based upon vehicle B being offered at \$15 000, \$17 000, \$19 000, and \$21 000.

The results shown in Figure CS5.1 include all n persons in the simulated survey. A least squares line has been drawn through the four points, f being the

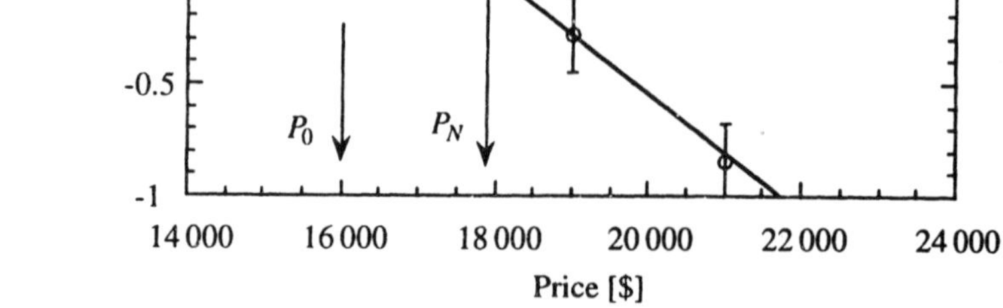

Figure CS5.1 A plot of the simulated results of the DV survey having 600 respondents for the name value of the vehicle manufacturer B versus A. A least squares line has been drawn through the four points, f being the fraction of respondents that chose the alternative relative to the baseline at each of four prices shown.

fraction of respondents that chose the alternative relative to the baseline at each of four prices shown. The price at the intercept of the least squares line with the horizontal line given by $4f-2=0$ is the neutral price. The difference between the neutral price and the baseline price equals the value of vehicle B less the value of vehicle A and represents the name value of B relative to A:

$$\Delta V = P_N - P_0 \tag{CS5.1}$$

The uncertainty in the intercept and the relative name value arises from the uncertainty in $4f-2$ as illustrated by the error bars. The relationship between $4f-2$ and price for the linear model is given by:

$$4f-2=\frac{3E_1}{2P_0}\{V-V_0-[P-P_0]\} \tag{CS5.2}$$

If the fraction of persons f selecting the alternative is greater than 0.75 or less than 0.25, then the function $4f-2$ on the left-hand side of Equation CS5.2 should be replaced by its logit form $\text{Ln}(f/[1-f])$.

Because a respondent selects either vehicle A or B for each price point, the frequency function for f should follow the binomial distribution, the standard deviation for f being given by:

$$s_f=\sqrt{\frac{f[1-f]}{n}} \tag{CS5.3}$$

where n is the number of responses to a given price and is assumed constant for each of the four prices. The standard deviation for $4f-2$ is four times s_f. The error bars shown in Figure CS5.1 are for two standard deviations of $4f-2$ or $8s_f$ equal to ± 0.1632 (at $f=1/2$) for the simulated sample size of $n=600$. The four points in the plot represent $k=4$ price subgroups.

The computations from the spreadsheet used to compute the uncertainty in the value difference are shown in Figure CS5.2. The first column lists the four price points which are followed by columns for f and $4f-2$. Each cell in the column labeled SSX contains the square of the difference between the price for its row and the average price which is computed in cell D54. Each cell in the column labeled SSY contains the square of the difference between the price for its row and the average price which is computed in cell F54. The sum of the squares in columns SSX and SSY are shown in cells G55 and H55, respectively.

The statistical properties of the plot of $4f-2$ versus price are shown in the block labeled Y versus X PLOT. These were computed using the function LINEST(D50:D53,F50:F53,1,1) which was entered into the three row by two column region G58:H60. The terms in the region E58:F60 identify the numerical results in region G58:H60.

The statistical properties of the plot of price versus $4f-2$ (the plot with the variables swapped in Figure CS5.1) is shown in the block labeled Y versus X PLOT. As above, these were computed using the function LINEST(D50:D53,F50:F53,1,1) which was entered into the three row by two

	C	D	E	F	G	H	I
46	n = 600			$4f-2$			
47	k = 4			Std Dev =	0.0816		
48							
49		Price	f	$4f$ -2	SSX	SSY	
50		15000	0.672	0.6870	9.00E+06	0.5008	
51		17000	0.588	0.3525	1.00E+06	0.1393	
52		19000	0.430	-0.2802	1.00E+06	0.0673	
53		21000	0.289	-0.8420	9.00E+06	0.6746	
54	Ave=	18000	0.495	-0.0207			
55				Sum=	2.00E+07	1.3820	
56							
57							
58		Y versus X	Slope	Const.	-0.000261	4.677	
59		PLOT	SE Slope	SE Const.	2.22E-05	0.403	
60			R^2	Y Std Dev	0.986	0.099	
61							
62		X versus Y	Slope	Const.	-3777	17922	
63		PLOT	SE Slope	SE Const.	321	189	
64			R^2	X Std Dev	0.986	378	313
65							
66				t (95%)=	1.96		
67				Value range @ intercept =	307		
68				Value difference=	1921		
69				Elasticity (E1) =	2.8		

Figure CS5.2 Worksheet used to analyze the error in value computed from the neutral price intercept.

column region G62:H64. The terms in the region E62:F64 to the left identify the numerical results two cells to the right in region G62:H64.

The quantity in cell I64 is the binomial theorem standard deviation of the price computed by dividing the standard deviation for $4f-2$ equal to $4s_f$ by the slope of Y versus X in cell G58. This theoretical result should be compared to the standard deviation for price (X Std Dev) in cell H64 using the F-ratio test for variances. For this problem, the Type I error for assuming that the variances had a different common population variance is large, 23% computed from the function FDIST(1.46,2,600). Statistical agreement between the two determinations of the standard deviation is support that the scatter about the line in Figure CS5.1 is primarily due to the sample size. (This was made rigorously so using computer simulation.)

The uncertainty in the neutral price listed in cell G67 was computed from the standard expression for the uncertainty in the linear regression average [1]:

$$\delta P_N = \left[\frac{\pm t_{\alpha/2} 4s_f}{SLOPE\ (\ Y \text{ vs } X\ )}\right]\sqrt{\frac{1}{k}+\frac{[Y-\bar{Y}]^2}{SSY}} \qquad \text{(CS5.4)}$$

for a Type I error of 5% ($\alpha = 0.05$), $k = 4$, $Y = 0$, and $\overline{Y} = \overline{4f-2}$ listed in cell F54. The spreadsheet expressions for this range are given in Table CS5.1 (cell G67) along with the expressions in cells for other quantities of interest.

The value difference is computed in cell G68. The price elasticity given by:

$$E_1 = \frac{\partial(4f-2)}{\partial P}\left[\frac{2P_0}{3}\right] \quad \text{(CS5.5)}$$

is listed in cell G69 and was computed by taking the derivative of $4f-2$ with respect to price in the paired choice expression given by Equation CS5.1.

When options are considered, the development of the uncertainty in the value of the options from a DV survey are similar to that given above except that the expression given by Equation CS5.1 is replaced by:

$$4f-2 = \frac{3E_{Opt}}{2V_{Opt}}\left[V_{Opt} - P_{Opt}\right] \quad \text{(CS5.6)}$$

The option form of the expression for $4f-2$ evaluates the option elasticity, E_{Opt}, at a price for the option equal to its value.

It is important to note that the uncertainty in the neutral price is influenced by both the sample size and the price elasticity (reflected in the slope of $4f-2$ versus price). In deciding how many respondents need to be surveyed, the term *SLOPE*(*Y* vs *X*) in Equation CS5.4 needs to be estimated using the expression:

$$SLOPE(Y \text{ vs } X) = \frac{3E_1}{2P_0} \quad \text{(CS5.7)}$$

In a paired choice survey, the elasticity E_1 will not be equal to the elasticity determined in a market where there are $N > 2$ competitors. If an estimate is not available for a prior survey, then a default value of 2 can be taken for E_1.

Reference

1. Walpole, R.E. and Myers, R.H. (1993) *Probability and Statistics for Engineers and Scientists*, 5th edn, Macmillan, New York, pp. 379–80.

Table CS5.1 Expressions in spreadsheet cells

Cell	*EXCEL formula*
I64 (Standard deviation for price)	=G47/G58
G66 (Student's *t* for 5% Type I error)	=TINV(1–0.95,600–2)
G67 (Value range for 5% Type I error)	=$G$66*(–G47/G58)*SQRT(1/4+(0–F54)^2/H55)
G68 (Value difference)	=–H58/G58–16000
G69 (Elasticity E_1)	=–G58*(16000)*2/3

Case study 6 Use of the direct value (DV) method when considering multiple attributes jointly

When multiple attributes are being considered in a consumer survey using the DV method described in Chapter 5, respondents, as a rule of thumb, will generally be able to handle the added cognitive stress of evaluating them jointly instead of one at a time when the number of attributes are seven or less. This means, for example, that an $L_8(2^7)$ orthogonal array (Appendix B and Chapter 8) could be a practical design if interactions were assumed negligible. The purpose of the case study is to demonstrate the use of the DV method using an $L_4(2^3)$ experimental design, for simplicity of illustration, to evaluate the value of binoculars as a function of two levels of style, 10× versus 7× magnification, and a provision for eyeglass wearers not to have to remove their glasses for viewing.

The $L_4(2^3)$ experimental design is shown in Table CS6.1 where the set point notations 11, 21, and 31 for the columns refer to the alternative style, 10× magnification, and the feature for eyeglass wearers, the baseline binoculars having the baseline style, 7× magnification, and no feature for eyeglass wearers. The baseline price is $110.

Trials 2, 3, and 4 generate three separate paired comparisons for the survey shown as Figures CS6.1, 6.2, and 6.3 respectively. The responses to this survey were simulated for a sample size of 600 and for style 2, 10× magnification, and the feature for eyeglass wearers adding $10, $30 and $20, respectively, in value

Table CS6.1 $L_4(2^3)$ **experimental design**

Trial	*0*	*11*	*21*	*31*
1	1	0	0	0
2	1	0	1	1
3	1	1	0	1
4	1	1	1	0

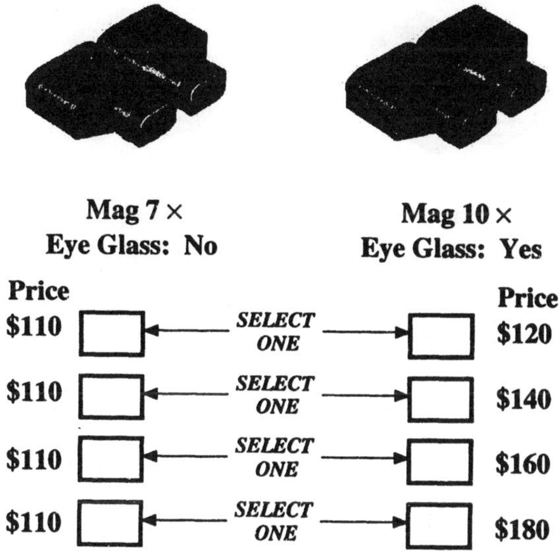

Figure CS6.1 Survey presentation for determining the neutral price of the binoculars in trial 2.

relative to the baseline binoculars. The simulated results for f are shown in Table CS6.2 and the logit transformation of f to $\text{Ln}(f/[1-f])$ is shown in Table CS6.3. The logit transformation is needed because the range of f is outside of the 0.25 to 0.75 limits for the linear model.

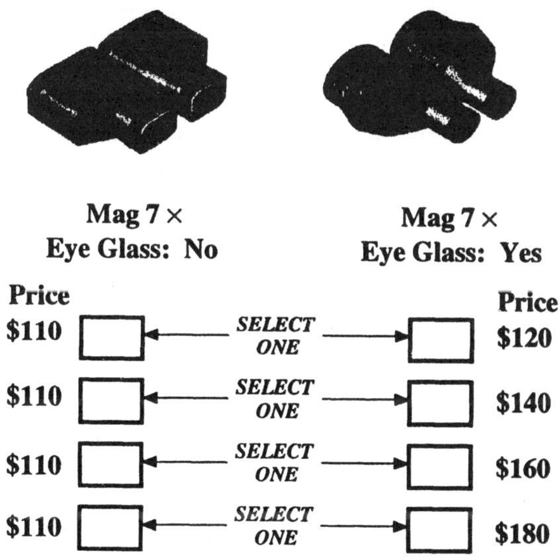

Figure CS6.2 Survey presentation for determining the neutral price of the binoculars in trial 3.

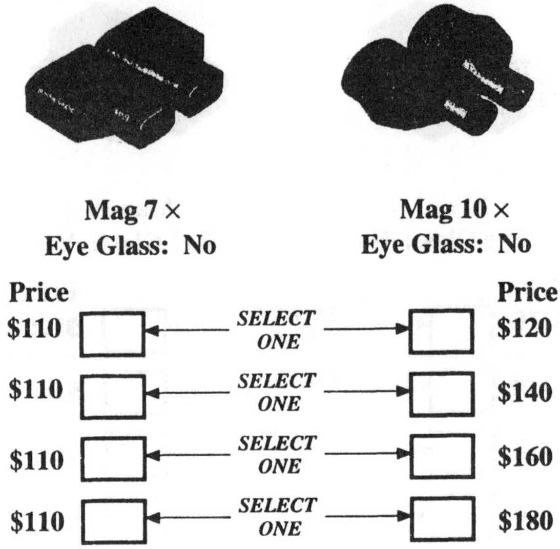

Figure CS6.3 Survey presentation for determining the neutral price of the binoculars in trial 4.

The Ln(f/ [1 − f]) results are plotted in Figure CS6.4 against price, the intercept at the zero horizontal line being the neutral price for each trial. The error analysis for the intercept is given in Table CS6.4. For a given trial in Table CS6.3, the difference between Ln(f/ [1 − f]) for each price and its average over the four prices is computed and then summed to compute SSY (sum of the squares Y). Shown below SSY in each column is the slope computed from the best straight line for the trial shown in Figure CS6.4. The error bars shown in Figure CS6.4 are for two standard deviations of Ln(f/ [1 − f]) The standard deviation computed using the linear model result as an approximation (Case Study 5) is $4s_f$ where s_f is the standard deviation given by binomial distribution taken as $\sqrt{[1/4]/600} = 0.02$ for this problem. The row shown as XSD in Table

Table CS6.2 Fraction f of respondents choosing the alternative versus the baseline binoculars for each trial

	Trial			
Price	*1*	*2*	*3*	*4*
120	*	0.935	0.773	0.889
140	*	0.789	0.462	0.617
160	*	0.534	0.207	0.351
180	*	0.208	0.061	0.122

Table CS6.3 Conversion of the results for f in Table CS6.2 to the $\mathrm{Ln}(f/[1-f])$ logit form and the neutral prices determined from the intercepts in Figure CS6.4

	Trial			
Price	*1*	*2*	*3*	*4*
120	*	2.67	1.23	2.08
140	*	1.32	−0.15	0.48
160	*	0.14	−1.35	−0.61
180	*	−1.34	−2.73	−1.98
Average =		0.70	−0.75	−0.01
$P_N =$		161	139	150
$P_N - P_0 =$		51	29	40

CS6.4 is the price standard deviation of the points about the best lines which was computed as 0.08/slope. The uncertainty in the neutral price was computed using Equation CS5.4 with t = 1.96 at 599 df for a 95% confidence level.

The results for the neutral prices are shown in column 3 of Table CS6.5. The baseline value of the binoculars of \$190 is shown in column 2 as determined from a (hypothetical) analysis of demand and price using Equation 3.11. Column 5 shows the total value for each trial. The last column shows the value that each attribute added to the baseline value of the binoculars computed using the relationship $\lambda = [X]^{-1}[V]$ (Appendix B). The style of alternative 2 (set point 11) added \$8.90 in value, the 10× magnification (21) added \$31, and the feature for eyeglass wearers added \$19.60. The 95% confidence range for the

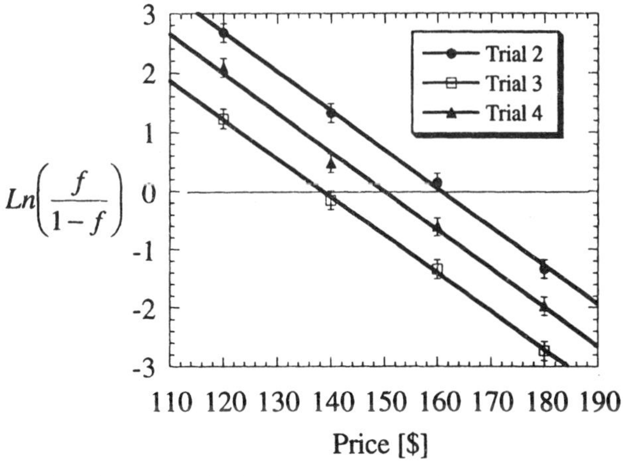

Figure CS6.4 Plot of $\mathrm{Ln}(f/[1-f])$ for determining the neutral price for trials 2, 3, and 4.

uncertainty in the added values can be taken (conservatively) as the largest range for the neutral price of ±$1.40.

Table CS6.4 Analysis of error in the neutral price for a 95% confidence range

	Trial			
Price	*1*	*2*	*3*	*4*
120	*	3.88	3.91	4.37
140	*	0.39	0.36	0.23
160	*	0.31	0.35	0.37
180	*	4.14	3.92	3.87
	SSY =	8.73	8.55	8.84
	Slope =	−0.066	−0.065	−0.066
	X SD =	1.2	1.3	1.2
	t =	1.96	1.96	1.96
P_N range ± 95%=		1.3	1.4	1.2

Table CS6.5 Computation of the total value for each trial. Total value for the baseline trial 1 was determined from (hypothetical) demand/price analysis. The value added to the baseline for each trial was determined from the difference between the neutral price for the trial and the baseline price. The added value for each of the three attributes is shown in the last column

Trial	*Base value, V_0*	*Neutral price, P_N*	*Added value $P_N - P_0$*	*Total value*	*ij*	λ_{ij}
1	$190	$110	$—	$190	0	190
2	$190	$161	$51	$241	11	8.9
3	$190	$139	$29	$219	21	31.0
4	$190	$150	$40	$230	31	19.6

Appendix A Probability and statistics

A.1 Concept of probability

When performing an experiment, several trials are made with the factors under evaluation being changed from trial to trial in a planned manner and one or more attributes of interest being measured at each trial. Each measurement of an attribute represents a single event, e, or outcome and the set of all possible outcomes defines the attribute's sample space [1]. Each possible outcome of a trial can have a probability assigned to it such that $n\wp(e)$ is the number of times that the outcome is expected to occur in n trials. An impossible outcome has a probability equal to zero and a definite outcome has a probability equal to one. No outcome has a negative probability and no outcome has a probability greater than one. Moreover, the summation of the probabilities for an attribute over its sample space must always be equal to unity.

Experiments based upon the toss of a fair coin, for example, have two single outcomes, heads or tails, with each having an equal probability of 1/2. The toss of a die with its sides numbered in sequence has six possible outcomes given by $i=1, 2, \ldots, 6$, and, if it is weighted uniformly, then $\wp(i)=1/6$ for $i=1, 2, \ldots, 6$. Two single outcomes from an experiment are mutually exclusive by definition; they cannot occur simultaneously. The coin toss is either heads or tails; the die lands and rests with just one side up; the distance for a single flight of the arrow can not be both 400 and 600 yards; and the bullseye cannot be both hit and missed on the same shot.

Single outcomes can, however, be linked to generate a combined outcome. Such combinations also have probabilities which again range only between zero and one. For example, one combined outcome is that either single outcome e_1 or single outcome e_2 will occur. The combining of two single outcomes, e_1 and e_5, out of nine possible single outcomes is shown schematically in Figure A.1 to form the combined probability for outcome C which is equal to the sum of the two single outcome probabilities for 1 and 5:

$$\wp(C) = \wp(e_1) + \wp(e_5)$$

For simplicity, we assume that each outcome has the same probability equal to 1/9. The interpretation of the combined outcome having the probability

$$\wp(C) = (1/9) + (1/9) = 2/9$$

is that C has occurred if either outcome 1 or outcome 5 has occurred.

When combined outcomes are themselves combined (Figure A.2) this probability is defined as their union, $\wp(A \cup B)$. There exists the possibility that certain single outcomes are common to both, such as outcomes 3 and 9 in Figure A.2 and they must not be counted twice in adding up the combined probabilities. The probability $\wp(A' \cup B')$ is equal to the sum of $\wp(A')$ and $\wp(B')$ less the joint probability that both can occur simultaneously, $\wp(A' \cap B')$, known as the intersection of A′ and B′:

$$\begin{aligned}\wp(C') &= \wp(A' \cup B') \\ &= \wp(A') + \wp(B') - \wp(A' \cap B')\end{aligned} \tag{A.1}$$

which for this example is given by:

$$\begin{aligned}\wp(C') &= \frac{1}{3} + \frac{1}{3} - \frac{2}{9} \\ &= \frac{4}{9}\end{aligned} \tag{A.2}$$

where $\wp(A' \cap B') = 2/9$ is equal to the double counting that would occur if $\wp(C')$ were taken simply as $\wp(A') + \wp(B')$.

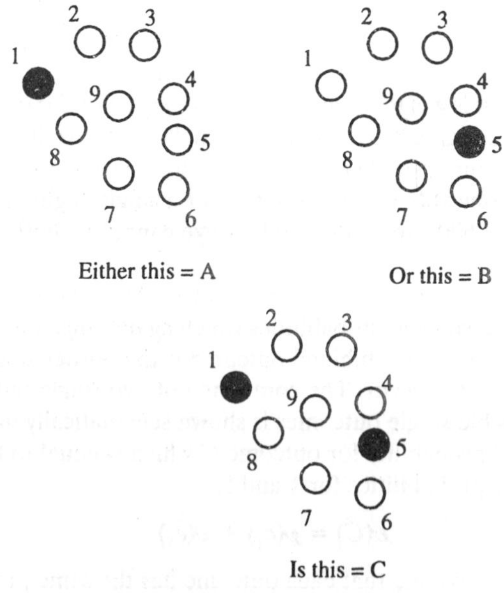

Figure A.1 Combining single outcome 1, representing A, and single outcome 5, representing B, to form the combined outcome C represented by the occurence of either 1 or 5.

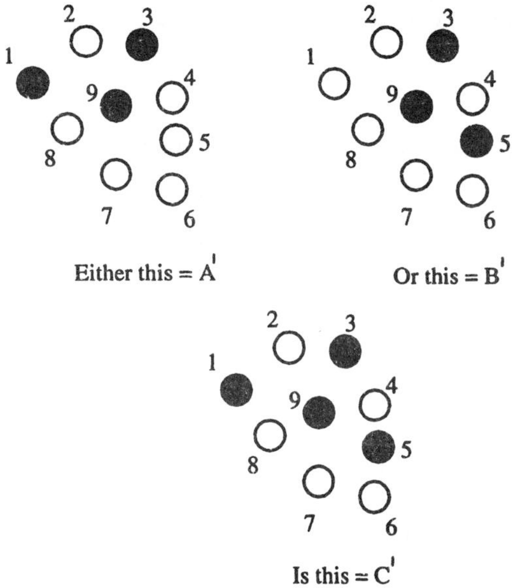

Figure A.2 Combining two already combined outcomes A' and B' to form C' in which two single outcomes are common between A' and B'.

If two outcomes, A and B, can occur, the general expression for the joint probability $\wp(A \cap B)$ can be written in terms of the product of the probability of the first outcome $\wp(A)$ times the conditional probability, $\wp(B \mid A)$, of the second outcome occurring given that the first outcome has occurred:

$$\begin{aligned}\wp(A \cap B) &= \wp(A)\,\wp(B \mid A) \\ &= \wp(B)\,\wp(A \mid B)\end{aligned} \quad \text{(general relation)} \qquad \text{(A.3)}$$

If the two outcomes are independent, then

$$\wp(A \cap B) = \wp(A)\,\wp(B) \quad \text{(if outcomes independent)} \qquad \text{(A.4)}$$

A.2 Frequency functions, expectations, and the normal distribution

The above example was for an experiment having discrete outcomes. In many situations, the outcomes can be represented by a continuous variable, x. The probability that the outcome lies between x and $x+\delta x$ is given by:

$$\wp(x) = f(x)\delta x, \qquad \text{(A.5)}$$

where $f(x) \geq 0$ is defined as the frequency function for x. The integral of $f(x')$ from $-\infty$ to x is known as the cumulative distribution of x:

(A.6)

and the integral of $f(x')$ over all x' is equal to one:

$$\int_{-\infty}^{+\infty} f(x')\,dx' = 1 \tag{A.7}$$

The expectation $E(x)$ for an arbitrary frequency function is the population mean given by:

$$\mu \equiv E(x) = \int_{-\infty}^{+\infty} x f(x)\,dx \tag{A.8}$$

The expectation $E([x-\mu]^2)$ is the population variance of a distribution defined by:

$$\sigma^2 \equiv E([x-\mu]^2) = \int_{-\infty}^{+\infty} (x-\mu)^2 f(x)\,dx. \tag{A.9}$$

The term, σ, equal to the square root of the variance, is known as the standard deviation of the distribution. For a finite sample of size having $y = 1,\ldots, n$ replications, the quantity given by:

$$s^2 = \frac{1}{n-1}\sum_{y=1}^{n}[x(y)-\bar{x}]^2 \tag{A.10}$$

is known as the unbiased sample variance whose expectation is equal to the population variance in that the average of a large number of unbiased sample variances taken from the population and having n samples each should equal the population variance. The expectation of the sample mean $\bar{x}$ determined by the expression:

$$\bar{x} = \frac{1}{n}\sum_{y=1}^{n} x(y) \tag{A.11}$$

is the population mean, μ.

The normal distribution whose frequency function is given by:

$$f_{norm}(x) = \left[\frac{1}{\sigma\sqrt{2\pi}}\right]\exp\left(-\left[\frac{x-\mu}{\sqrt{2}\sigma}\right]^2\right) \tag{A.12}$$

is one of the most important functions in statistics and is shown in Figure A.3 as a function of the standard normal variable, $Z \equiv [x-\mu]/\sigma$.

Three separate ranges under the distribution are shown in Table A.1 along with the area for each as determined from Equation (A.7) on transforming the differential dx to dx/σ and changing the limits of integration to the ranges noted.

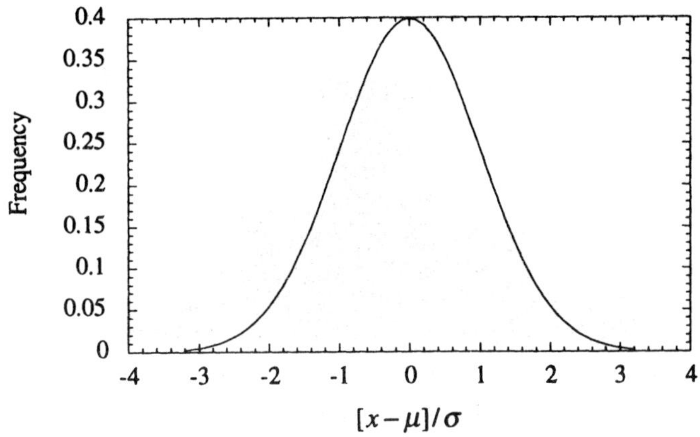

Figure A.3 Frequency function for standard nominal distribution.

A.3 Central limit theorem

The importance of the normal distribution is a result of the central limit theorem which states that if the variable x has a distribution with mean μ and standard deviation σ, then the variable $\{[\bar{x} - \mu] / [\sigma/\sqrt{n}]\}$ computed from the sample average $\bar{x}$ has a distribution that approaches the standard normal distribution as n becomes infinite. In other words, the distribution for the sample averages $\bar{x}$ computed from a series of samples of size n taken at random from a much larger distribution having a mean μ and a standard deviation σ will become normal as the sample size becomes large even though the distribution for x itself with standard deviation σ may not be normal. The standard deviation for the distribution of $\bar{x}$ will be $\sigma/\sqrt{n}$. A demonstration of the central limit theorem is given in Chapter 10 in the discussion of statistical tolerancing. The experimental frequency function as revealed by the histogram of the statistical tolerancing data is shown in Figure A.4 as a function of its standard normal variable. The histogram is seen to closely resemble the standard normal distribution shown above. The area enclosed by the histogram is also seen to be equal to unity.

Table A.1 Area enclosed under the frequency distribution over ±1 to ±3 standard normal ranges

Z range	*Area percentage*
±1	68.26
±2	95.44
±3	99.73

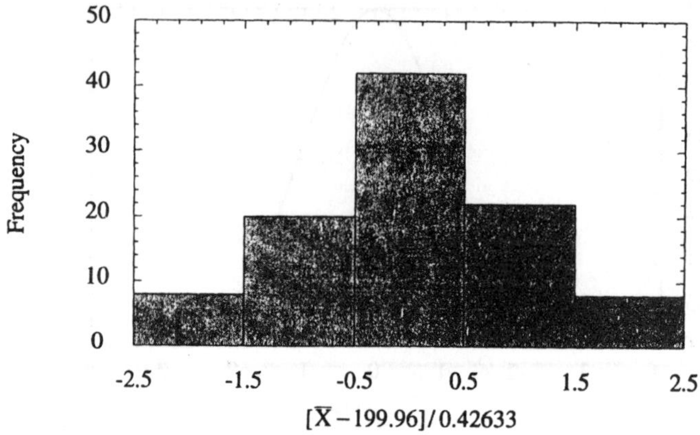

Figure A.4 Histogram determined from statistical tolerance data in Table 11.5.

A.4 Normal probability plots

It is often useful to plot the sample data from an experiment on normal probability paper shown as Figure A.5a. (Note that the percent scale is not logarithmic.) This is done in a series of steps which are illustrated for a sample calculation in Table A.2 with the resulting distribution plotted in Figure A.5b. First the twenty data in column two of Table A.2 are arranged in ascending order, shown in column three. The cumulative probabilities for the points are then estimated for each outcome x_i (i = 1, 2, 3, ... n) using the relationship [2]:

$$\wp_C(x_i) = 100[i - 0.5]/n. \tag{A.13}$$

The resulting points for the sample calculation are shown in column four of Table A.2. The probability paper is designed so that a straight line will be the best fit to normally distributed data. Deviation from normality is generally indicated by curvature at the ends of the distribution. The median of the distribution is the point on the outcome axis at which the line through the points intersects the 50% cumulative probability line. The standard deviation can be estimated by subtracting the mean (equal to the median for a normal distribution) from the point on the outcomes axis where the line intersects the cumulative probability of 84.13%.

A.5 Hypothesis testing

Developing new products and making improvements to existing products often requires experimentation to validate that a proposed new design or process innovation actually improves the performance or reduces the cost in the manner intended. Typically it comes down to comparing the performance of the

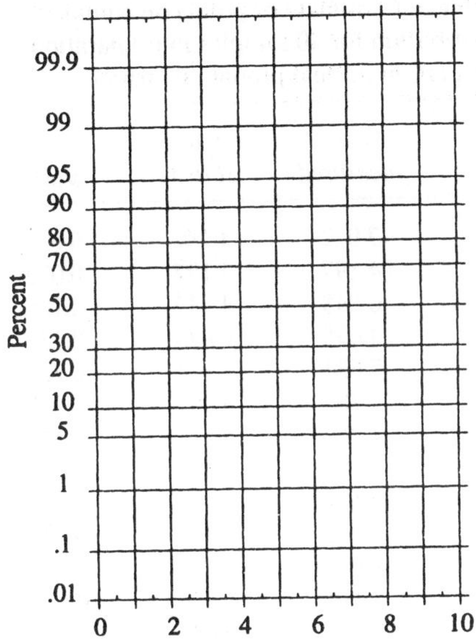

Figure A.5a Normal probability paper.

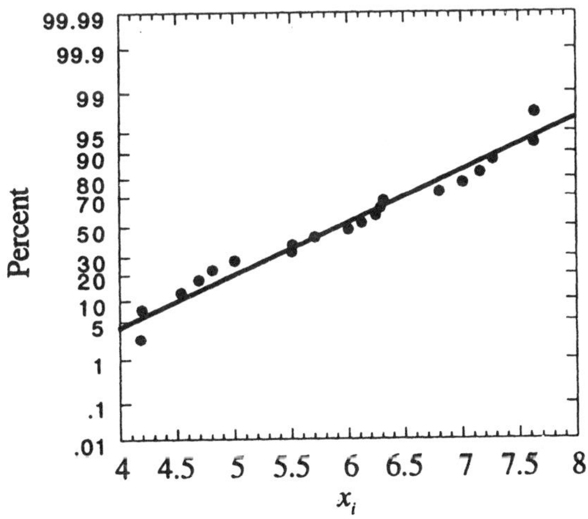

Figure A.5b Normal probability plot of data in Table A.2.

current product with the performance of a prototype of the proposed product. Shown in Figure A.6 are two hypothetical performance curves. Assume that the measurement is for a larger-is-better attribute, that the function x_0 peaking at 9 is for the current production, and that the function x_1 peaking at 10.5 is for the prototype.

Table A.2 Estimate of points on cumulative distribution for 20 samples in preparation for plot on normal probability paper

i	x_i *as measured*	x_i *as sorted*	$\wp_c(x_i)$
1	7.012	4.182	0.025
2	4.197	4.197	0.075
3	6.248	4.539	0.125
4	4.695	4.695	0.175
5	7.634	4.814	0.225
6	5.708	5.007	0.275
7	7.280	5.511	0.325
8	7.164	5.515	0.375
9	6.125	5.708	0.425
10	5.007	6.006	0.475
11	4.182	6.125	0.525
12	4.814	6.248	0.575
13	6.291	6.291	0.625
14	6.321	6.321	0.675
15	6.805	6.805	0.725
16	5.515	7.012	0.775
17	5.511	7.164	0.825
18	6.006	7.280	0.875
19	7.639	7.634	0.925
20	4.539	7.639	0.975

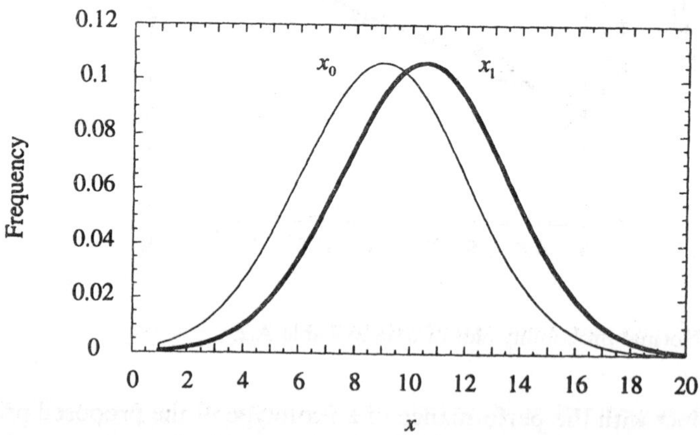

Figure A.6 Hypothetical performance curves for current production and for a prototype planned for future production.

The prototype peaks at a slightly higher point and, because larger is better, the decision may seem to be obvious, put the prototype into production. The stumbling block is that this may require a very large investment. How much confidence can you place in the prototype being better than the current product based upon the above results? Your baseline product may have been in production for some time with satisfactory performance. Millions may have been produced and sold. Are you going to bet the company on the results from a limited number of prototypes? This, of course, is done every day in the marketplace which requires that companies interpret their limited test results using good statistical techniques.

In a more general fashion, we define the hypothesis $H2_0$ that the population mean of the distribution of x_1 is equal to the population mean of the baseline distribution, x_0, the observed shift in the averages from the sample distributions being no more than what might be expected from random fluctuations. The hypothesis defined as $H2_1$ is that the population mean for the x_1 distribution is different from that for the baseline. Hypothesis testing is made by first defining a critical region on either side of the sample mean of the baseline distribution x_0 such that a decision will be made that $H2_1$ is true if the average for the new sample distribution x_1 falls in this region. A point representing a sample distribution falling in the critical region is shown in Figure A.7. The possible outcomes in the decision matrix are shown in Table A.3.

The critical region is generally chosen such that the so-called Type I error is minimized because there is reluctance to depart from a known baseline if there is not high confidence that the proposed alternative is better. The area under the frequency distribution curve that lies in the critical region defines the size of Type I error expected. In most of our considerations, we will be interested in whether or not an alternative under consideration is better than baseline which focuses our attention on a single tail of the distribution, the side of the distribution indicating a product improvement.

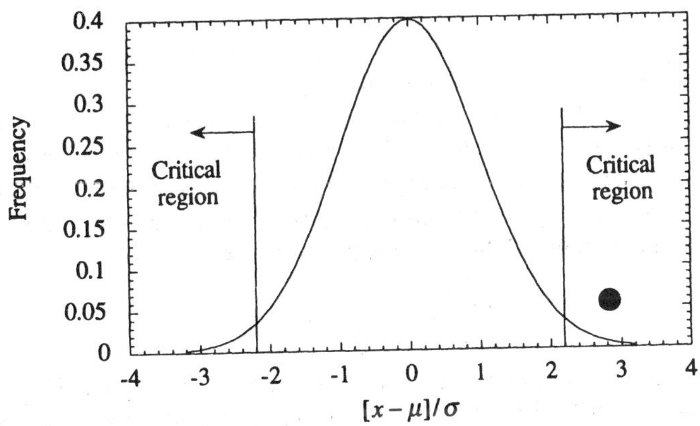

Figure A.7 Critical region for hypothesis testing.

Table A.3 Outcomes for hypotheses in regard to Type I and Type II errors

Condition	*Action*	*If $H2_0$ true*	*If $H2_1$ true*
Point falls in critical region	Accept $H2_1$	Type I error	Correct decision
Point falls outside critical region	Reject $H2_1$	Correct decision	Type II error

The error expected in rejecting $H2_0$ for a given critical region can be determined using Student's t test when the populations are normally distributed. This test of significance recognizes that small sample sizes are generally used to arrive at the experimentally measured frequency distributions with the unknown population variance being taken as the unbiased sample variance. Student's t distribution is a function of the degrees of freedom (df) equal to ν and is given by:

$$f_s(t) = \left\{ \frac{\Gamma\left(\frac{\nu+1}{2}\right)}{\sqrt{\nu\pi}\,\Gamma\left(\frac{\nu}{2}\right)} \right\} \left\{ 1 + \left[\frac{t^2}{\nu}\right]^{-\left[\frac{\nu+1}{2}\right]} \right\}$$

where Γ is the gamma function. The mean of the t distribution is zero and its variance is $\nu/[\nu - 2]$ for $\nu > 2$. The df is the number of sample measurements less one, $n - 1$, used to compute the sample variances which appear in the definition of the t statistic given by [3]:

$$t = \frac{[\bar{x}_2 - \bar{x}_1] - [\mu_2 - \mu_1]}{\{[s_2^2 + s_1^2]/n\}^{1/2}} \tag{A.14}$$

where $\bar{x}_1$ and $\bar{x}_2$ are the averages of the sample distributions x_1 and x_2, respectively, μ_1 and μ_2 are the two population means (assumed to be equal for hypothesis $H2_0$), and s_1^2 and s_2^2 are the unbiased sample variances of the two distributions being compared. For simplicity, the two sample sizes are assumed to be the same here (the same number of replications, n). The sample variances must be from populations having a common variance σ^2 which can be taken as the expectation of the average of the two sample variances:

$$E\left(\frac{s_1^2 + s_2^2}{2}\right) \equiv \sigma^2. \tag{A.15}$$

The hypothesis of a common population variance, noted here as $H1_0$, can be tested using the F distribution, described shortly, the statistic given by the ratio of the two variances. The df for the pooled population variance given by Equation A.15 is equal to $2(n - 1)$.

A comparison of Student's distribution for ν = 1, 3, and 5 with the normal distribution is shown in Figure A.8. When the df are small, it is seen that the distribution lies well above the normal distribution in the regions far from the mean. The basic assumption for using Student's test is that the sample is taken from a large population whose distribution is normal. The integral for the area under Student's distribution from t to infinity (the single-tailed distribution) is the error expected for assuming that the mean of a distribution is greater than baseline. The value of this integral is shown for several values of t and ν in Table A.4.

Table A.4 is interpreted as follows: If the outcome of the experiment for $\nu = 6$ were $t = 2$, for example, then you should be wrong 4.62% of the time if you accept that the mean of the distribution x_1 will be greater than the mean for the baseline, i.e. you choose hypothesis $H2_1$ over $H2_0$. If t were equal to 0.75 for $\nu = 4$, then, by interpolation, the choice of the hypothesis $H2_1$ should be incorrect 25% of the time. Table A.4 was computed using the function TDIST(t,v,1) in the spreadsheet. (The availability of this and other statistical functions in the spreadsheet makes it unnecessary to use look-up tables for most statistical quantities.)

A.6 The chi-squared distribution

The frequency distribution for the variance of a normally distributed variable follows a chi-squared distribution (Figure A.9) which is defined by the relation:

$$f(x,\nu) = \begin{cases} \dfrac{1}{2^{\nu/2}\Gamma(\nu/2)} x^{(\nu/2)-1} e^{-x/2}, & x > 0 \\ 0, & x \le 0 \end{cases} \tag{A.16}$$

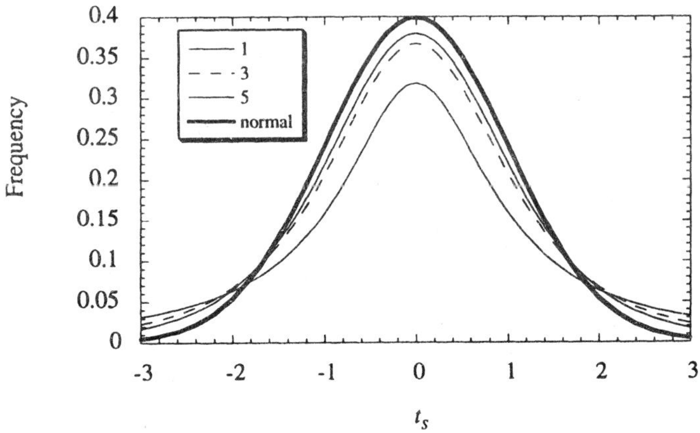

Figure A.8 Student's t distribution for three different degrees of freedom compared to the normal distribution.

Table A.4 Integral from t to infinity of single-tailed t distribution for several degrees of freedom

ν	t = 0.5	1	1.5	2	2.5	3	3.5	4
2	0.3333	0.2113	0.1362	0.0918	0.0648	0.0477	0.0364	0.0286
4	0.3217	0.1870	0.1040	0.0581	0.0334	0.0200	0.0124	0.0081
6	0.3174	0.1780	0.0921	0.0462	0.0233	0.0120	0.0064	0.0036
8	0.3153	0.1733	0.0860	0.0403	0.0185	0.0085	0.0040	0.0020
10	0.3139	0.1704	0.0823	0.0367	0.0157	0.0067	0.0029	0.0013
12	0.3131	0.1685	0.0797	0.0343	0.0140	0.0055	0.0022	0.0009
14	0.3124	0.1671	0.0779	0.0326	0.0127	0.0048	0.0018	0.0007
16	0.3119	0.1661	0.0765	0.0314	0.0118	0.0042	0.0015	0.0005
20	0.3113	0.1646	0.0746	0.0296	0.0106	0.0035	0.0011	0.0004
25	0.3107	0.1634	0.0731	0.0282	0.0097	0.0030	0.0009	0.0002
30	0.3104	0.1627	0.0720	0.0273	0.0091	0.0027	0.0007	0.0002

where

$$x \equiv \chi^2 = \frac{\nu s^2}{\sigma^2} \tag{A.17}$$

in which ν is equal to the number of samples or replications less one, s^2 is the measured unbiased sample variance, and σ^2 is the true population variance.

The integral of the frequency distribution from $\chi^2_{2.5}$ to $\chi^2_{97.5}$ includes 95% of the observations and the integral from $\chi^2_{0.135}$ to $\chi^2_{99.865}$ includes 99.7% (same as the amount included in $\pm 3\sigma$ for a normal distribution). The 99.7% range for the sample variance s^2 from a normal population with a known variance of σ^2 is given by:

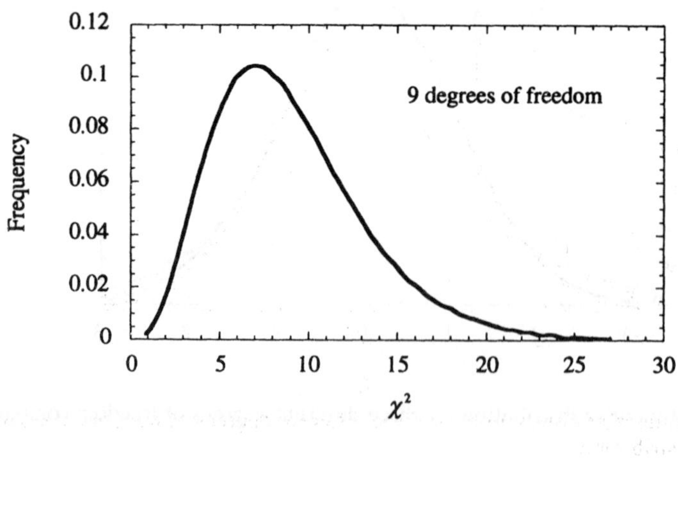

$$\frac{\chi^2_{0.135}\sigma^2}{\nu} \le s^2 \le \frac{\chi^2_{99.865}\sigma^2}{\nu}. \tag{A.18}$$

The limits on a variance chart for statistical process control are constructed from the two limits given by Equation A.18. With a measured sample variance of $s^2(q)$ for an experimental trial (q), the unknown population variance lies within the range given by:

$$\frac{\nu s^2(q)}{\chi^2_{99.865}} \le \sigma^2(q) \le \frac{\nu s^2(q)}{\chi^2_{0.135}}$$

at 99.7% confidence. The mean of the χ^2 distribution is ν and its variance is 2ν. Because the expected value of $1/\chi^2$ is $1/[\nu-2]$, it follows that the variance, $s^2(q)$, of a sample taken from a normal population having a variance, $\sigma^2(q)$, is less than $\sigma^2(q)$ more often than not. The variance of $1/\chi^2$ is $[\nu-2]^{-1}\{[\nu-4]^{-1}-[\nu-2]^{-1}\}$.

Although the chi-squared distribution approaches a normal distribution as the sample size becomes large, it has a significant tail on the upside of the distribution. Moreover, the low side of the distribution never falls below zero because the variance can never be negative. These properties are seen in Figure A.9. When a process is placed under statistical process control, the chi-squared distribution can be used to set the upper and lower control limits on the measured sample variances. For simplicity in practice, control limits are often placed on the recorded sample ranges whose expected value is linearly related to the population standard deviation (Equation 11.4).

Table A.5 shows the ranges and sample standard deviations for 25 subgroups each having ten samples taken from a normal population having a mean of 10 and a variance of 1. Charts displaying the two results are shown in Figure A.10. The limits shown are for the standard deviation and were computed from the square roots of Equation A.18 for $\chi^2_{0.135} = 0.371$ and $\chi^2_{99.865} = 1.73$. The average (approximately equal to the median) of the 25 ranges in Table A.5 is near the theoretical value of 3.078 from Table 11.2 shown by the vertical line. The distribution of the ranges is shown as a normal probability plot in Figure A.11.

A.7 The *F* distribution (significance between two measured sample variances)

Although the *t* statistic given by Equation A.14 can be used to test the significance of the difference between two averages using the *t* distribution, there are many occasions when it is also important to test the significance between two experimentally measured variances. In fact, the conditions for homogeneity of variance should be verified before pooling the sample variances and computing the *t* statistic in Equation A.15. Instead of actually taking the difference between the two variances under consideration, the ratio defined as *F* is determined:

Table A.5 Range, variance, and standard deviation of a normal variable

Sample	*Range*	*Variance*	*Std dev*
1	2.862	1.186	1.089
2	4.169	2.029	1.425
3	3.737	1.178	1.085
4	2.368	0.668	0.817
5	3.459	1.102	1.050
6	2.516	0.882	0.939
7	2.794	0.647	0.804
8	3.067	0.781	0.884
9	3.256	1.152	1.073
10	3.025	0.957	0.978
11	2.932	0.788	0.888
12	3.341	1.338	1.157
13	3.918	1.334	1.155
14	2.829	0.930	0.964
15	2.445	0.579	0.761
16	3.633	1.082	1.040
17	2.330	0.519	0.721
18	3.528	1.012	1.006
19	3.120	0.752	0.867
20	2.333	0.715	0.845
21	2.767	0.605	0.778
22	1.490	0.321	0.567
23	3.493	1.234	1.111
24	2.476	0.827	0.909
25	2.795	0.556	0.746
Ave.=	2.987	0.927	

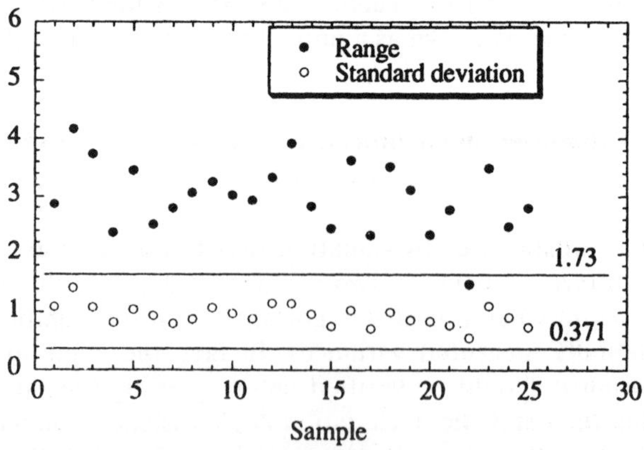

Figure A.10 Display of range and sample standard deviation for 25 simulated subgroups taken from a normal population having a standard deviation of 1 for a sample size of $n=10$ for each. The control limits on the standard deviation were determined from the chi-squared distribution at 99.7% confidence.

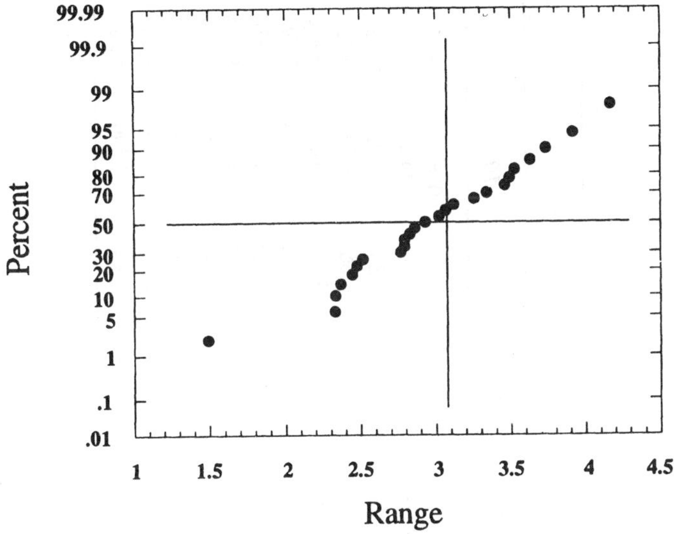

Figure A.11 The scatter of the range (25 points) about its theoretical average of 3.078 for a population standard deviation of 1 with $n=10$.

$$F \equiv \frac{s_1^2 / \sigma_1^2}{s_2^2 / \sigma_2^2} \tag{A.19}$$

When the F statistic is used to test the confidence level for the null hypothesis $H1_0$ that the two sample variances are from the same population, $\sigma_1^2 = \sigma_2^2$, the larger sample variance, shown above as s_1^2, is placed in the numerator by convention. If the variances under evaluation are independent and are from normally distributed populations, the F statistic will follow the F distribution given by:

$$f(F) = cF^{\frac{1}{2}[\nu_1 - 2]}\left[\nu_2 + \nu_1 F\right]^{\frac{1}{2}[\nu_1 + \nu_2]}$$

The coefficient c is given by:

$$c = \frac{\nu_1^{\frac{\nu_1}{2}} \nu_1^{\frac{\nu_2}{2}} \Gamma\left(\frac{\nu_1 + \nu_2}{2}\right)}{\Gamma\left(\frac{\nu_1}{2}\right)\Gamma\left(\frac{\nu_2}{2}\right)}$$

where ν_1 and ν_2 are the df associated with the measurements of s_1^2 and s_2^2, respectively. Generally, significance testing for variance using the F distribution employs a critical region of 1–5% that ensures a small Type I error. The resulting ratios for 0.01 and 0.05 error levels are shown in Tables A.6a and A.6b, respectively, as a function of ν_1 and ν_2.

Table A.6a Critical values of F ratio for an error level of 0.01

ν_2	ν_1: 1	2	3	4	5	6	7	8	9	10	11	12
1	4052	4999	5404	5624	5764	5859	5928	5981	6022	6056	6083	6107
2	98.5	99	99.16	99.25	99.3	99.33	99.36	99.38	99.39	99.4	99.41	99.42
3	34.12	30.82	29.46	28.71	28.24	27.91	27.67	27.49	27.34	27.23	27.13	27.05
4	21.2	18	16.69	15.98	15.52	15.21	14.98	14.8	14.66	14.55	14.45	14.37
5	16.26	13.27	12.06	11.39	10.97	10.67	10.46	10.29	10.16	10.05	9.963	9.888
6	13.75	10.92	9.78	9.148	8.746	8.466	8.26	8.102	7.976	7.874	7.79	7.718
7	12.25	9.547	8.451	7.847	7.46	7.191	6.993	6.84	6.719	6.62	6.538	6.469
8	11.26	8.649	7.591	7.006	6.632	6.371	6.178	6.029	5.911	5.814	5.734	5.667
9	10.56	8.022	6.992	6.422	6.057	5.802	5.613	5.467	5.351	5.257	5.178	5.111
10	10.04	7.559	6.552	5.994	5.636	5.386	5.2	5.057	4.942	4.849	4.772	4.706
11	9.646	7.206	6.217	5.668	5.316	5.069	4.886	4.744	4.632	4.539	4.462	4.397
12	9.33	6.927	5.953	5.412	5.064	4.821	4.64	4.499	4.388	4.296	4.22	4.155

Table A.6b Critical values of F ratio for an error level of 0.05

ν_2	ν_1 = *1*	*2*	*3*	*4*	*5*	*6*	*7*	*8*	*9*	*10*	*11*	*12*
1	161.4	199.5	215.7	224.6	230.2	234	236.8	238.9	240.5	241.9	243	243.9
2	18.51	19	19.16	19.25	19.3	19.33	19.35	19.37	19.38	19.4	19.4	19.41
3	10.13	9.552	9.277	9.117	9.013	8.941	8.887	8.845	8.812	8.785	8.763	8.745
4	7.709	6.944	6.591	6.388	6.256	6.163	6.094	6.041	5.999	5.964	5.936	5.912
5	6.608	5.786	5.409	5.192	5.05	4.95	4.876	4.818	4.772	4.735	4.704	4.678
6	5.987	5.143	4.757	4.534	4.387	4.284	4.207	4.147	4.099	4.06	4.027	4
7	5.591	4.737	4.347	4.12	3.972	3.866	3.787	3.726	3.677	3.637	3.603	3.575
8	5.318	4.459	4.066	3.838	3.688	3.581	3.5	3.438	3.388	3.347	3.313	3.284
9	5.117	4.256	3.863	3.633	3.482	3.374	3.293	3.23	3.179	3.137	3.102	3.073
10	4.965	4.103	3.708	3.478	3.326	3.217	3.135	3.072	3.02	2.978	2.943	2.913
11	4.844	3.982	3.587	3.357	3.204	3.095	3.012	2.948	2.896	2.854	2.818	2.788
12	4.747	3.885	3.49	3.259	3.106	2.996	2.913	2.849	2.796	2.753	2.717	2.687

A.8 Selecting points at random from distributions

Often we will want to determine the significance of an experimentally measured sample average of a complicated function which does not have a well-characterized statistical distribution. For example, the expressions for the CIQ for a proposed new product in Table 8.2 are given in terms of the population average and variance. For a normal population both the average and variance are well-characterized functions. However, the CIQ formed by their combinations are not well characterized statistically. Moreover, the resulting expressions for profit and market share are even further removed from being well characterized statistically. However, in almost all instances, we can build up these more complex distributions of interest through combinations of standard distributions and study their statistics using Monte Carlo computer simulation methods which can be carried out in a straightforward manner using spreadsheet programs.

A.8.1 SIMULATING t AND CHI-SQUARED DISTRIBUTIONS

Because small sample sizes are used in many experiments, the frequency distribution for Student's t statistic given by:

$$t = \frac{\bar{x}(q) - \mu(q)}{s(q)/\sqrt{n}}$$

for experimental trial (q) is of great interest in which $\bar{x}(q)$ is the sample average of the measured basic variables $x(q,y)$ for $y = 1, 2, \ldots, n$ replications. The term $\mu(q)$ is the population mean and $s(q)$ is the sample standard deviation of the measured basic variables. If the sample measurements for an attribute of interest are $g(q,y)$, then the basic variables for smaller is better (SIB), nominal is best (NIB), and larger is better (LIB) are defined in Table 8.1. If variances from the experimental trials have not been pooled, ν is equal to $n-1$.

When the population mean $\mu(q)$ is known for the conditions represented by trial (q), the range expected for the sample averages is determined from the t statistic using the relation:

$$\mu(q) + \frac{t_{99.865}s(q)}{\sqrt{n}} \le \bar{x}(q) \le \mu(q) + \frac{t_{0.135}s(q)}{\sqrt{n}}.$$

The more important expression for our purposes is for the range of uncertainty in the population mean:

$$\bar{x}(q) + \frac{t_{1-\alpha}s(q)}{\sqrt{n}} \le \mu(q) \le \bar{x}(q) + \frac{t_{\alpha}s(q)}{\sqrt{n}}$$

based upon a measured sample average and sample standard deviation. The above expression is interpreted to mean that there is $1-2\alpha$ probability that $\mu(q)$

lies between the two limits for the t statistic. The range can also be interpreted in the manner of a cumulative probability $\wp_\mu(\alpha)=1-\alpha$ that the true population mean lies between $-\infty$ and $\bar{x}(q) + t_\alpha s(q)/\sqrt{n}$. The uncertainty in the population mean can be simulated by randomly selecting values, t_S, from Student's distribution and inserting them into the expression:

$$\mu_S(q) = \bar{x}(q) + \frac{t_S s(q)}{\sqrt{n}} \qquad \text{(random variable)} \quad \text{(A.20)}$$

The cumulative distribution formed by many random selections of $\mu_S(q)$ using Equation A.20 will be defined as $\wp_\mu(\alpha)$. Based upon sample measurements of $\bar{x}(q)$ and $s(q)$ for a normally distributed variable, the uncertainty in the position of the population mean, $\mu(q)$, is given by this distribution. Likewise, the uncertainty in the population variance for a measured sample variance having ν df can be simulated by randomly selecting χ_S^2 values from the chi-squared distribution and inserting them into the expression:

$$\sigma_S^2(q) = \frac{\nu_S^2(q)}{\chi_S^2} \qquad \text{(random variable)} \quad \text{(A.21)}$$

Based upon a measurement of s^2 with ν degrees of freedom for a normally distributed variable, the possibilities for the population variance, $\sigma^2(q)$, are given by Equation A.21. In other words, if many points are selected at random from the chi-squared distribution and substituted into Equation A.21 for χ^2, the resulting cumulative distribution expresses the uncertainty in our knowledge of the exact value for $\sigma^2(q)$.

The random selection of points t_S from Student's t distribution can be made by using the function RAND() to select random numbers uniformly distributed between zero and one, setting each equal to a value of $\wp_C$ for the cumulative probability (equal to the integral of the frequency distribution from minus infinity to t, and using the t inverse function, which is given by TINV(1–$\wp_C$,ν), to obtain a random selection of a point from the t distribution for ν degrees of freedom. Because the TINV() function yields a value for t for a given value for one minus the cumulative distribution for t positive only, it follows that the equation for the t statistic for a randomly selected point from the cumulative distribution $\wp_C$ = RAND() is given by:

$$t_S = \text{IF}(\wp_C > 0.5,1,-1) \times \text{TINV}(\text{IF}(\wp_C > 0.5,\ 2 \times (1-\wp_C),\ 2 \times \wp_C),\ \nu) \quad \text{(A.22)}$$

Simulation of the range in population variance uses the inverse chi-squared function, CHIINV and a second randomly selected point on the cumulative distribution $\wp_C'$ = RAND():

$$\chi_S^2 = \text{CHIINV}(1-\wp_C',\ \nu)$$

with the result inserted into the denominator of Equation A.21. Note: the df are often pooled, $\nu = \nu_p$, in using the above expressions.

On some computers, however, the computations of the TINV and CHIINV functions are relatively slow or the functions may not be available with certain spreadsheet programs. Fortunately, the RAND() function (a random number generator between 0 and 1) is generally available and computes relatively quickly. The RAND() may be used to generate a random selection of points that obey the t or chi-squared distributions as follows (this process is used in Case Study 4).

From Equation A.17 we see that:

$$\chi^2(\nu) \approx \nu s_\zeta^2(\nu)$$

where:

$$s_\zeta^2(\nu) \equiv \left[\frac{1}{\nu}\right]\sum_1^{\nu+1}\zeta^2 \tag{A.24}$$

is the sum, divided by ν, of $\nu + 1$ standard normal random variables (SNRV) noted as ζ having zero mean and unit variance. The ζ are obtained using computer simulation to evaluate the SNRV approximation given by:

$$\zeta \cong \frac{1}{\sqrt{2}}\sum_{j=1}^{24}[\text{RAND}() - 0.5]_j \tag{A.25}$$

which follows from the central limit theorem and the fact that the variance of a uniform distribution from –0.5 to +0.5 is equal to 1/12. Likewise it follows from the t distribution that

$$t_s(\nu) \approx \frac{\bar{\zeta}(\nu+1)}{\sqrt{s_\zeta^2(\nu+1)/[\nu+1]}}$$

where

$$\bar{\zeta}(\nu+1) \equiv \left[\frac{1}{\nu+1}\right]\sum_1^{\nu+1}\zeta. \tag{A.26}$$

When results are not pooled in estimating variance, $\nu = n - 1$. It follows from Equation A.21 and the above that the simulated uncertainty in the population variance which obeys the chi-squared distribution is given by the ratio of the measured sample variance to the simulated variance of the SNRV when each have the same number of degrees of freedom and replications:

$$\sigma_S^2(q) = \frac{s^2(q,\nu)}{s_\zeta^2(\nu)} \quad \text{(a random variable)} \tag{A.27}$$

It follows from Equation A.20 and the above that the simulated t distribution for the uncertainty in the population mean is given by:

$$\mu_s(q) = \bar{x}(q) + \bar{\zeta}(\nu+1)\sqrt{\frac{s^2(q,\nu)/n}{s_\zeta^2(\nu)/[\nu+1]}} \qquad \text{(a random variable)} \quad \text{(A.28)}$$

Note that in using Equations A.27 and A.28, the df may have been pooled ($\nu = \nu_P$) and thus not equal to $n-1$. When ν is large (ν Æ 30), the t distribution can be replaced by the standard normal distribution and the χ^2 distribution can be replaced by a normal distribution of mean ν and standard deviation $\sqrt{2\nu}$. Thus for large ν, Equation A.27 becomes:

$$\sigma_s^2(q) = \frac{\nu s^2(q)}{\nu + \sqrt{2\nu\zeta}} \qquad \text{(a random variable, } \nu \text{ Æ 30)} \quad \text{(A.29)}$$

and Equation A.28 becomes:

$$\mu_s(q) = \bar{x}(q) + \frac{\zeta s(q)}{\sqrt{n}} \qquad \text{(a random variable, } \nu \text{ Æ 30)} \quad \text{(A.30)}$$

A.8.2 SIMULATING THE LOGNORMAL DISTRIBUTION

Situations arise when the basic variable is not normally distributed and a transformation is required to make it so. For example, the lognormal transformation should be made for the SIB and LIB basic variables unless the distribution is known to be something else (Table 8.1):

$$\ell(q,y) = \begin{cases} \text{SIB:} & \text{Ln}(g(q,y)) \\ \text{LIB:} & \text{Ln}(1/g(q,y)) \end{cases} \qquad \text{(A.31)}$$

The average $\bar{\ell}(q)$ and variance $s_\ell^2(q)$ of the lognormal distribution can be used in Equations A.20 and A.21 (or Equations A.28 and A.27) to simulate the uncertainty in the population mean $\mu_\ell(q)$ and population variance $\sigma_\ell^2(q)$, respectively, of the lognormal variable for SIB and LIB. Each of these simulated variables can then be transformed back to physical space from lognormal space using the expressions [4]:

$$\mu_s(q) = \exp\left(\mu_\ell(q) + \frac{\sigma_\ell^2(q)}{2}\right) \qquad \text{(A.32)}$$

$$\sigma_S^2(q) = \exp(2\mu_\ell(q))\{\exp(2\sigma_\ell^2(q)) - \exp(\sigma_\ell^2(q))\} \qquad \text{(A.33)}$$

The cost of inferior quality for SIB and LIB basic variables that display log normal statistics is given in terms of the natural log transformed population statistics by:

$$\Omega(q) / k_T = \exp(2\mu_\ell(q) + 2\sigma_\ell^2(q)) \qquad \text{(A.34)}$$

A.9 Significance of ξ'_{ij} coefficients

Although the F statistic can be used to test if a pair of sample variances are unlikely to have had a common population variance, the problem posed by an experiment involves several to many variances (one for each trial). Bartlett's test addresses the issue of assessing whether a distribution of variances (beyond a pair) can be considered as having a common population variance [5]. However, it does not treat directly the quantities of interest here related to variance, the ξ'_{ij} coefficients defined by Equation 9.7. These measure how much a given off-baseline set point changes −10 times the logarithm of the variance. Moreover, with Taguchi's robust design concepts, we are not so much interested in the homogeneity of variance, which Bartlett's test was designed to address, but in the inhomogeneity of variance. Specifically, we want to know which set points (factors) are significant in reducing the variance and which likely are not.

We test the null hypothesis, noted as $H1_0$, that the sample measurements for each trial had a common population variance by computing the scatter distribution of the variance coefficients ξ'_{ij} in accord with this hypothesis and comparing it against the experimentally measured ξ'_{ij}. A condition for using the ξ'_{ij} scatter distribution is that the variances must be from a normally distributed population (or approximately so).

Fortunately, it does not matter what we take for the common population variance in computing the scatter distribution for the null hypothesis, only that it be constant over all of the trials. This follows because the logarithm of the ratio of the sample variance to the population variance as determined from Equation A.17 depends only upon the chi-squared distribution divided by the degrees of freedom:

$$-10 \text{ Log}\left(\frac{s^2(q)}{\sigma^2}\right) = -10 \text{ Log}\left(\frac{\chi^2}{\nu}\right) \tag{A.35}$$

On unfolding the left-hand side of the above expression into two separate logarithms, we have

$$-10 \text{ Log}(s^2(q)) = -10 \text{ Log}\left(\frac{\chi^2}{\nu}\right) - 10 \text{ Log}(\sigma^2) \tag{A.36}$$

Now a property of the model for the coefficients, Equation B.4, is that off-baseline coefficients are not influenced by the addition of a constant to each element of the outcome column vector **[Y]**. As a result, the off-baseline ξ'_{ij} scatter distribution for the null hypothesis depends upon the df for the trial as well as the **[X]** array defining the experimental design but it is completely independent of the numerical value of the assumed common population variance.

The distribution of the ξ'_{ij} coefficients will be symmetric about zero for the null hypothesis if for each positive element, $\omega_{ij}(q) > 0$, in each row ij of the solution matrix given by $[\mathbf{X}^T\mathbf{X}]^{-1}[\mathbf{X}^T]$ (Appendix B) there is a negative element,

$\omega_{ij}(q') < 0$, equal to it in absolute value, $\omega_{ij}(q') = -\omega_{ij}(q)$. This condition is satisfied for all orthogonal arrays when expressed in $[\mathbf{X}]_\lambda$ form and for two-level orthogonal arrays expressed in the $[\mathbf{X}]_\varphi$ form (see Appendix B for definitions). In what follows, we restrict our considerations of the test for the significance of the ξ'_{ij} coefficients to [X] arrays which yield ξ'_{ij} coefficients symmetric about zero when the null hypothesis of a common population variance is satisfied.

The scatter distribution of the measured ξ'_{ij} coefficients for the null hypothesis is expected to follow a new statistic given by:

$$Z^*_{ij} \equiv \frac{\xi'_{ij}}{\mathrm{E}([\xi'_{ij} - \mu_\xi]^2)}.$$

For the null hypothesis, the symmetrically distributed ξ'_{ij} coefficient, are a random variable given by:

$$\begin{aligned}\xi'_{ij} &= -10 \sum_{q=1,k} \omega_{ij}(q)\log\left(\frac{\chi^2(q)}{\nu}\right) \\ &= -10 \sum_{\omega_{ij}(q)>0} \omega_{ij}(q)\left\{\log\left(\frac{\chi^2(q)}{\nu}\right) - \log\left(\frac{\chi^2(q')}{\nu}\right)\right\}\end{aligned} \tag{A.37}$$

It follows from the above expression that the mean of the ξ'_{ij} distribution is zero:

$$\mu_\xi = 0$$

and that the variance of ξ'_{ij} is given by:

$$\begin{aligned}E([\xi'_{ij} - \mu_\xi]^2) &= E([\xi'_{ij}]^2) \\ &= [100/2]\sigma_\xi^2 \sum_{q=1,k} \omega_{ij}^2(q) \\ &= 50\sigma_\xi^2 \operatorname{cov}_{ijij}\end{aligned} \tag{A.38}$$

A factor of 1/2 was introduced in Equation A.38 to compensate for the summation of $\omega_{ij}^2(q)$ being written over all q instead of just the q for which $\omega_{ij}(q) > 0$ as in Equation A.37. The summation over the $q = 1, ..., k$ trials in Equation A.38 was then replaced in the last line by $\operatorname{cov}_{ijij}$, the diagonal element for set point ij in the covariance matrix $[\mathbf{X}^T\mathbf{X}]^{-1}$ which is defined in Appendix B. The variance σ_ξ^2 is given by:

$$\sigma_\xi^2 \equiv E\left(\left[\log\left(\frac{\chi^2(q)}{\nu}\right) - \log\left(\frac{\chi^2(q')}{\nu}\right)\right]^2\right).$$

Thus the new statistic can be written as :

$$Z^*_{ij} = \frac{\xi'_{ij}}{\sqrt{50\sigma_\xi^2 \operatorname{cov}_{ijij}}} \tag{A.39}$$

Computation of the new statistic as well as the variance σ^2_ξ can be avoided by first considering the pairwise problem which has two trials and two unknowns, the baseline coefficient and the single off-baseline coefficient $\xi'(=\xi'_{11})$. The off-baseline coefficient is given by:

$$\xi' = 10 \text{ Log}\left(\frac{s^2(\nu)_1}{s^2(\nu)_2}\right)$$

where the assumption has been made that the sample variance of trial 1 is larger than the variance for trial 2. The F ratio is given by:

$$F \equiv \frac{s_1^2}{s_2^2} = 10^{[|\xi'|/10]}$$

For the pairwise problem $\text{cov}_{ijij} = 2$ and it follows that the F ratio in the above expression can be written in terms of the new statistic for the pairwise problem, Z^*, as:

$$F = 10^{\left[|Z^*|\sqrt{\sigma^2_\xi}\right]} \tag{A.40}$$

At this point we can generalize Equation A.40 by substituting the general expression given by Equation A.39 for Z^* after placing subscripts ij on the F ratio yielding:

$$F_{ij} \equiv 10^{|G_{ij}|} \tag{A.41}$$

where

$$G_{ij} = \frac{\xi'_{ij}}{\sqrt{50\,\text{cov}_{ijij}}} \tag{A.42}$$

The absolute value is used for the exponent in Equations A.40 and A.41 to make the F ratio for coefficient ij always one or greater. In evaluating the significance of a measured ξ'_{ij}, we are simply transforming the measurement into the F_{ij} statistic and seeing how likely that is under the hypothesis $H1_0$. The result for F_{ij} can be substituted into the expression FDIST(F_{ij}, ν, ν) to arrive at an estimate of the pairwise error. This error needs to be multiplied by the number of off-baseline set point coefficients to assess the overall experiment-wise error (Chapter 9).

References

1. Walpole, R.E. and Myers, R.H. (1993) *Probability and Statistics for Engineers and Scientists*, 5th edn, Macmillan, New York, p. 9.
2. DeVor, R.E., Chang, T., and Sutherland, J. (1992) *Statistical Quality Design and Control*, Macmillan, New York, pp. 90–8.

3. Walpole, R.E. and Myers, R.H. (1993) *Probability and Statistics for Engineers and Scientists*, 5th edn, Macmillan, New York, p. 259.
4. Logothetis, N. and Wynn, H.P. (1989) *Quality Through Design*, Clarendon Press, Oxford, U.K., p. 436.
5. Bartlett, M.S. and Kendall, D.G. (1949) The statistical analysis of variance – heterogeneity and the logarithmic transformation. *Journal of the Royal Statistical Society*, Suppl. 8, 128–38.

Appendix B Design of experiments

B.1 Introduction

A major tool for improving product quality is the planning and execution of well designed experiments. Mathematical models are used to reduce the number of experiments that need to be run and to aid in interpreting the results. The models yield phenomenological coefficients which couple the independent variables (control factors) to the dependent, outcome variables such as quality, cost, value, investment, profit, etc. measured for each of the experimental trials.

The independent variables can differ widely in their nature and some may be continuous and others may be discrete. For example, the speed of a sailboat will depend on the speed of the wind, a continuous variable, and whether or not it has its spinnaker in place, a yes or no discrete variable. The coefficient that couples (linear) wind speed to boat speed would be dimensionless, whereas the coefficient that couples yes or no for the spinnaker would have the units of speed per first spinnaker in place. Two spinnakers in place would introduce a new variable and a new coefficient. If the wind speed increased infinitesimally, we would expect a very small increment of speed. However, if we added or deleted the spinnaker we would expect a discontinuous change in speed. The effect of the spinnaker should also be strongly coupled to wind speed resulting in a large interaction between the two variables.

In all problems that we will study, the variables will be defined at discrete **set points** whether they are continuous or not. One reason for this is that most of the variables related to new product development will be discrete either rigorously or effectively because even potentially continuous variables often come in standard sizes. For the automotive engineer, it's either going to be the 3.3 or the 3.6 liter engine; your choice is between an aluminum or magnesium housing; brake lining XZ33 from supplier XZ or CB29 from supplier CB are the alternatives before you.

B.2 The lambda set point series

The lambda set point series for the general outcomes $Y(q)$ of trial q is given by:

$$\lambda_0 + \sum_{ij} \lambda_{ij} X_{ij}(q) + \frac{1}{2!} \sum_{\substack{ijkl \\ k \neq i}} \lambda_{ijkl} X_{ij}(q) X_{kl}(q) +$$

$$\frac{1}{3!} \sum_{\substack{ijklmn \\ k \neq i \\ m \neq k,i}} \lambda_{ijklmn} X_{ij}(q) X_{kl}(q) X_{mn}(q) + \ldots + \text{err}(q) = Y(q) \tag{B.1}$$

in which the generalized set point variables $X_{ij}(q)$ are binary (0 or 1) and dimensionless. The random error term, err(q) for trial q, although always present, will not be carried along explicitly in what follows or in the body of the text. However, in keeping with Taguchi's concept of robust design, the variance of this term will not necessarily be considered as common across the trials as is often done in the traditional approaches to experimental design.

The first summation is over the paired subscripts ij denoting the type i and level j the of set point. Summations beyond the first contain pair, triplet, etc. interactions but exclude higher-order self interactions, the summation over $ijkl$ being for $k \neq i$ and the summation over $ijklmn$ being for $k \neq i$ and $m \neq k, i$. With this form, all of the **coupling coefficients** – λ_{ij}, λ_{ijkl}, λ_{ijklmn}, etc. – have the same units as the outcome variables, $Y(q)$. This makes it easy to see immediately which set points are most important because the coupling coefficients, which are the unknowns, do not need to be multiplied by another factor once they have been determined to sense the size of their impact.

The 0 or 1 value of the binary variable $X_{ij}(q)$ is determined by the conditions:

$$X_{ij}(q) = \begin{cases} 1 \text{ if set point } ij \text{ is `on' for trial } q. \\ 0 \text{ if set point } ij \text{ is `off' for trial } q. \end{cases}$$

The baseline condition, noted as level j=0 for each variable i, will often have historical significance as the point of departure, the as-is condition from which improvements are sought. The notations for variable type i and level j are only convenient numerical representations for the names of the variable and level and do not necessarily represent addresses in an ordered set; level 1 could, for example, be higher on the scale for the variable j than level 3 and level 3 could be higher than level 6 or below level 1 if desired.

The set point series can be used to model the change of any process from its baseline condition to a new state (Figure B.1). Changes in the six set points represent changes in input, controls, and tools. For each set of changes, a new outcome is measured. Instead of modeling a process, a change in a product could be modeled as shown schematically for six set points in Figure B.2. Here the set point changes could refer to proposed new features for the new product versus the current production model. The outcome could be measured in terms of a customer ranking from a market survey that evaluated combinations of these set points. This type of example is considered in the binoculars conjoint analysis discussed in Chapter 5. The outcomes could also be measured in terms of the performance of a device, an example being the design changes of the automobile described for the SQD methodology example described in Chapter 7.

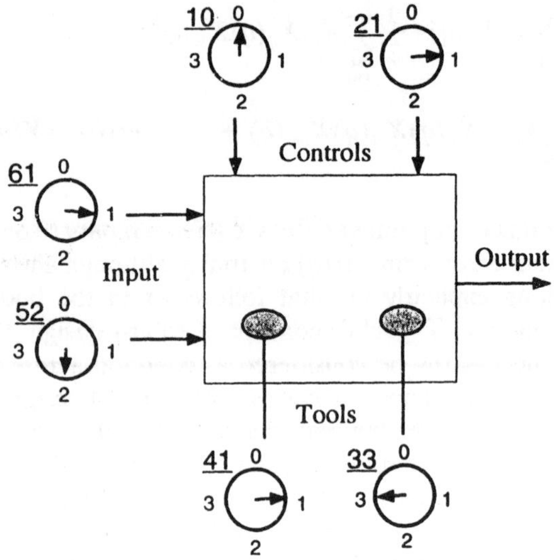

Figure B.1 Schematic of a set point model for a process showing input, output, controls, and tools.

B.2.1 NUMBER OF UNKNOWNS EQUALS THE NUMBER OF TRIALS

The general equations for the measured outcomes $Y(q)$ given by Equation B.1 are linear in terms of the unknown lambda coefficients λ_0, λ_{ij}, λ_{iklj}, λ_{ijklmn}, ... With the series on the left-hand side cut off at w unknowns, a set of $k = w$ linearly independent, simultaneous equations are needed which can be written in matrix form as:

$$[\mathbf{X}][\lambda]=[\mathbf{Y}] \tag{B.2}$$

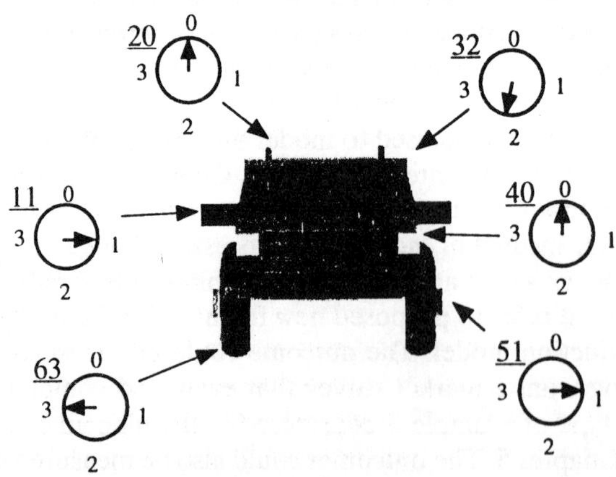

Figure B.2 Schematic of set point model for product design modifications.

where **[λ]** is the column matrix having coefficients $\lambda_0, \lambda_{10}, \lambda_{11}, \ldots \lambda_{1010}, \ldots \lambda_{1110}, \ldots$. The order of the coefficients (0, 10, 11, ... , 1010, 1110, ...) down the column is the same as the order of columns (0, 10, 11, ... , 1010, 1110, ...) across the matrix **[X]**. The elements of the design array **[X]** are equal to either 0 or 1 depending on whether or not a particular set point (0, *ij*, *ijkl*, *ijklmn*,...) is 'on' (= 1) or 'off' (= 0) for a specific trial (row). The strength of the 'main effects' is represented by coefficients λ_{ij}. The others having multiple pair subscripts represent different levels of interaction. The λ_{ijkl} represent pairwise interactions and the λ_{ijklmn} represent triplet interactions, respectively, between the set points. The variable $Y(q)$ in Equation B.1 represents a general outcome. In robust parameter design, the $Y(q)$ represent the signal-to-noise ratios for the trials [1].

The unknown coefficients can be solved using standard linear algebra methods provided that the series is cut off at a finite level. Before this is done, however, the relationships between the λ coefficients described in what follows are used to simplify the expression. When the number of linearly independent trials is equal to the number of unknowns, the coefficients in the **[λ]** matrix can be solved for using the relation:

$$[\boldsymbol{\lambda}] = [\mathbf{X}]^{-1}[\mathbf{Y}] \tag{B.3}$$

The matrix $[\mathbf{X}]^{-1}$ is the inverse of the design array, **[X]**.

B.2.2 NUMBER OF TRIALS IS GREATER THAN THE NUMBER OF UNKNOWNS

When the number of trials, *k*, exceed the number of unknowns, *w*, the transpose of **[X]**, written as $[\mathbf{X}^T]$, is multiplied against both sides of Equation B.2 to reduce the number of equations to *w* from $k > w$:

$$[\mathbf{X}^T][\mathbf{X}][\boldsymbol{\lambda}] = [\mathbf{X}^T][\mathbf{Y}] \ .$$

The solution for the λ coefficients given by:

$$[\boldsymbol{\lambda}] = [\mathbf{X}^T\mathbf{X}]^{-1}[\mathbf{X}^T][\mathbf{Y}] \tag{B.4}$$

provides a least squares fit to the outcomes in determining the λ coefficients. The quantity $[\mathbf{X}^T\mathbf{X}]^{-1}$ is the variance–covariance matrix for a variance of unity. It is always a $w \times w$ square matrix. This matrix is generally shown in textbooks as being multiplied by σ^2, the assumed common variance of the random error for each trial. As already stated, we will not make the automatic assumption that the random trial errors have a common variance. Thus, instead of $\sigma^2[\mathbf{X}^T\mathbf{X}]^{-1}$ we use the matrix $[\mathbf{X}^T\mathbf{X}]^{-1}$ which we will refer to simply as the **covariance matrix**. It is necessary that there be at least *w* linearly independent equations within the matrix **[X]** for the covariance matrix to exist and therefore for a solution to be found. The solution for a specific coefficient in Equation B.4 can also be written as:

$$\lambda_{ij} = \sum_{q=1,k} \omega_{ij}(q) Y(q) \tag{B.5}$$

where the term $\omega_{ij}(q)$ is the element in column q and row ij of the **solution** matrix defined by $[\mathbf{X}^T\mathbf{X}]^{-1}[\mathbf{X}^T]$. Each trial q will often be replicated n times (two replications means that two runs are made for the given trial) and $Y(q)$ represents the average outcome measured for trial q.

B.3 Comparison to Taylor's series

For a continuous variable, the set point notation can be compared to a Taylor's expansion about the baseline position given by:

$$Y(q) = Y_0 + \sum_{ij} Y_{,i}\Delta x_i(j)X_{ij}(q) + \frac{1}{2!}\sum_{ijkl} Y_{,ik}\Delta x_i(j)\Delta x_k(l)X_{ij}(q)X_{kl}(q) + \frac{1}{3!}\sum_{ijklmn} Y_{,ikm}\Delta x_i(j)\Delta x_k(l)\Delta x_m(n)X_{ij}(q)X_{kl}(q)X_{mn}(q) + \ldots \quad \text{(B.6)}$$

where

$$Y_{,i} = \partial Y/\partial x_i\,, \quad \text{(B.7)}$$

$$Y_{,ik} = \partial^2 Y/\partial x_i \partial x_k\,, \quad \text{(B.8)}$$

and

$$Y_{,ik} = \partial^3 Y/\partial x_i \partial x_k \partial x_l\,. \quad \text{(B.9)}$$

The set point notation $X_{ij}(q)$ is related to the continuous variable Δx_i measured from the baseline position. The relation $\Delta x_i = \Delta x_i(j)X_{ij}(q)$ defines the continuous variable Δx_i in terms of the discrete integer level j. In the summations over the level j shown in Equation B.6, the set point variable $X_{ij}(q)$ is zero except when level j is present for the trial q. On comparing Equation B.6 with Equation B.1, we see that

$$\lambda_0 = Y_0\,, \quad \text{(B.10)}$$

$$\lambda_{ij} = Y_{,i}\Delta x_i(j) + \frac{1}{2!}Y_{,ii}\Delta x_i^2(j) + \frac{1}{3!}Y_{,iii}\Delta x_i^3(j)\,, \quad \text{(B.11)}$$

and

$$\lambda_{ijkl} = Y_{,ik}\Delta x_i(j)\Delta x_k(l) + \frac{1}{2}\left\{Y_{,ikk}\Delta x_i(j)\Delta x_k^2(l) + Y_{,iik}\Delta x_i^2(j)\Delta x_k(l)\right\}. \quad \text{(B.12)}$$

The baseline condition is given by $\Delta x_i = \Delta x_i(0) = 0$ in the continuous variable notation and by $X_{ij} = X_{i0}$ in the set point notation. It follows therefore that:

$$\lambda_{i0} = 0 \quad \text{(B.13)}$$

and

$$\lambda_{i0kl} = \lambda_{ijk0} = \lambda_{i0k0} = 0\,. \tag{B.14}$$

The generalization of Equation B.14 to the triplet interactions yields:

$$\lambda_{i0klmn} = \lambda_{ijk0mn} = \lambda_{ijklm0} = \lambda_{i0k0m0} = 0\,. \tag{B.15}$$

Thus, any time a variable is at its baseline level with respect to itself alone or involving interactions, the corresponding coupling coefficient λ is zero.

The advantage in using the set point series as a model for the experiments over a Taylor's expansion expression is that (1) the set point variables are binary and dimensionless, (2) the complexity of powers are eliminated, (3) the λ coefficients have the same units as the outcomes Y, and (4) the possible combinations of pairwise, triplet, etc. interactions which could influence (confound) the outcome of a given trial can readily be identified and measured with additional experiments if analysis of preliminary results indicate that interactions may be present.

B.4 The phi set point series

When orthogonal experimental designs are used, it is common to use a model baseline different from a specific trial as described above for the λ coefficients. The phi set point model has the same form as the lambda series (terms beyond the triplet being ignored):

$$\begin{aligned}&\varphi_0 + \sum_{ij}\varphi_{ij}X_{ij}(q) + \frac{1}{2!}\sum_{\substack{ijkl\\k\neq i}}\varphi_{ijkl}X_{ij}(q)X_{kl}(q) + \\ &\frac{1}{3!}\sum_{\substack{ijklmn\\k\neq i\\m\neq k,i}}\varphi_{ijklmn}X_{ij}(q)X_{kl}(q)X_{mn}(q) = Y(q)\end{aligned} \tag{B.16}$$

The difference arises from the special summation relationships. Those for the main effect coefficients are given by:

$$\sum_j \varphi_{ij} = 0 \tag{B.17}$$

which for two levels j = 0, 1 for any variable (i) yields:

$$\varphi_{i0} = -\varphi_{i1} \tag{B.18}$$

The relations between the pairwise interaction coefficients for any i, k are given by:

$$\sum_j \varphi_{ijkl} = \sum_l \varphi_{ijkl} = 0, \tag{B.19}$$

which for two levels yields:

$$\varphi_{i1k1} = \varphi_{i0k0} = -\varphi_{i1k0} = -\varphi_{i0k1} \tag{B.20}$$

The least square solution for the column matrix of φ coefficients is given by the same form as Equation B.4.

B.5 Baseline considerations

When the φ coefficients are used, we will write the array of coefficients X_{ij} as the matrix $[\mathbf{X}]_\varphi$ and when the baseline is one of the trials, we will write the array as $[\mathbf{X}]_\lambda$. However, for simplicity in writing matrix operations, the arrays will be often written simply as $[\mathbf{X}]$, it being understood which form applies. For an orthogonal experimental design, the columns of the $[\mathbf{X}]_\varphi$ array are explicitly orthogonal to each other but the columns of the $[\mathbf{X}]_\lambda$ array for the same experimental design will not be. The two underlying experimental designs, however, are exactly the same, nevertheless. The conclusions regarding the significance of the coefficients will, in general, be different because the $[\mathbf{X}]_\lambda$ array measures the coefficients relative to a specific trial and the $[\mathbf{X}]_\varphi$ array measures the coefficients relative to the average of all the trials. When an orthogonal experimental design is used, we will refer to both the $[\mathbf{X}]_\varphi$ array and the $[\mathbf{X}]_\lambda$ array as orthogonal arrays meaning that the experimental design is orthogonal even though the columns of the $[\mathbf{X}]_\lambda$ array will not be in the strict mathematical sense.

In addition to the model baseline, an experimental design can also have a production baseline. The production baseline is the combination of set points that represent the product currently in production and this will always be made to be trial 1, if possible, when using either the φ coefficients or the λ coefficients. The model baseline defined by the condition $Y(B) = \lambda_0$ for the $[\mathbf{X}]_\lambda$ array will also be trial 1 making the model and production baselines one and the same. For the $[\mathbf{X}]_\varphi$ array, the model baseline condition $Y(B) = \varphi_0$ is not a specific trial but an average of all the trials and each φ coefficient is measured about this average. For the λ coefficients, we use the convention that level 0 is where the λ_{i0} coefficients equal zero (Equations B.13 and B.14).

B.6 [X] array construction rules

The rules for constructing $[\mathbf{X}]_\lambda$ and $[\mathbf{X}]_\varphi$ arrays are straightforward:

1. The elements in column 0 are all 1s.
2. Each main effect variable ij has a column for each of its levels j except for the baseline level, $j = 0$ (column $i0$) of the main effect which is always absent.
3. For a set point series using the λ coefficients, column ij has $X_{ij} = 1$s for rows (trials) where level j is present and 0s for rows where the level is different from j. Such columns are always combinations of 0s and 1s.
4. For set point series using the φ coefficients, column ij has $X_{ij} = -1$s for rows where the set point for the main effect i is at its baseline level $j = 0$ (the level not shown explicitly with a column heading in the array), $X_{ij} = 1$s where it is

at level j, and X_{ij} = 0s when at a level different from j other than baseline. Such columns for two-level factors are always combinations of 1s and –1s. For three- and higher-level factors, the columns will have 1s, –1s, and 0s.

5. The rows of the interaction columns *ijkl*, etc. are filled by products of the elements $X_{ij}X_{kl}$ etc. in the row under consideration.

In contrast to the more common representations of experimental design matrices by 1s, 2s, 3s, etc. or pluses and minuses, the set point description using either the $[\mathbf{X}]_\varphi$ or $[\mathbf{X}]_\lambda$ array can readily be used in spreadsheet programs for efficiently performing the required matrix operations and yet maintain the clarity of the common representations as to which set points are 'on' and 'off' for a given trial.

B.6.1 CONSTRUCTION OF [X] FOR AN ORTHOGONAL ARRAY

Let's now consider an example to illuminate the straightforward rules given above for constructing [X] arrays. The standard form noted here as $L_9(3^3)$ in Table B.1a is taken from the first three main effect columns of the $L_9(3^4)$ orthogonal array given by Taguchi and Konishi [2]. Although the standard forms are good road maps for setting up experimental trials, they need to be converted if they are to be used for matrix operations with a spreadsheet. The first step in conversion is to generate the base 0 form shown to the right of the standard form by subtracting one from each of its elements.

In Table B.1b, the base 0 form of the $L_9(3^3)$ orthogonal array in Table B.1a has been converted into the $[\mathbf{X}]_\varphi$ form which can be used in the matrix operations and which has the relations between main effect coefficients given by Equation B.17. Column 0 denotes the baseline set point which is equal to unity in every row because φ_0 appears on the left-hand side of Equation B.16 for each trial, the contribution from the baseline being present in every trial. The largest number y for each factor (column) in the base 0 form generates y columns for the factor in both the $[\mathbf{X}]_\varphi$ and $[\mathbf{X}]_\lambda$ forms. Columns 11 and 12 denote the elements of the $[\mathbf{X}]_\varphi$ matrix that multiply φ_{11} and φ_{12}, respectively, representing variable 1 at levels 2 and 3 in the standard form. The columns for level 0 in the standard form are eliminated through the use of Equation B.16 which places the –1s in columns $i1$ and $i2$. Table B.1c shows the $[\mathbf{X}]_\lambda$ form for this array which uses trial 1 as the model baseline. It becomes straightforward to key in the $[\mathbf{X}]_\lambda$ and $[\mathbf{X}]_\varphi$ forms directly by inspection of the standard form after using the base zero intermediate form once or twice.

B.7 Comparisons with full factorial arrays

Let's now consider several orthogonal arrays and full factorial arrays to further illuminate the construction rules described above. A full factorial array measures

Table B.1a $L_9(3^3)$ orthoganol array in standard form and base zero form. (source of standard form: G. Taguchi and S. Konishi (1987) *Taguchi Methods Orthogonal Arrays and Linear Graphs*, © Copyright, American Supplier Institute, Allen Park, MI; reproduced by permission under License No. 961201)

Standard form				*Base zero form*			
Trial	*1*	*2*	*3*	*Trial*	*1*	*2*	*3*
1	1	1	1	1	0	0	0
2	1	2	2	2	0	1	1
3	1	3	3	3	0	2	2
4	2	1	2	4	1	0	1
5	2	2	3	5	1	1	2
6	2	3	1	6	1	2	0
7	3	1	3	7	2	0	2
8	3	2	1	8	2	1	0
9	3	3	2	9	2	2	1

Table B.1b $L_9(3^3)$ experimental design as $[\mathbf{X}]_\varphi$ array

Trial	*0*	*11*	*12*	*21*	*22*	*31*	*32*
1	1	–1	–1	–1	–1	–1	–1
2	1	–1	–1	1	0	1	0
3	1	–1	–1	0	1	0	1
4	1	1	0	–1	–1	1	0
5	1	1	0	1	0	0	1
6	1	1	0	0	1	–1	–1
7	1	0	1	–1	–1	0	1
8	1	0	1	1	0	–1	–1
9	1	0	1	0	1	1	0

Set Points 10, 20, 30 are not shown explicitly but are included implicitly in the base column, column 0.
When any set point is at its baseline level for a specific trial, –1s appear under each of the columns for that set point.

the effect of each main effect on the outcomes as well as all possible interactions between the main effects. The $[\mathbf{X}]_\varphi$ array shown in Table B.2 represents an $L_4(2^3)$ design [2]. It is known as a fractional factorial array. For it to generate a successful experimental design, interactions between the main effect variables must be weak because this array is obtained from the two-level two-main-effect full factorial array shown in Table B.3 (noted here as the 2^2 array) by replacing its assumed weak 1121 interaction with another main effect variable noted as 3 in Table B.2.

Table B.1c $L_9(3^3)$ experimental design as $[X]_\lambda$ array

Trial	*0*	*11*	*12*	*21*	*22*	*31*	*32*
1	1	0	0	0	0	0	0
2	1	0	0	1	0	1	0
3	1	0	0	0	1	0	1
4	1	1	0	0	0	1	0
5	1	1	0	1	0	0	1
6	1	1	0	0	1	0	0
7	1	0	1	0	0	0	1
8	1	0	1	1	0	0	0
9	1	0	1	0	1	1	0

Set Points 10, 20, 30 are not shown explicitly but are included implicitly in the base column, column 0.
When any set point is at its baseline level for a specific trial, 0s appear under each of the columns for that set point.

Table B.2 $L_4(2^3)$ experimental design as $[X]_\varphi$ array

Trial	*0*	*11*	*21*	*31*
1	1	–1	–1	–1
2	1	–1	1	1
3	1	1	–1	1
4	1	1	1	–1

Table B.3 2^2 experimental design as $[X]_\varphi$ full factorial array

Trial	*0*	*11*	*21*	*1121*
1	1	–1	–1	1
2	1	–1	1	–1
3	1	1	–1	–1
4	1	1	1	1

The full factorial array for three variables at two levels each (2^3) is shown in Table B.4. Again if the interactions are weak, the four interaction columns can be replaced by main effect variables yielding the $L_8(2^7)$ design in the $[X]_\varphi$ form (Table B.5). The full factorial array for two variables with three levels each is shown in Table B.6. The four interaction columns in Table B.6 can be replaced by two additional main effect variables if interactions are weak yielding the $L_9(3^4)$ design in the $[X]_\varphi$ form (Table B.7).

Table B.4 2^3 experimental design as $[\mathbf{X}]_\varphi$ full factorial array

Trial	*0*	*11*	*21*	*31*	*1121*	*1131*	*2131*	*112131*
1	1	–1	–1	–1	1	1	1	–1
2	1	–1	–1	1	1	–1	–1	1
3	1	–1	1	–1	–1	1	–1	1
4	1	–1	1	1	–1	–1	1	–1
5	1	1	–1	–1	–1	–1	1	1
6	1	1	–1	1	–1	1	–1	–1
7	1	1	1	–1	1	–1	–1	–1
8	1	1	1	1	1	1	1	1

Table B.5 $L_8(2^7)$ experimental design as $[\mathbf{X}]_\varphi$ array

Trial	*0*	*11*	*21*	*31*	*41*	*51*	*61*	*71*
1	1	–1	–1	–1	–1	–1	–1	–1
2	1	–1	–1	1	–1	1	1	1
3	1	–1	1	–1	1	–1	1	1
4	1	–1	1	1	1	1	–1	–1
5	1	1	–1	–1	1	1	–1	1
6	1	1	–1	1	1	–1	1	–1
7	1	1	1	–1	–1	1	1	–1
8	1	1	1	1	–1	–1	–1	1

Table B.6 3^2 experimental design as $[\mathbf{X}]_\varphi$ full factorial array

Trial	*0*	*11*	*12*	*21*	*22*	*1121*	*1122*	*1221*	*1222*
1	1	–1	–1	–1	–1	1	1	1	1
2	1	–1	–1	1	0	–1	0	–1	0
3	1	–1	–1	0	1	0	–1	0	–1
4	1	1	0	–1	–1	–1	–1	0	0
5	1	1	0	1	0	1	0	0	0
6	1	1	0	0	1	0	1	0	0
7	1	0	1	–1	–1	0	0	–1	–1
8	1	0	1	1	0	0	0	1	0
9	1	0	1	0	1	0	0	0	1

Conversion to an $[\mathbf{X}]_\lambda$ array (a set point series with trial 1 as baseline) is generated by replacing the –1s in the main effect columns by 0s and then multiplying the appropriate main effects columns together to obtain the columns for the interactions. The transformations of the last two arrays above to the $[\mathbf{X}]_\lambda$ form are shown as Tables B.8 and B.9.

Table B.7 $L_9(3^4)$ experimental design as $[X]_\varphi$ array

Trial	*0*	*11*	*12*	*21*	*22*	*31*	*32*	*41*	*42*
1	1	–1	–1	–1	–1	–1	–1	–1	–1
2	1	–1	–1	1	0	1	0	1	0
3	1	–1	–1	0	1	0	1	0	1
4	1	1	0	–1	–1	1	0	0	1
5	1	1	0	1	0	0	1	–1	–1
6	1	1	0	0	1	–1	–1	1	0
7	1	0	1	–1	–1	0	1	1	0
8	1	0	1	1	0	–1	–1	0	1
9	1	0	1	0	1	1	0	–1	–1

Table B.8 3^2 experimental design as $[X]_\lambda$ full factorial array

Trial	*0*	*11*	*12*	*21*	*22*	*1121*	*1122*	*1221*	*1222*
1	1	0	0	0	0	0	0	0	0
2	1	0	0	1	0	0	0	0	0
3	1	0	0	0	1	0	0	0	0
4	1	1	0	0	0	0	0	0	0
5	1	1	0	1	0	1	0	0	0
6	1	1	0	0	1	0	1	0	0
7	1	0	1	0	0	0	0	0	0
8	1	0	1	1	0	0	0	1	0
9	1	0	1	0	1	0	0	0	1

Table B.9 $L_9(3^4)$ experimental design as $[X]_\lambda$ array

Trial	*0*	*11*	*12*	*21*	*22*	*31*	*32*	*41*	*42*
1	1	0	0	0	0	0	0	0	0
2	1	0	0	1	0	1	0	1	0
3	1	0	0	0	1	0	1	0	1
4	1	1	0	0	0	1	0	0	1
5	1	1	0	1	0	0	1	0	0
6	1	1	0	0	1	0	0	1	0
7	1	0	1	0	0	0	1	1	0
8	1	0	1	1	0	0	0	0	1
9	1	0	1	0	1	1	0	0	0

B.8 Designing an experiment

In designing an experiment you generally want to do the following:

1. Run no more experiments than necessary.

2. Handle data effectively and efficiently.
3. Avoid making errors and blunders.
4. Minimize wasted efforts.
5. Find a satisfactory answer as quickly as possible.
6. Confirm the finding (recommended by Taguchi).

Although dedicated software can be purchased to assist in handling the experimental data and in interpreting the results, a spreadsheet program is used throughout here because of its greater flexibility and its ability to perform the Monte Carlo simulations used in significance testing as described in Chapter 9.

B.8.1 EFFICIENT EXPERIMENTAL DESIGN

The first concern in designing an experiment to solve for w unknowns is to have at least w linearly independent trials out of the total array of $k \geq w$ trials. Although a variety of **[X]** arrays can be used for an experiment, some will yield less error for the coefficients for a given number of n replications than others. An important second criterion, therefore, is to choose the **[X]** array so that the labor needed to gain the desired accuracy is minimized.

Orthogonal arrays are usually the most efficient designs for determining the coefficients to a given level of accuracy. Thus, if a specific orthogonal array is suitable for the experiments when factoring in all other considerations, it should be used. Examples of orthogonal arrays are described in Chapter 8 as well as in what follows here. In fact, the orthogonal array may require k_{OA} trials greater than the minimum number w trials needed to solve for the unknowns for the specific problem at hand but it will, nevertheless, be more efficient in that the number of replications n_{OA} required by it for a specific error level (standard deviation) will be less than the n required for the same error using the minimum of w trials which can be expressed as $k_{OA}n_{OA} < wn$.

B.8.2 CONFOUNDING

When interactions are assumed weak in an experiment but are actually strong, confounding of main effects with interactions (placing a main effect variable in the column for the interaction between two other main effects) can lead to erroneous conclusions. The simplest problem for illustrating this issue involves an orthogonal array of three main variables at two levels each; the $[\mathbf{X}]_\lambda$ matrix for this design is shown in Table B.10a.

If set points 11, 21, and 31 interact significantly, then the full factorial experiment would be required to evaluate the three set points (factors) including their interactions. If it is believed that this is not the case and that the interactions are probably weak, it is also of interest, nevertheless, to understand, in case one or more interactions did turn out to be strong, how the main effect measurements in an arbitrary design, **[X]**, would be confounded by the interaction. This is readily determined by using the spreadsheet to construct the **confounding** matrix given by:

Table B.10a $L_4(2^3)$ orthagonal array as $[X]_A$ array

Trial	*0*	*11*	*21*	*31*
1	1	0	0	0
2	1	0	1	1
3	1	1	0	1
4	1	1	1	0

$$[X_C]=[X^TX]^{-1}[X^T][X^*_{FF}]$$

The array $[X^*_{FF}]$ is an increase in the columns but not the rows of the original $[X]$ showing the interactions between main effects not included $[X]$.

Consider, for illustrative purposes, the possible confounding between the design in Table B.10a and the array for the same four trials when all possible interactions are considered among the main effects (Table B.10b). The resulting confounding array computed from the above expression is given in Table B.10c. The 1121 interaction, for example, is seen to confound with main effects 11 and 21 contributing 1/2 of its value to each and also to confound with main effect 31 contributing –1/2 of its value to it. The 112131 triplet interaction is seen not to confound because the three main-effect set points were not on at the same time in the four trials shown in Table B.10a. Each main effect, of course, interacts with itself giving the diagonal of ones in Table B.10c.

When you have assumed incorrectly at the outset that an interaction is absent or weak, your discovery that the interaction is, in fact, significant will often be made on running a confirmation trial made at the end of the experiment to verify your findings. For example, it may be found after running the first four trials shown in Table B.10a needed to evaluate the three main effect variables in the absence of interactions that the confirmation trial does not agree with the expected result. If a significant interaction between variables 1 and 2 is suspected to be the cause, a fifth trial and column 1121 can be added to the experimental design to measure this interaction as shown in Table B.11.

If the suspicion of an interaction between variables 1 and 2 is indeed correct, only this one additional trial has to be run as the outcomes measured for the

Table B.10b Array for the four trials in Table B.10a when all possible interactions are considered

Trial	*0*	*11*	*21*	*31*	*1121*	*1131*	*2131*	*112131*
1	1	0	0	0	0	0	0	0
2	1	0	1	1	0	0	1	0
3	1	1	0	1	0	1	0	0
4	1	1	1	0	1	0	0	0

Table B.10c Confounding array between the arrays in Tables B.10a and B.10b

Trial	*0*	*11*	*21*	*31*	*1121*	*1131*	*2131*	*112131*
0	1	0	0	0	0	0	0	0
11	0	1	0	0	0.5	0.5	–0.5	0
21	0	0	1	0	0.5	–0.5	0.5	0
31	0	0	0	1	–0.5	0.5	0.5	0

original four trials remain valid with the new design in place. The resulting experiment no longer represents an orthogonal design but it may yield an acceptable error and, therefore, an expedient design versus the use of the full factorial array for the problem which would require eight trials as shown by the $[\mathbf{X}]_\lambda$ form in Table B.12 and by the $[\mathbf{X}]_\varphi$ form in Table B.13. Column 31 in Table B.13 is derived from column 4 of the standard Taguchi form of the $L_8(2^7)$ orthogonal array shown in Figure 8.9. Column 4 is used instead of column 3 for factor *c* to avoid confounding with the interaction between factors *a* and *b*. The 1121 interaction column for factors *a* and *b* is derived from column 3 in Figure 8.9.

Now consider what happens if the third main effect variable were placed in the third column of the standard $L_8(2^7)$ orthogonal array, the resulting $[\mathbf{X}]_\lambda$ matrix with the 1121 interaction is shown in Table B.14. By observation, this array is seen to have only four unique trials (1, 3, 5, and 7) which are replicated once (2, 4, 6, and 8). It is therefore insufficient to solve for the five unknowns. At least one additional unique trial is needed to solve for the three main effects and the pairwise interaction. Apart from showing replications explicitly, the above matrix is no different from the $L_4(2^3)$ matrix for the confounding problem considered above and resolution of the problem is no different in that at least one additional unique trial must be run.

B.8.3 THE NOT-QUITE-ORTHOGONAL ARRAY

In running an experiment using an orthogonal array, equipment difficulties or a blunder may cause one or more set points not to follow the orthogonal pat-

Table B.11 $[\mathbf{X}]_\lambda$ array for three main effect variables at two levels each and 11 × 12 pairwise interaction

Trial	*0*	*11*	*21*	*31*	*1121*
1	1	0	0	0	0
2	1	0	1	1	0
3	1	1	0	1	0
4	1	1	1	0	1
5	1	0	0	1	0

Table B.12 Three main effect variables at two levels each and 11 × 12 pairwise interaction extracted from the $[\mathbf{X}]_\lambda$ form of an $L_8(2^7)$ orthogonal array

Trial	*0*	*11*	*21*	*31*	*1121*
1	1	0	0	0	0
2	1	0	0	1	0
3	1	0	1	0	0
4	1	0	1	1	0
5	1	1	0	0	0
6	1	1	0	1	0
7	1	1	1	0	1
8	1	1	1	1	1

Table B.13 Three main effect variables at two levels each and 11 × 12 pairwise interaction extracted from the $[\mathbf{X}]_\varphi$ form of an $L_8(2^7)$ experimental design

Trial	*0*	*11*	*21*	*31*	*1121*
1	1	–1	–1	–1	1
2	1	–1	–1	1	1
3	1	–1	1	–1	–1
4	1	–1	1	1	–1
5	1	1	–1	–1	–1
6	1	1	–1	1	–1
7	1	1	1	–1	1
8	1	1	1	1	1

Table B.14 Example of replication of rows replacing confounding of columns

Trial	*0*	*11*	*21*	*31*	*1121*
1	1	0	0	0	0
2	1	0	0	0	0
3	1	0	1	1	0
4	1	0	1	1	0
5	1	1	0	1	0
6	1	1	0	1	0
7	1	1	1	0	1
8	1	1	1	0	1

tern for a given trial. The work done using the non-orthogonal form can usually be salvaged, however. An example is shown in Table B.15 for the $[\mathbf{X}]_\lambda$ form of what would be an $L_9(3^4)$ orthogonal array if not for mistakes in the columns 11, 21 and 22 of trial 6 where the highlighted 0s and 1 should be 1s and 0, respectively, for the orthogonal design. Because the resulting $[\mathbf{X}]_\lambda$ matrix still inverts, no additional trials need to be run unless more accuracy were required or unless interactions were found to be significant. If a mistake were such that the matrix did not invert, one or more additional trials would be needed.

B.9 Student's *t* for λ coefficients

Orthogonal arrays are encouraged for experimental designs because of their efficiency, which arises from a balanced selection of control factor (set point) combinations (Chapter 8) from the array of all possible combinations given by the full factorial array. The manner in which the [X] matrix influences the efficiency of the experiment can be understood by evaluating the uncertainty in the measured coefficients λ_{ij} (or φ_{ij}).

The lambda coefficients (Equation B.5) represent a weighted sample average of the form:

$$\lambda_{ij} \equiv \bar{\lambda}_{ij} = \sum_{q=1,k} \omega_{ij}(q) Y(q) \qquad \text{(B.21)}$$

for an experiment where $Y(q)$ is average of the outcomes measured over the $y=1 \rightarrow n$ replications for trial (q) given by:

$$Y(q) = \frac{1}{n} \sum_{y=1,n} Y(q,y). \qquad \text{(B.22)}$$

Table B.15 A not orthogonal experimental design in the $[\mathbf{X}]_\lambda$ form obtained by modifying an $L_9(3^4)$ orthogonal array

Trial	*0*	*11*	*12*	*21*	*22*	*31*	*32*	*41*	*42*
1	1	0	0	0	0	0	0	0	0
2	1	0	0	1	0	1	0	1	0
3	1	0	0	0	1	0	1	0	1
4	1	1	0	0	0	1	0	0	1
5	1	1	0	1	0	0	1	0	0
6	1	**0**	0	**1**	**0**	0	0	1	0
7	1	0	1	0	0	0	1	1	0
8	1	0	1	1	0	0	0	0	1
9	1	0	1	0	1	1	0	0	0

The outcomes $Y(q, y)$ are assumed to be normally distributed having a variance equal to $\sigma^2(q)$, the variance of $Y(q)$ being $\sigma^2(q)/n$. The standard normal statistic for set point ij is given by:

$$Z_{ij} = \frac{\lambda_{ij} - \mu_{ij}}{\sqrt{\sigma_{ij}^2 / n}} \tag{B.23}$$

We are interested in this statistic only for off-baseline set points (all set points but λ_0) and the hypothesis that the outcomes are taken from normal populations having the same mean. Because $\sum_{q=1,k} \omega_{ij}(q) = 0$ for all off-baseline set points, it follows that $\mu_{ij} = 0$ for this hypothesis and that:

$$\sigma_{ij}^2 = \sum_q \omega_{ij}^2(q)\sigma^2(q). \tag{B.24}$$

On replacing the population variances by the sample variances, it follows that the normal statistic becomes a t statistic for set point ij given by:

$$t_{ij} = \frac{\lambda_{ij}\sqrt{n}}{\sqrt{\sum_{q=1,k}[\omega_{ij}^2(q)s^2(q)]}} \tag{B.25}$$

having $\nu_P = k(n-1)$ df if the individual trial population variances given by $\sigma^2(q)$ are equal in which case the trial variances $s^2(q)$ can be replaced by their pooled variance s_P^2 given by:

$$s_P^2 = \frac{1}{k}\sum_{q=1\to k} s^2(q) \tag{B.26}$$

The test of whether the individual trial variances in Equation B.25 can reasonably be replaced by their pooled or average variance can be made using the F_{ij} statistic as described in Appendix A and also in Chapter 9. If all of the variances can be pooled, Equation B.25 can be written as:

$$t_{ij} = \frac{\lambda_{ij}\sqrt{n}}{s_P\sqrt{\sum_{q=1,k}[\omega_{ij}^2(q)]}} = \frac{\lambda_{ij}\sqrt{n}}{s_P\sqrt{\text{cov}_{ijij}}} \tag{B.27}$$

which follows because the summation over the square of the elements $\omega_{ij}(q)$ in the solution matrix given by $[\mathbf{X}^T\mathbf{X}]^{-1}[\mathbf{X}^T]$ is equal to the element cov_{ijij} on the diagonal of the covariance matrix $[\mathbf{X}^T\mathbf{X}]^{-1}$. For example, if several $[\mathbf{X}]_\lambda$ design matrices are considered for an experiment, the one having the smallest cov_{ijij} will be most sensitive for measuring the coefficient λ_{ij}.

If the test for homogeneity of variance is not satisfied, Equation B.25 can be used as an approximate t statistic which we write in the form:

$$t_{ij} = \frac{\lambda_{ij}}{\sqrt{s_{ij}^2 / n}} \tag{B.28}$$

where s_{ij}^2 is an effective sample variance for the coefficient λ_{ij} given by:

$$s_{ij}^2 = \sum_{q=1}^{k} s^2(q)\omega_{ij}^2(q) \tag{B.29}$$

having ν_{ij} effective df given by:

$$\nu_{ij} = \frac{[n-1]\left[\sum_{q=1}^{k} s^2(q)\omega_{ij}^2(q)\right]^2}{\sum_{q=1}^{k} s^4(q)\omega_{ij}^4(q)} = \frac{[n-1]s_{ij}^4}{\sum_{q=1}^{k} s^4(q)\omega_{ij}^4(q)} \tag{B.30}$$

The above expression for the effective df (written for the case we are considering in which the sample sizes are the same for each trial) is a generalization (by inspection) of the standard pairwise formula for computing the effective df for the approximate t statistic when the population variances are not the same [3]. Both the effective variance and the effective df are readily computed using the spreadsheet once the solution matrix and the column matrix of the measured sample variances are in the spreadsheet.

The t statistic for the sum (or difference) of two coefficients with the pooled variance is given by:

$$t_{ij\pm kl} = \frac{\{\lambda_{ij} \pm \lambda_{kl}\}\sqrt{n}}{s_P\sqrt{\mathrm{cov}_{ijij} + \mathrm{cov}_{klkl} \pm 2\,\mathrm{cov}_{ijkl}}} \tag{B.31}$$

B.9.1 ORTHOGONAL ARRAYS (OA) VERSUS ONE AT A TIME (OAAT)

Using the $L_8(2^7)$ design in Table B.16, the following analysis is designed to illustrate the superior properties of an OA, versus the OaaT designs (Table B.17). The solution matrices for the OA and OaaT designs in the $[\mathbf{X}]_\lambda$ form are shown in Tables B.18 and B.19, respectively. The respective arrays with elements $\omega_{ij}^2(q)$ are shown in Tables B.20 and B.21. The sum of the $\omega_{ij}^2(q)$ along any row is seen to be 2 for the OaaT array but only 1/2 for the OA. For completeness, the covariance matrix, $[\mathbf{X}^T\mathbf{X}]^{-1}$, for the $L_8(2^7)$ OA in the $[\mathbf{X}]_\lambda$ form is shown in Table B.22 where the sum of the $\omega_{ij}^2(q)$ along a row in Table B.20 can be read off as the element cov_{ijij} of the array in Table B.22. Thus, we see from Equation B.27 and Tables B.20 and B.21 that the OaaT design needs four times the number of replications per trial to achieve the same value for t (level of experimental error) as the orthogonal array, assuming that the trial variances are the same. When persons are being surveyed, however, as opposed to the evaluations of machines, it may be more effective to use the simpler OaaT design which is easier for respondents to follow. The number of persons surveyed is increased to gain the needed sampling accuracy. This is discussed in Chapter 5.

B.10 $[X]_\lambda$ versus $[X]_\varphi$ for significance testing

It is instructive to consider the t_{ij} for the set points constructed from the $[X]_\lambda$ array versus those constructed from the $[X]_\varphi$ array using the example of the $L_9(3^4)$ experimental design. The $[X]_\lambda$ matrix for this design is the top array shown in Table B.23 and the $[X]_\varphi$ matrix is shown as the top array in Table B.24. The same outcome vector [Y] is shown for the trials for the two experiments. The middle arrays in Tables B.23 and B.24 are the covariance matrices corresponding to the $[X]_\lambda$ and the $[X]_\varphi$ matrices, respectively. The last array in each table is the solution matrix. The resulting λ and φ coefficients divided by their respective $\sqrt{\text{cov}_{ijij}}$ are shown in Tables B.25 and B.26, respectively.

Table B.16 $L_8(2^7)$ experimental design in the $[X]_\lambda$ form

Trial	*0*	*11*	*21*	*31*	*41*	*51*	*61*	*71*
1	1	0	0	0	0	0	0	0
2	1	0	0	0	1	1	1	1
3	1	0	1	1	0	0	1	1
4	1	0	1	1	1	1	0	0
5	1	1	0	1	0	1	0	1
6	1	1	0	1	1	0	1	0
7	1	1	1	0	0	1	1	0
8	1	1	1	0	1	0	0	1

Table B.17 One-at-a-time array in the $[X]_\lambda$ form

Trial	*0*	*11*	*21*	*31*	*41*	*51*	*61*	*71*
1	1	0	0	0	0	0	0	0
2	1	1	0	0	0	0	0	0
3	1	0	1	0	0	0	0	0
4	1	0	0	1	0	0	0	0
5	1	0	0	0	1	0	0	0
6	1	0	0	0	0	1	0	0
7	1	0	0	0	0	0	1	0
8	1	0	0	0	0	0	0	1

Table B.18 The solution matrix with elements ω_{ij} (q) for the $L_8(2^7)$ experimental design written in the $[\mathbf{X}]_\lambda$ form

	q							
ij	*1*	*2*	*3*	*4*	*5*	*6*	*7*	*8*
0	1	0	0	0	0	0	0	0
11	–0.25	–0.25	–0.25	–0.25	0.25	0.25	0.25	0.25
21	–0.25	–0.25	0.25	0.25	–0.25	–0.25	0.25	0.25
31	–0.25	–0.25	0.25	0.25	0.25	0.25	–0.25	–0.25
41	–0.25	0.25	–0.25	0.25	–0.25	0.25	–0.25	0.25
51	–0.25	0.25	–0.25	0.25	0.25	–0.25	0.25	–0.25
61	–0.25	0.25	0.25	–0.25	–0.25	0.25	0.25	–0.25
71	–0.25	0.25	0.25	–0.25	0.25	–0.25	–0.25	0.25

Table B.19 The solution matrix with elements ω_{ij} (q) for the one-at-a-time experimental design written in the $[\mathbf{X}]_\lambda$ form

	q							
ij	*1*	*2*	*3*	*4*	*5*	*6*	*7*	*8*
0	1	0	0	0	0	0	0	0
11	–1	1	0	0	0	0	0	0
21	–1	0	1	0	0	0	0	0
31	–1	0	0	1	0	0	0	0
41	–1	0	0	0	1	0	0	0
51	–1	0	0	0	0	1	0	0
61	–1	0	0	0	0	0	1	0
71	–1	0	0	0	0	0	0	1

Table B.20 The matrix with elements ω_{ij}^2 (q) for the $L_8(2^7)$ experimental design in the $[\mathbf{X}]_\lambda$ form

	q								$\sum_{q=1,8}\omega_{ij}^2(q)$
ij	*1*	*2*	*3*	*4*	*5*	*6*	*7*	*8*	
0	1	0	0	0	0	0	0	0	1
11	0.0625	0.0625	0.0625	0.0625	0.0625	0.0625	0.0625	0.0625	0.5
21	0.0625	0.0625	0.0625	0.0625	0.0625	0.0625	0.0625	0.0625	0.5
31	0.0625	0.0625	0.0625	0.0625	0.0625	0.0625	0.0625	0.0625	0.5
41	0.0625	0.0625	0.0625	0.0625	0.0625	0.0625	0.0625	0.0625	0.5
51	0.0625	0.0625	0.0625	0.0625	0.0625	0.0625	0.0625	0.0625	0.5
61	0.0625	0.0625	0.0625	0.0625	0.0625	0.0625	0.0625	0.0625	0.5
71	0.0625	0.0625	0.0625	0.0625	0.0625	0.0625	0.0625	0.0625	0.5

Table B.21 The matrix with elements $\omega_{ij}^2\,(q)$ for the one-at-a-time experimental design in the $[\mathbf{X}]_\lambda$ form

	q								$\sum_{q=1,8} \omega_{ij}^2(q)$
ij	*1*	*2*	*3*	*4*	*5*	*6*	*7*	*8*	
0	1	0	0	0	0	0	0	0	1
11	1	1	0	0	0	0	0	0	2
21	1	0	1	0	0	0	0	0	2
31	1	0	0	1	0	0	0	0	2
41	1	0	0	0	1	0	0	0	2
51	1	0	0	0	0	1	0	0	2
61	1	0	0	0	0	0	1	0	2
71	1	0	0	0	0	0	0	1	2

The interpretation of the λ coefficients is that, when either set point 11 or 12 is 'on' and the others are 'off,' the output is increased by 10 over the model baseline outcome of 20 represented by trial 1. Similarly for the φ coefficients, when either 11 or 12 is 'on', the outcome is increased by 3.33 over their model baseline outcome of 26.66 which is equal to the average of the trials. When set point 10 is 'on,' the outcome is decreased by 6.66 (equal to minus one times the sum of the other two coefficients (Equation B.16)).

For a given critical region of size α (Appendix A) which fixes the t_{ij} coefficients for a fixed pooled variance, it can be seen by inspection of Tables B.25 and B.26 that the number of replications n can be chosen such that $\lambda_{ij}/\sqrt{\text{cov}_{ijij}} \geq t_{ij}\sqrt{s_p^2/n}$ and that $\varphi_{ij}/\sqrt{\text{cov}_{ijij}} \leq t_{ij}\sqrt{s_p^2/n}$. For a LIB (larger-is-better) condition, for example, λ_{12} would be significant and favorable when these inequalities are satisfied. However, φ_{11} and φ_{12}, although favorable, would not be significant and φ_{10} would be significant but unfavorable. Using the $[\mathbf{X}]_\lambda$ array, we would conclude that either 11 or 12 should be turned 'on' and arrive at a new output of thirty but using the $[\mathbf{X}]_\varphi$ array we might decide (incorrectly) to stay at the model

Table B.22 Covariance matrix for the $L_8(2^7)$ experimental design in the $[\mathbf{X}]_\lambda$ form

	0	*11*	*21*	*31*	*41*	*51*	*61*	*71*
0	1	–0.25	–0.25	–0.25	–0.25	–0.25	–0.25	–0.25
11	–0.25	0.5	0	0	0	0	0	0
21	–0.25	0	0.5	0	0	0	0	0
31	–0.25	0	0	0.5	0	0	0	0
41	–0.25	0	0	0	0.5	0	0	0
51	–0.25	0	0	0	0	0.5	0	0
61	–0.25	0	0	0	0	0	0.5	0
71	–0.25	0	0	0	0	0	0	0.5

Table B.23 $L_9(3^4)$ experimental design in $[\mathbf{X}]_\lambda$ form and resulting covariance and solution matrices

$[\mathbf{X}]_\lambda$

Trial	*0*	*11*	*12*	*21*	*22*	*31*	*32*	*41*	*42*	*Y*
1	1	0	0	0	0	0	0	0	0	20
2	1	0	0	1	0	1	0	1	0	20
3	1	0	0	0	1	0	1	0	1	20
4	1	1	0	0	0	1	0	0	1	30
5	1	1	0	1	0	0	1	0	0	30
6	1	1	0	0	1	0	0	1	0	30
7	1	0	1	0	0	0	1	1	0	30
8	1	0	1	1	0	0	0	0	1	30
9	1	0	1	0	1	1	0	0	0	30

$[\mathbf{X}^T\mathbf{X}]^{-1}$

	0	*11*	*12*	*21*	*22*	*31*	*32*	*41*	*42*
0	1	−0.333	−0.333	−0.333	−0.333	−0.333	−0.333	−0.333	−0.333
11	−0.333	0.667	0.333	0	0	0	0	0	0
12	−0.333	0.333	0.667	0	0	0	0	0	0
21	−0.333	0	0	0.667	0.333	0	0	0	0
22	−0.333	0	0	0.333	0.667	0	0	0	0
31	−0.333	0	0	0	0	0.667	0.333	0	0
32	−0.333	0	0	0	0	0.333	0.667	0	0
41	−0.333	0	0	0	0	0	0	0.667	0.333
42	−0.333	0	0	0	0	0	0	0.333	0.667

$[\mathbf{X}^T\mathbf{X}]^{-1}[\mathbf{X}^T]$

ij	*1*	*2*	*3*	*4*	*5*	*6*	*7*	*8*	*9*
0	1	1E–16	1E–16	1E–16	1E–16	1E–16	1E–16	1E–16	1E–16
11	−0.333	−0.333	−0.333	0.333	0.333	0.333	0	0	0
12	−0.333	−0.333	−0.333	0	0	0	0.333	0.333	0.333
21	−0.333	0.333	0	−0.333	0.333	0	−0.333	0.333	0
22	−0.333	0	0.333	−0.333	0	0.333	−0.333	0	0.333
31	−0.333	0.333	0	0.333	0	−0.333	0	−0.333	0.333
32	−0.333	0	0.333	0	0.333	−0.333	0.333	−0.333	0
41	−0.333	0.333	0	0	−0.333	0.333	0.333	0	−0.333
42	−0.333	0	0.333	0.333	−0.333	0	0	0.333	−0.333

Table B.24 $L_9(3^4)$ experimental design in $[X]_\varphi$ form and resulting covariance and solution matrices

$[X]_\varphi$

Trial	*0*	*11*	*12*	*21*	*22*	*31*	*32*	*41*	*42*	*Y*
1	1	–1	–1	–1	–1	–1	–1	–1	–1	20
2	1	–1	–1	1	0	1	0	1	0	20
3	1	–1	–1	0	1	0	1	0	1	20
4	1	1	0	–1	–1	1	0	0	1	30
5	1	1	0	1	0	0	1	–1	–1	30
6	1	1	0	0	1	–1	–1	1	0	30
7	1	0	1	–1	–1	0	1	1	0	30
8	1	0	1	1	0	–1	–1	0	1	30
9	1	0	1	0	1	1	0	–1	–1	30

$[X^TX]^{-1}$

	0	*11*	*12*	*21*	*22*	*31*	*32*	*41*	*42*
0	0.111	0	0	0	0	0	0	0	0
11	0	0.222	–0.111	0	0	0	0	0	0
12	0	–0.111	0.222	0	0	0	0	0	0
21	0	0	0	0.222	–0.111	0	0	0	0
22	0	0	0	–0.111	0.222	0	0	0	0
31	0	0	0	0	0	0.222	–0.111	0	0
32	0	0	0	0	0	–0.111	0.222	0	0
41	0	0	0	0	0	0	0	0.222	–0.111
42	0	0	0	0	0	0	0	–0.111	0.222

$[X^TX]^{-1}[X^T]$

ij	*1*	*2*	*3*	*4*	*5*	*6*	*7*	*8*	*9*
0	0.111	0.111	0.111	0.111	0.111	0.111	0.111	0.111	0.111
11	–0.111	–0.111	–0.111	0.222	0.222	0.222	–0.111	–0.111	–0.111
12	–0.111	–0.111	–0.111	–0.111	–0.111	–0.111	0.222	0.222	0.222
21	–0.111	0.222	–0.111	–0.111	0.222	–0.111	–0.111	0.222	–0.111
22	–0.111	–0.111	0.222	–0.111	–0.111	0.222	–0.111	–0.111	0.222
31	–0.111	0.222	–0.111	0.222	–0.111	–0.111	–0.111	–0.111	0.222
32	–0.111	–0.111	0.222	–0.111	0.222	–0.111	0.222	–0.111	–0.111
41	–0.111	0.222	–0.111	–0.111	–0.111	0.222	0.222	–0.111	–0.111
42	–0.111	–0.111	0.222	0.222	–0.111	–0.111	–0.111	0.222	–0.111

Table B.25 $\lambda_{ij} / \sqrt{\text{cov}_{ijij}}$ **computed from the** $[X]_\lambda$ **form of the** $L_9(3^4)$ **experimental design**

ij	λ_{ij}	$\lambda_{ij} / \sqrt{\text{cov}_{ijij}}$
0	20	
11	10	12.2
12	10	12.2
21	0	0
22	2E–15	2.2E–15
31	0	0
32	0	0
41	0	0
42	0	0

baseline of 26.66. But the baseline is not physically accessible for any combination of the set points.

This contradiction is resolved as follows: if we first turn on all of the set points φ_{i0} (only set point φ_{10} is non-zero for the example being considered), we arrive at trial 1, this being an unfavorable but significant change relative to the model baseline. Next, we turn set point φ_{12} 'on.' (We could have also chosen set point φ_{11}.) The test for significance for this second change is based upon the set point difference given by $\varphi_{12} - \varphi_{10}$ which is equal to $2\varphi_{12} + \varphi_{11}$ yielding the t test inequality given by:

$$t_{12-10}\sqrt{s_P^2 / n} \le \frac{2\varphi_{12} + \varphi_{11}}{\sqrt{4\,\text{cov}_{1212} + 4\,\text{cov}_{1112} + \text{cov}_{1111}}} \qquad \text{(B.32)}$$

Table B.26 $\varphi_{ij} / \sqrt{\text{cov}_{ijij}}$ **computed from the** $[X]_\varphi$ **form of the** $L_9(3^4)$ **experimental design**

ij	φ_{ij}	$\varphi_{ijij} / \sqrt{\text{cov}_{ijij}}$	ij	φ_{i0}	$\varphi_{i0} / \sqrt{\text{cov}_{i0i0}}$	Convert to Trial 1 baseline: ij	φ_{ij}	$\varphi_{ijij} / \sqrt{\text{cov}_{ijij}}$
0	26.67					0	20.00	
11	3.33	7.07				11	10	12.2
12	3.33	7.07	10	–6.67	–14.14	12	10	12.2
21	0	0				21	0	
22	0	0	20	0		22	0	
31	0	0				31	0	
32	0	0	30	0		32	0	
41	0	0				41	0	
42	0	0	40	0		42	0	

On substituting for the parameters on the right-hand side of Equation B.32, we find that it is equal to 12.2 and the above inequality is satisfied. The resulting significance is identical (as it should be) to the level of significance of the λ coefficients in Table B.25. Thus, when we compute significance relative to a specific trial, the $[\mathbf{X}]_{\lambda}$ and $[\mathbf{X}]_{\varphi}$ matrix forms give the same sensitivity.

References

1. Phadke, M.S. (1989) *Quality Engineering Using Robust Designs*, Prentice Hall, Englewood Cliffs, NJ.
2. Taguchi, G. and Konishi, S. (1987) *Taguchi Methods Orthogonal Arrays and Linear Graphs*, American Supplier Institute, Allen Park, MI.
3. Walpole, R.E. and Myers, R.H. (1993) *Probability and Statistics for Engineers and Scientists*, Macmillan, pp. 258–9.

Index